Plant Genetics and Genomics: Crops and Models

Volume 23

Series Editor

Richard A. Jorgensen
University of Arizona
Tucson, AZ, USA

Plant Genetics and Genomics: Crops and Models book series provides current overviews and summaries of the state of the art in genetics and genomics for each of the important crop plants and genetic models for which such a volume does not now exist or is out of date.

Most volumes will focus on a single crop, species, or group of close relatives, including especially those plants that already have advanced genomic resources developed and preferably complete or advancing genome sequences. Potential targets include sorghum, maize, poplar, the Brassicaceae, rice, Medicago and Mimulus. Nearly all of these genomes have been sequenced or are currently being sequenced and have substantial communities of biologists and biotechnologists working on them. Priorities are set according to importance of the crop or model species, size of community, and stage of development of the field.

Other volumes take a disciplinary or technological focus, including the currently planned volume on plant cytogenetics and potential volumes on topics such as comparative genomics, epigenetics, and functional genomics.

More information about this series at http://www.springer.com/series/7397

Christopher A. Cullis
Editor

Genetics and Genomics of Linum

 Springer

Editor
Christopher A. Cullis
Department of Biology
Case Western Reserve University
Cleveland, OH, USA

ISSN 2363-9601   ISSN 2363-961X   (electronic)
Plant Genetics and Genomics: Crops and Models
ISBN 978-3-030-23963-3   ISBN 978-3-030-23964-0   (eBook)
https://doi.org/10.1007/978-3-030-23964-0

This Springer imprint is published by the registered company Springer Nature Switzerland AG
The registered company address is: Gewerbestrasse 11, 6330 Cham, Switzerland

Preface

Flax (*Linum usitatissimum* L.) is one of the founding agricultural crops that was domesticated from the wild progenitor, commonly known as pale flax, by hunter-gatherers some 30,000 years ago. It has been selected either for the production of oil or for fiber. Fiber flax has been bred for its long, mainly unbranched stem containing long fibers. Linseed (or oil-seed flax) has been selected for short and highly branched plants to increase the number of flowers to maximize seed production. The health-related properties of flax in human and animal nutrition have opened an additional avenue for flax improvement. Modern flax breeding has the overall objectives to develop cultivars with increased fiber or seed yields, or improved health-related properties. Among the important characteristics are an improved adaptation and disease resistance to augment yields of the commercially important commodities. More than 40,000 accessions representing 54 *Linum* species are conserved in the gene banks around the world. These accessions provide an essential genetic resource for flax breeding and research, and this genetic resource has been partially characterized from cytogenetic, genetic, and genomic aspects. The new molecular genetic tools will make the flax germplasm more accessible to flax. The Canadian breeding program is presented as a model for how variety improvement has proceeded and the choice of which important characteristics need to be considered.

The initial molecular characterization of the flax germplasm was through molecular markers including isozymes, random amplified polymorphic DNAs (RAPDs), and restriction fragment length polymorphisms. The TUFGEN (Total Utilization of Flax Genomics) project has been a turning point for flax genomics resources with the development of the whole-genome sequence of CDC Bethune and the underpinning resources such as the datasets for BAC-end sequences and large-scale ESTs. The initial genome assembly has been refined by the inclusion of a range of techniques, including flax BNG optical map, the BAC physical map, and consensus genetic maps, resulting in a high-quality integrated genome. The assembly has resulted in chromosomes ranging from 15.6 to 29.4 Mb in size. However, the 15 pairs of chromosomes appear to be indistinguishable cytologically in size in spite of a twofold difference in DNA content.

The assembled genome sequence has been augmented by other molecular resources including the characterization of both coding and small RNA molecules. The transcriptomic analysis of the flax seed development will further the improvement of oil production and health-modulating phytochemicals of flax, which, along with the bast fiber, are arguably the most economically important traits in flax. The assembled genome sequence has also facilitated the theoretical identification of small RNA molecules as well as their physical identification.

Although there is more variation in the wild flax progenitor germplasm than in the cultivated varieties, chemical mutagenesis-derived mutant populations have been successfully used to identify important genes involved in cell wall formation among others. Flax is also amenable to transformation, and the development of the floral dip procedure for this species may facilitate other approaches to mutant production and increasing available variation. However, the issues with the transgenic flax, Triffid, and its unregulated escape into the commercial seed provide a cautionary tale for the development of genetically modified flax.

Some flax varieties have a particular characteristic, which is not known in other plant species, namely, the rapid modulation of its genome under certain growth conditions. The variation, with the generation of altered stable lines termed genotrophs, has been shown to occur within a specific defined subset of the genome. Understanding the mechanisms and characteristics of this genome compartment that appears to have the function of modifying the phenotype without the deleterious effects associated with a random mutagenesis makes it an interesting evolutionary mechanism. The plethora of molecular tools available allows this phenomenon to be understood at the molecular level, and potentially capable of manipulation.

The molecular resources available, the commercially important, and the health-related characteristics make flax an accessible interesting model system for understanding complex pathways while also providing the basis for improving the commercial attractiveness of the crop.

The development of the molecular resources for flax is directly attributable to the investment in the TUFGEN program and to the collaborations that were fostered, and continue to thrive, through the program.

Cleveland, OH, USA   Christopher A. Cullis

Contents

Chapter 1
A Taxonomic View on Genetic Resources in the Genus *Linum* L. for Flax Breeding

Axel Diederichsen

1.1 Introduction

The most comprehensive taxonomic review of the genus *Linum* mentioned that there are about 200 botanical species (Winkler 1931). This conspectus is, however, outdated. The lack of a recent review has caused the treatments used for *Linum* taxa in local, national or regional floras to apply scientific names and classifications that are not harmonized. Inconsistencies among floras in assigning a taxonomic rank to a given taxon occur. Many synonyms exist, and they are often used without cross-referencing. A conspectus is needed to compare *Linum* species described from different areas with different names that in fact may be closely related or even identical. A coherent taxonomic review is required to understand and communicate information on the species and infraspecific diversity in the genus *Linum*. Such taxonomic inconsistencies are quite common in many genera and have great implications on managing crop genepools in genebanks. For example, similar scenarios exists in sunflower (*Helianthus annuus* L.; Atalgić and Teryić 2016) and chickpea (*Cicer arietinum* L.; Diederichsen et al. 2009).

Since 1998, the Canadian national genebank for plant genetic resources for food and agriculture, Plant Gene Resources of Canada, has put major efforts into the conservation and research of flax diversity. Classical taxonomic approaches were

A. Diederichsen (✉)
Plant Gene Resources of Canada, Saskatoon Research and Development Centre,
Agriculture and Agri-Food Canada, Saskatoon, SK, Canada
e-mail: axel.diederichsen@agr.gc.ca

© Springer Nature Switzerland AG 2019
C. A. Cullis (ed.), *Genetics and Genomics of Linum*, Plant Genetics and
Genomics: Crops and Models 23, https://doi.org/10.1007/978-3-030-23964-0_1

part of these efforts to group and assess the diversity of the Canadian genebank holdings of 3500 accessions of cultivated flax and more than 100 accessions of 18 other *Linum* species (Diederichsen and Fu 2008; PGRC 2016). This book chapter uses flax as a case study to describe the role of taxonomy and systematics in the conservation and utilization of plant genetic resources. The objective is to review contributions made by taxonomists and crop plant researchers to find a broadly accepted way to describe and communicate the diversity in the genus *Linum* with emphasis on the conservation and utilization of this diversity in the breeding of cultivated flax.

Rational utilization and conservation of genetic diversity, as well as assessing this diversity, requires a tool to communicate unambiguously about such diversity. The taxonomical unit of the species is generally used without much thought. The loss of a species is of great concern and makes the news in public media. Species extinction mobilizes policymakers to take action. However, when we speak of genetic resources, we deal mostly with the diversity within a species, which is the genetic diversity. Loss of such infraspecific diversity is much less spectacular, but to lose such diversity in crop plants and their wild relatives may be of critical importance due to the implications on food security.

It is very much disputed whether systematics and taxonomy, the categorizing and naming of groups of similar elements using scientific methods, are appropriate tools for distinguishing genetic diversity within a species, i.e. on the infraspecific level. Moreover, the relevance of taxonomy for utilization of plant genetic resources is sometimes doubted, as are the concepts of taxonomy in general. Some taxonomists have questioned the Linnaean species concept as such (Bachmann 1998). The main argument is that taxonomy is not considered an objective science, but a subjective view on diversity. Following the path initiated by Linnaeus and well-described in its unfolding over time by Stearn (1986), taxonomy has in particular shied away from applying established taxonomic principles to genepools of crop plants. Consequently, the rules of the International Code for Botanical Nomenclature of Cultivated Plants (Brickell et al. 2009) are a useful tool for naming modern cultivars, but are not useful for orientation in the wider genepool of a cultivated species, which is urgently required for genebank management (Diederichsen 2004).

Classical taxonomy was strictly based on morphological features distinguishable with the naked eye. Later, micro-morphological features, chemical features and, more recently, molecular technologies culminating in sequencing the nuclear DNA have become available and are today dominating the debate on biodiversity. Interestingly, research in flax that started in the 1950s has shown that the amount of nuclear DNA within the species varies and can vary from generation to generation (Durrant 1962; Cullis 2005). We may have reached a point in time where we need to acknowledge that the molecular descriptions only represent another approach that contributes to the understanding of a species but do not allow us to capture fully a species as it unfolds in real life. As previously articulated (see review by Small 1989), the author also feels that each approach to a species will contribute to the understanding of that species. However, we must acknowledge that any definition of a botanical species or other taxon is only an approximation of our thinking to the whole and complex truth and is based on the tools for perception we use for making our observations. Without tools we return to what classical taxonomists relied on: the bare eye, the senses given to us as humans to perceive the world. The highly sophisticated tools used in phenomics applied today are nothing but a major refinement of what our eyes can observe. In addition, molecular methods, including all kind of markers and sequencing of plastome and nuclear DNA, are nothing but observations using other tools and are essentially of a descriptive nature. Slight modifications happen constantly in living organisms. The recently emerging discipline of epigenetics points at genomic plasticity, a phenomenon observed in flax 50 years ago (Durrant 1962; Cullis 2005). It is remarkable that flax has directed research towards epigenetics but at the same time has been the species that guided research to the gene-for-gene interaction between a plant and a pathogen based on very static genetic principles (Flor 1955).

It may be impossible to capture the idea of a biological entity with terms that suggest a static and unchangeable state because a living being will always change while alive and even more so over generations. Mansfeld (1962) clearly described this dilemma or pseudo-problem (he used the term *Scheinproblem*) of systematics. We need to find a convention for communication that comes closest to this ever-changing biological entity and, at the same time, enhances our ability to work with it in plant breeding, in agriculture, in trade, in conservation efforts or in research.

Classical concepts may be currently falling out of favour, but to overview trends over time, it is important to understand the earlier publications that were built on the classical concepts for assessing the diversity of crop genepools. This method was essential to the Vavilov school in crop plant research, which shaped genetic resources exploration in the former Soviet Union and influenced in particular the eastern European countries and Japan (Vavilov 1931; Loskutov 1999). The phenotype of a plant is immediately accessible to our senses, and many phenotypic characters are very relevant for the utilization of that plant. The functionality of many morphological features allows interpretations in the larger ecological or evolutionary context of a species. For efficient communication, the species name is relevant. It is for good reason that common names for plant groups, which are congruent with what later was recognized by cartesian science as a botanical species, have been developed in all cultures. In addition, many infraspecific groups for particular usages based on distinct differences in phenotype have received common names for distinguishing such types within a cultivated plant species (Diederichsen 2004). For farmers and trade, and for interaction among humans, such naming and language have been instrumental for thousands of years. Scientific conventions, such as the taxonomy used in the genebank database system GRIN-Global, maintained by the United States Department of Agriculture to communicate the content of the National Plant Germplasm System, show how relevant this still is (Wiersema and Leon 2016). The International Seed Testing Association (ISTA) also relies on consistent taxonomy for supporting trade and standardization of testing methods (Wiersema 2013). Small (1993) emphasized the great economic impact plant systematics have on agriculture. Other crop plant researchers such as Hanelt (1988) stressed the relevance of taxonomy for managing genebanks. Unfortunately, most modern genebanks lack a close association with taxonomists. As a result, material is sometimes stored and passed on to genebank clients without stringent protocols for botanical identification. This is of great concern and negatively affects our ability to conserve and utilize this diversity. Concerns regarding the lack of taxonomic expertise for effective conservation of biodiversity have been expressed elsewhere (Small et al. 1995), but how this impacts the operations of genebanks preserving plant genetic resources for food and agriculture is rarely articulated.

1.2   The Primary Genepool of Genetic Resources for Flax Breeding

The only agricultural crop in the genus *Linum* is the species *Linum usitatissimum* L., and it has two distinct usages: linseed is used for seed oil and fibre flax is used for extracting the long stem fibre for textile production. In many languages, distinct names exist for flax plant types used for seed oil extraction and for types used for fibre (Diederichsen and Richards 2003). Obviously, common sense guided people in various language families in creating these distinct names to facilitate efficient communication of genetic diversity below the species rank. Taxonomists, concerned about the correct scientific naming of plants, have become reluctant over the years to be as precise as common language in assigning scientific names to these distinct usage groups in flax. Diederichsen and Richards (2003) supported the grouping of cultivated flax into four major groups and using the scientific names for them proposed by Kulpa and Danert (1962). A wide range of diversity in agronomically important characters exists in cultivated flax (Diederichsen et al. 2013). Research has relied on identifying infraspecific groups in flax to interpret archaeological findings and make conclusions on its evolution under domestication (Herbig and Maier 2011). Even flax workers reluctant to apply a detailed infraspecific classification may see the practical evidence in the grouping of cultivated flax into these four principal groups (convarieties).

1.3   Key for Determination of the Four Convarieties of Cultivated Flax, *L. usitatissimum*

A	Capsules open completely septicidaly and loculicidaly during ripening (Fig. 1.1). Seeds are shattered. Later, empty capsules are dropped.	**1.** convar. *crepitans*
A*	Capsules not opening during ripening or only slightly septicidaly separating from each other (Fig. 1.1). Seeds are not easily shattered. Empty capsules are not dropped.	**B**
B	Plant height more than 70 cm and only the upper 1/3 or less of the entire stem length with side branches; if less than 70 cm, then stem branches only in the upper 1/5 of the entire stem length.	**2.** convar. *elongatum*
B*	Plant height usually less than 70 cm; more than 1/5 of the entire stem length with side branches.	**C**
C	Weight of 1000 seeds more than 9 g; plants usually without basal branches.	**3.** convar. *mediterraneum*
C*	Weight of 1000 seeds less than 9 g; plants often with basal branches.	**4.** convar. *usitatissimum*

1. ***L. usitatissimum*** convar. ***crepitans*** (Boeninningh.) Kulpa et Danert, Kulturpflanze, Beih. 3, (1962) 374.

The convar. *crepitans* (Fig. 1.2) refers to dehiscent flax. It has been used in Central and Southeast Europe as a fibre plant (Hegi 1925). Seed shattering makes it difficult to harvest the seeds. This type of flax is no longer cultivated, and only the germplasm collections conducted early in the twentieth century facilitated conservation of this type in genebanks. Its range of variation is limited. With the exception of the dehiscence of the capsules, the plants are phenotypically similar to those of the convar. *usitatissimum*.

Fig. 1.1 Different degrees of dehiscence in mature flax capsules. The complete dehiscence characterizes the primitive type of cultivated flax, L. *usitatissimum* convar. *crepitans*. (Photo: R. Underwood, Agriculture and Agri-Food Canada)

Fig. 1.2 Flowering/maturing plant of *L. usitatissimum* convar. *crepitans*. This type of flax has only been conserved in genebanks. PGRC accession CN 100852, landrace from Portugal. (Photograph: Z. Bainas, Agriculture and Agri-Food Canada)

2. ***L. usitatissimum*** convar. ***elongatum*** Vav. et Ell. in Kul't. Fl. SSSR 5, 1 (1940) 153, pro prole sub *L. indehiscens* Vav. et Ell. subsp. *eurasiaticum* Vav. ex Ell.

The convar. *elongatum* (Fig. 1.3) refers to typical fibre flax. It has long stems, which are only branched at the top. This type of flax has been of great importance in the temperate and northern areas of Europe and in particular Eastern Europe (Vavilov 1926). China is also a centre of diversity for fibre flax. Fibre flax has a shorter vegetative period than the large-seeded flax. This group is identical with the fibre flax as defined by Dillman (1953).

3. ***L. usitatissimum*** convar. ***mediterraneum*** (Vav. ex Ell.) Kulpa et Danert, Kulturpflanze, Beih. 3, (1962) 376.

The convar. *mediterraneum* (Fig. 1.4) refers to large-seeded flax with large flowers and capsules and branched stems. It is only used for seed production. Flax of this type originates from the Mediterranean area and has a long vegetative period. This group is identical with the Mediterranean seed flax defined by Dillman (1953).

4. ***L. usitatissimum*** convar. ***usitatissimum***.

The convar. *usitatissimum* (Fig. 1.5) refers to the intermediate flax, or dual purpose flax. This is the most common type of flax in the world. Within this convariety, further segregation into several different morphotypes is possible. This group covers the spring-type seed flax, winter-type seed flax and Indian and Ethiopian (Abyssinian) flax, as

Fig. 1.3 Flowering/maturing plant of *L. usitatissimum* convar. *elongatum*, the typical plant type for using the long stem fibre that was common in Europe. PGRC accession CN 18991, cultivar Nike from Poland. (Photograph: Z. Bainas, Agriculture and Agri-Food Canada)

Fig. 1.4 Flowering/maturing plant of *L. usitatissimum* convar. *mediterraneum*. This type is only used for seed production and most flax from India and the Mediterranean area belongs to this group. PGRC accession CN 98566, landrace Neelum (3/2) from India. (Photograph: Z. Bainas, Agriculture and Agri-Food Canada)

Fig. 1.5 Flowering/maturing
plant of *L. usitatissimum*
convar. *usitatissimum.* This
intermediate flax covers the
majority of flax types and is
common in all regions.
PGRC accession CN 19017,
cultivar CDC Normandy
from Canada. (Photograph:
Z. Bainas, Agriculture and
Agri-Food Canada)

defined by Dillman (1953). All Canadian oilseed cultivars belong to this group because their weight of 1000 seeds does not reach the size of the large-seeded flax types.

On the next lower taxonomic level, Kulpa and Danert (1962) proposed the distinction of 28 botanical varieties within these 4 convarieties of cultivated flax. This classification system was used to compare the genebank holdings of the national genebanks of Germany (1606 accessions) and Canada (2748 accessions) (Diederichsen 2009). The results showed a similar concentration towards certain phenotypes in both genebanks. This grouping was also instrumental when assembling a core collection of the Canadian flax collection (Diederichsen et al. 2012). However, the utilization of formal taxonomic names for describing genetic diversity within a species is not widely accepted, and it will probably remain a tool only used by genebank curators specializing in particular crops.

Other formal infraspecific groupings of cultivated flax into several botanical varieties exist. Alefeld (1866) described 11 botanical varieties. Howard (1924) suggested a grouping of Indian flax into 26 botanical varieties based on flower and seed colour. Elladi (1940) expanded the classification to 119 botanical varieties. The most recent formal classification of cultivated flax was proposed by Černomorskaja and Stankevič (1987) and distinguished five subspecies (Diederichsen and Richards 2003). None of these classifications has been applied to a genebank collection. In this context, it is also important to note Dillman's (1953) very comprehensive description and categorization of diversity of cultivated flax. This is not a formal taxonomic grouping, but a grouping into cultivar groups. In contrast to the formal taxonomic groupings, it considered agronomically relevant traits such as the need for vernalization as a distinguishing feature. Dillman's system was applied to the world flax collection of the United States Department of Agriculture (USDA) by grouping the genepool into six plant types: (1) fibre flax; (2) spring-type seed flax; (3) winter-type seed flax; (4) short, large-seeded, Indian seed flax; (5) Ethiopian forage-type flax; and (6) Mediterranean or Argentine seed flax.

1.4   The Wild Progenitor of Cultivated Flax Belonging to the Primary Genepool

Linum bienne Mill., Gard. dict. ed. 8 (1768) n. 8 – *L. usitatissimum subsp. angustifolium* (Huds.) Thell., Fl. adv. Montp. (1912) 361; other synonyms are *L. ambiguum* Jord., *L. hohenhackeri* Boiss., *L. usitatissimum* subsp. *hispanicum* Thell., *L. dehiscens* Vav. et Ell. subsp. *angustifolium* (Huds.) Vav. et Ell. and *L. angustifolium* Huds. (Hammer 2001). The English common name of the wild progenitor is pale flax.

Pale flax (Fig. 1.6) has a biennial or perennial growth habit, i.e. it needs a vernalization to induce flowering. Heer (1872) was the first to identify pale flax as the wild

Fig. 1.6 The middle row are plants of *L. bienne* Mill., the wild progenitor of cultivated flax. The many basal branches, bushy growth habit and the dehiscence and shattering of capsules are typical for this species. (Photograph: A. Diederichsen, Agriculture and Agri-Food Canada)

progenitor of cultivated flax. The flowers are homostylous and self-pollinated. The capsules open spontaneously and the seeds shatter. The species occurs in the Mediterranean area and in Western Europe. Tammes (1928) demonstrated that this species is interfertile with cultivated flax. However, no reports exist about its usage in flax breeding.

A significant contribution towards the conservation and understanding of *L. bienne* was made by scientists collecting germplasm of this species in Turkey and depositing it in a genebank (Uysal et al. 2012). This germplasm originated from areas of flax domestication that were until recently not represented in world gene-banks, and it is in particular useful for domestication research in flax. Investigation of the morphology of pale flax showed that some characters have a wider range of variation in this taxon than in cultivated flax (Diederichsen and Hammer 1995). Recent molecular studies have confirmed the relationship with the cultivated species (Fu and Allaby 2010; Soto-Cerda et al. 2014). Additional germplasm of this species from the Balkans has recently been deposited in genebanks (Gutaker 2014).

From the biological point of view, pale flax should be placed in the same botanical species as cultivated flax since the two types can freely intercross. Harlan and de Wet (1971) proposed the genepool concept as a basis for rational classification of crop genepools and strongly suggested that formal taxonomy reflected such biological and evolutionary relationships. Based on this principle, the correct name for pale flax is *L. usistatissimum* subsp. *angustifolium* (Huds.) Thell., and all cultivated flax would fall in the other subspecies, L. *usitatissimum* subsp. *usitatissimum*. Hammer (2001) followed this principle. However, lengthy names are often cumbersome to use. Even worse, if someone is lax and omits the subspecies name, the cultivated form and wild progenitor can be easily confused. For that reason, and for convenience, it seems the species rank is mostly still applied to distinguish pale flax, and the correct name for pale flax at the botanical species rank is *L. bienne* Mill. All other *Linum* species will not produce fertile offspring when crossed with cultivated flax and, therefore, belong to the tertiary genepool from the perspective of plant breeding. Various reports about crossing cultivated flax with taxa other than its wild progenitor are very questionable, as the taxonomic identification of the material was not presented convincingly (Diederichsen 2007). This exemplifies how important it is to verify the botanical identity of material based on solid taxonomy. When scientists specializing in physiology or genomics use misidentified material, errors in communication occur.

1.5   The Secondary and Tertiary Genepools of Species in the Genus *Linum*

Studies of the relationships among *Linum* taxa have not changed the views that were based on the earlier morphological and cytological results (Fu and Allaby 2010). The review of Winkler (1931) is still the baseline. In recent years, descriptions of new wild species in the genus *Linum* have been provided for Southern Italy (Peruzzi 2011), Turkey (Tugay et al. 2010; Yılmaz 2010; Yılmaz and Kaynak 2008), Greece (Iatrou, 1989; Christodoulakis 1999) and Mexico (Rogers 1982). The species composition in the genus *Linum* has not been subject to many studies.

Diederichsen (2007) reported that 33 world genebanks preserved about 600 accessions of 52 *Linum* species belonging to the secondary and tertiary gene-pools for breeding of cultivated flax. A recent inspection of the database at the World Information and Early Warning System (WIEWS) on Plant Genetic Resources for Food and Agriculture of the Food and Agricultural Organization of the United Nations (FAO 2016) showed that 82 institutions around the world maintain a total of 685 *Linum* accessions that are not listed as *L. usitatissimum*, *L. bienne*, *L. angustifolium* or *L. usitatissimum* subsp. *angustifolium*. Assuming all these botanical identifications are correct, these 685 accessions would belong to *Linum* taxa that are neither cultivated flax nor its wild progenitor, pale flax. At the same time, the WIEWS database lists 3479 accessions as *Linum* sp., i.e. the botanical identification has not been conducted or at least not been reported to the database. Thus, the situation of *Linum* germplasm conservation has not changed much since Diederichsen's (2007) review. It is obvious, however, that the botanical identification of material even to the species level is a bottleneck in many genebank collections.

The described situation highlights the wide margin of error we must accept when considering the global status of both in situ and ex situ conservation of plant genetic resources in the genus *Linum*, especially when decisions are made that directly affect the conservation of these genetic resources. Without a consistent taxonomy and lacking reliable species identification, we can only make very approximate estimates of the diversity and conservation status of *Linum* germplasm. Databases that are meant to support decision making for conservation and utilization of genetic resources suffer tremendously due to taxonomic insecurity. For example, the Global Biodiversity Information Facility lists only 26 accepted species names in the genus *Linum* (GBIF 2016). This list does not include cultivated flax or the wild progenitor species of the primary genepool. However, older, well-reputed floras for Turkey (Davis 1967) and Europe (Ockendon and Walters 1968) list more species than that. The database "Plant List" (Royal Botanic Gardens Kew and Missouri Botanical Garden 2016) lists scientific names of species and infraspecific taxa and includes 169 accepted names and 220 names that are unresolved in the genus *Linum*. The species delimitations remain vague for many species, in particular in the *L. flavum* and *L. perenne* groups (Ockendon and Walters 1968). Ornamental plants have been selected in these two groups. These examples show that a great deal of taxonomic confusion continues to exist in the genus *Linum* and that even on the species level we have not improved our ability to communicate *Linum* diversity using taxonomy since Winkler (1931) despite enormous advances in technology. For field botanists, it is essential to have botanical keys that allow identification of plants at least to the species level in the field. In Fig. 1.7, a situation encountered by the author on the Crimean Peninsula south of Sevastopol with three *Linum* species growing on the same meadow in close proximity to each other illustrates this need (Diederichsen et al. 2012). This illustrates the usefulness of species delimitations based on morphological characters that can be recognized by the field botanist, germplasm collector, genebank curator or plant breeder.

Fig. 1.7 Three *Linum* species encountered at the same location in close proximity on the Crimean peninsula south of Sevastopol in 2009. Left to right: *Linum tenuifolium* L., *L. corymbulosum* Reichenb. and *L. austriacum* L. (Photograph: A. Diederichsen, see also Diederichsen et al. 2012).

1.6   Conclusions

A taxonomic revision of the genus *Linum* is urgently required to allow progress in understanding the diversity in this genus and provide an essential tool for conservation and utilization of this diversity. In flax, classical taxonomic concepts can be used to communicate distinct intraspecific diversity. The classical approaches of systematics and taxonomy remain useful tools for those encountering diversity during collecting missions or when dealing with phenotypic diversity during regeneration of diverse ex situ collections of cultivated plants. Results from additional observations using modern molecular methods can be interpreted based on the classical phenotypic groupings and add considerably to our understanding of crops and crop wild relatives.

Acknowledgements Very helpful comments on the manuscript were made by Y.-B. Fu and R.K. Gugel from Agriculture and Agri-Food Canada.

References

Alefeld F (1866) Landwirthschaftliche Flora. [Agricultural flora]. Wiegandt and Hempel, Berlin

Atalgić J, Teryić S (2016) The challenges of maintaining a collection of wild sunflower (*Helianthus*) species. Genet Resour Crop Evol 63:1219–1236

Bachmann K (1998) Species as units of diversity: an outdated concept. Theor Biosci 117:213–230

Brickell CD, Alexander C, David JC, Hetterscheid WLA, Leslie AC, Malecot V, Jin X, Cubey JJ (2009) International code of botanical nomenclature for cultivated plants (ICNP or cultivated plant code) 8th ed. *Scripta Horticulturae*, International Society of Horticultural Science 10:1–184

Černomorskaja NM, Stankevič AK (1987) K voprosu o vnutryvidodvoj klassifikacii l'na obyknovennogo (*Linum usitatissimum* L.). [On the problem of intraspecific classification of common flax (*Linum usitatissimum* L)]. Selekcija I genetika techničeskych kul'tur 113:53–63

Christodoulakis (1999) *Linum phitosianum* spec. nova (linaceae) aus Griechenland. Phyton 33: 289–294

Cullis CA (2005) Mechanisms and control of rapid genomic changes in flax. Ann Bot 95:201–206

Davis PH (1967) *Linum* L. In: Davis PH, Cullen J, Coode MJE (eds) Flora of Turkey, vol 2. Edinburgh University Press, Edinburgh, pp 425–450

Diederichsen A (2004) Case studies for the use of infraspecific classifications in managing germplasm collections of cultivated plants. Acta Hortic (634):127–139

Diederichsen A (2007) *Ex-situ* collections of cultivated flax (*Linum usitatissimum* L.) and other species of the genus *Linum* L. Genet Resour Crop Evol 54:661–678

Diederichsen A (2009) Taxonomical and morphological assessments of infraspecific diversity in cultivated flax (*Linum usitatissimum* L.). J Agr Rural Dev Trop Subtrop Supplement 92:33–51

Diederichsen A, Fu YB (2008) Flax genetic diversity as the raw material for future success. Proceedings, the 2008 international conference on flax and other Bast plants, Saskatoon, Canada, July 21–23, 2008, pp 270–280

Diederichsen A, Hammer K (1995) Variation of cultivated flax (*Linum usitatissimum* L. subsp. *usitatissimum*) and its wild progenitor pale flax (subsp. *angustifolium* [Huds.] Thell.). Genet Resour Crop Evol 42:263–272

Diederichsen A, Richards KW (2003) Cultivated flax and the genus *Linum* L. – taxonomy and germplasm conservation. In: Muir A, Westcott N (eds) Flax, the genus *Linum*. Taylor & Francis, London, UK, pp 22–54

Diederichsen A, Seferova IV, van der Maesen LJG, Lülsdorf M (2009) Challenges in preserving, communicating and utilizing the diversity of wild perennial species of the genus *Cicer* L. proceedings international conference ПРОБЛЕМЫ ЭВОЛЮЦИИ И СИСТЕМАТИКИ КУЛЬТУРНЫХ РАСТЕНИЙ St. Peterburg, Russia, Dec 9–11, 2009, pp 273–278

Diederichsen A, Rozhkov RV, Korzhenevsky VV, Boguslavsky RL (2012) Collecting genetic resources of crop wild relatives in Crimea, Ukraine, in 2009. Crop Wild Relative 8:34–38

Diederichsen A, Kusters PM, Kessler D, Bainas Z, Gugel RK (2013) Assembling a core collection from the flax world collection maintained by Plant Gene Resources of Canada. Plant Genetic Res Crop Evol 60:1479–1485

Dillman AC (1953) Classification of flax varieties 1946. United States Department of Agriculture Technical Bulletin no. 1054. United States Department of Agriculture, Washington, 56 p

Durrant A (1962) The environmental induction of heritable change in *Linum*. Heredity 17:27–61

Elladi VN (1940) *Linum usitatissimum* (L.) Vav. Consp. Nov. – Len. (Russ.). In: Vul'f EV, Vavilov NI (eds) *Kul'turnaja flora SSSR, fibre plants*, Sel'chozgiz, Moskva, Leningrad, vol. 5, Part 1, pp 109–207

FAO (2016) World Information and Early Warning System (WIEWS) on Plant Genetic Resources for Food and Agriculture (PGRFA). http://www.fao.org/wiews-archive/wiews.jsp

Flor HH (1955) Host-parasite interaction in flax rust – its genetics and other implications. Phytopathology 45:680–685

Fu YB, Allaby RG (2010) Phylogenetic network of *Linum* species as revealed by non-coding chloroplast DNA sequences. Genet Resour Crop Evol 57:667–677

GBIF (2016) Global biodiversity information facility. http://www.gbif.org/

Gutaker R (2014) The genetic variation of cultivated flax (*Linum usitatissimum*) and the role of its wild ancestor (*Linum bienne* Mill.) in its evolution. Ph.D. Thesis, University of Warwick, 186 p

Hammer K (2001) Linaceae. In: Hanelt P. and Institute of Plant Genetics and Crop Plant Research (eds.). In: Mansfeld's encyclopedia of agricultural and horticultural crops (except ornamentals), vol 2. Springer, Berlin, Heidelberg, pp 1106–1108

Hanelt P (1988) Taxonomy as a tool for studying plant genetic resources. Kulturpflanze 36:169–187

Harlan JR, de Wet JMJ (1971) Toward a rational classification of cultivated plants. Taxon 20:509–517

Heer O (1872) Über den Flachs und die Flachskultur im Altertum. Neujahrsblatt der Naturforschenden Gesellschaft Zurich 74:1–26

Hegi G (1925) Ilustrierte Flora von Mitteleuropa. Lehmanns Verlag, München, vol. 5, Part 1, pp 3–38

Herbig C, Maier U (2011) Flax for oil or fibre? Morphometric analysis of flax seeds and new aspects of flax cultivation in late Neolithic wetland settlements in Southwest Germany. Veg Hist Archaeobotany 20:527–533

Howard GLC (1924) Studies in Indian oil seeds no. 2 linseed. India Dept. Agriculture, Memoirs. Botanical Series 12:135–183

Iatrou GA. (1989) *Linum hellenicum* (Linaceae), a new species from Peloponnesos, Greece. Wildenovia 19: 69–73

Kulpa W, Danert S (1962) Zur Systematik von *Linum usitatissimum* L. [On the systematics of *Linum usitatissimum* L.]. (in German). Kulturpflanze, Beiheft 3:341–388

Loskutov I (1999) Vavilov and his institute, a history of the world collection of plant genetic resources in Russia. International Plant Genetic Resources Institute, Rome, p 188p

Mansfeld R (1962) Über "alte" und "neue" Systematik der Pflanzen. Kulturpflanze Beiheft 3:26–46

Ockendon DJ, Walters SM (1968) *Linum* L. In: Tutin TG, Heywood VH (eds) Flora Europaea, vol 2. Cambridge University Press, Cambridge, pp 206–211

Peruzzi L (2011) A new species of *Linum perenne* group (Linaceae) from Calabria (S Italy). Plant Biosys 145:938–944

PGRC (2016) Plant Gene Resources of Canada – Ressources phytogénétiques du Canada. http://pgrc.agr.gc.ca/

Rogers CM (1982) *Linum mgvaughii*, a new species from Mexico. Contr. Univ. Mich. Herb. 15:205–207

Royal Botanic Gardens Kew and Missouri Botanical Garden (2016). The Plant List, a working list of all plant species. http://www.theplantlist.org/

Small E (1989) Systematics of biological systematics (or, taxonomy of taxonomy). Taxon 38:335–356

Small E (1993) The economic value of plant systematics in Canadian agriculture. Can J Bot 71:1537–1551

Small E, Cayouette J, Catling PM, Brookes B (1995) An opinion survey of priorities for plant systematic and phytogeography in Canada. Bull Canadian Botanical Association 28:19–22

Soto-Cerda B, Diederichsen A, Duguid S, Booker H, Rowland G, Cloutier S (2014) The potential of pale flax as a source of useful genetic variation for cultivated flax revealed through molecular diversity and association analyses. Mol Breed 34:2091–2107

Stearn WT (1986) Historical survey of the naming of cultivated plants. Acta Hortic (182):19–28

Tammes T (1928) The genetics of the genus *Linum*. Bibliographica Genetica 4:1–36

Tugay O, Bağci Y, Uysal T (2010) *Linum ertugrulii* (Linaceae), a new species from Central Anatolia, Turkey. Ann Bot Fenn 47(2):135–138

Uysal H, Kurt O, Fu Y-B, Diederichsen A, Kusters P (2012) Variation in morphological and phenotypic characters of pale flax (*Linum bienne* Mill) from Turkey. Genetic Res Crop Evol 59:19–30

Vavilov NI (1926) Studies on the origin of cultivated plants. Bull Appl Botany 16(2):3–248

Vavilov NI (1931) Linneevskij vid kak sistema. [The Linnaean species as concept]. (in Russian). Tr po prikl bot, gen i sel 26(3):109–134

Wiersema JH (2013) ISTA list of stabilized plant names, ed 6. http://www.ars-grin.gov/~sbmljw/istaintrod.html

Wiersema JH, Leon B (2016) The GRIN-taxonomy crop wild relative inventory. In Maxted N, Mulloo ME, Ford-Lloyd BV (eds) Enhancing crop genepool use: capturing wild relative and landrace diversity for crop improvement. Council of Applied Biology International, Oxford, UK, pp 453–457

Winkler H (1931) Linaceae, Trib. r. 3. Linoideae-Eulineae. In Engler A. (ed) Die natürlichen Pflanzenfamilien nebst ihren Gattungen und wichtigeren Arten, insbesondere den Nutzpflanzen, 2nd edn. W. Engelmann, Leipzig, vol. 19a, pp 111–120.

Yılmaz Ö, Kaynak G (2008) A new species of *Linum* (Linaceae) from west Anatolia, Turkey. Bot J Linn Soc 156:459–462

Yilmaz Ö (2010) *Linum kaynakiae* sp. nov. (sect. Syllinum, Linaceae) from Turkey. Nordic Journal of Botany 28: 605–612

Chapter 2
A Molecular View of Flax Gene Pool

Yong-Bi Fu

2.1 Introduction

The concept of the crop gene pool was developed by Harlan and de Wet (1971) to assist the access of crop genetic resources for plant breeding and to assess the extent of gene flow among populations of a crop and related taxa. They classified germ-plasm resources into primary, secondary, and tertiary gene pools according to their genetic accessibility to the target cultigen. The primary gene pool is comprised of related species that can be directly mated with the crop to produce fertile progeny. The secondary gene pool consists of those closely related species which can be hybridized with the crop to give a partially fertile hybrid. The tertiary gene pool includes those distantly related species which produce sterile or inviable hybrids with the crop only through exploitation with advanced techniques such as those of genetic engineering. Accordingly, the classification of these gene pools has been made for many crops such as barley, wheat, and common bean (Spillane and Gepts 2001). However, practical issues associated with the use of this gene pool classification are not lacking (Smartt 1984; Hammer 1998; Spillane and Gepts 2001; Gladis and Hammer 2002), and a modification of the original gene pool model was also made to include the quaternary gene pool for genetically engineered crops (Gepts 2000; Gepts and Papa 2003). Traditionally, plant breeders have emphasized and/or utilized closely related, well-adapted domesticated materials within the primary gene pool as sources of genetic diversity (Kannenberg and Falk 1995; Kelly et al. 1998), but technical advances such as plant transformation and genomics have made

Y.-B. Fu (✉)
Plant Gene Resources of Canada, Saskatoon Research and Development Centre, Agriculture and Agri-Food Canada, Saskatoon, SK, Canada
e-mail: yong-bi.fu@canada.ca

© Springer Nature Switzerland AG 2019
C. A. Cullis (ed.), *Genetics and Genomics of Linum*, Plant Genetics and
Genomics: Crops and Models 23, https://doi.org/10.1007/978-3-030-23964-0_2

the utilization of the secondary and tertiary gene pools more feasible. Efforts have been increasingly made to expand existing crop gene pools for plant breeding through introgressing traits by pre-breeding (e.g., see Sharma et al. 2013) and collecting crop wild relatives as genetic resources (e.g., Maxted et al. 1997, 2012; Castañeda-Álvarez et al. 2016). Thus, it is important to classify the gene pools for a crop species as genetic resources to guide germplasm exploration and conservation (Ford-Lloyd et al. 2011) and to evaluate them for plant breeding (Henry 2014). However, little research is done to characterize the classified gene pools (Spillane and Gepts 2001; Brozynska et al. 2016).

Flax (*Linum usitatissimum* L.) is one of the founding agricultural crops in the Near East (Zohary and Hopf 2000) and was domesticated for production of either oil or for fiber by hunter-gatherers some 30,000 years ago (Kvavadze et al. 2009). Fiber flax is bred for its long stem containing long fibers and is mainly grown in Russia, China, Egypt, and near the Northwestern European coast, whereas linseed was cultivated for short and highly branched plants to increase the number of flowers for seed production in Canada, China, USA, India, and Russia (Vromans 2006). Modern flax breeding has faced many challenges with the overall objectives to develop flax cultivars with increased fiber or seed yields, better adaptation, and disease resistance for changing market needs (Soto-Cerda et al. 2014a). Climate change is also expected to impact flax breeding and production (Hall et al. 2016). The health-related properties of flax in human and animal nutrition will stimulate the search for new traits in flax germplasm collections and their incorporation into breeding schemes (Muir and Wescott 2003). Meeting these challenges requires an accelerated access to diverse flax gene pools and an effective search for useful genetic variability (Brozynska et al. 2016), as natural variation is the raw material for any crop improvement and constitutes a critical part of any long-term strategy to enhance the productivity, sustainability, and resilience of crop varieties and agricultural systems (Godfray et al. 2010; Henry and Nevo 2014).

Currently, more than 40,000 flax accessions representing 54 *Linum* species are conserved in 81 genebanks around the world (Diederichsen 2007; FAO 2010). These accessions provide an essential genetic resource for flax breeding and research (Diederichsen and Fu 2008). Efforts have been made around the world to evaluate and characterize this genetic resource (e.g., Fu 2005; Smýkal et al. 2011; Soto-Cerda et al. 2013). Molecular genetic tools have been developed and applied to enhance germplasm characterization, making flax germplasm more accessible to flax breeding (e.g., Soto-Cerda et al. 2013, 2014b). These applications also provide an opportunity to characterize the poorly understood flax gene pool. In this chapter, we attempt to classify the flax gene pool mainly using the concept of the crop gene pool developed by Harlan and de Wet (1971) and to identify its molecular features (or patterns of genetic variation derived from molecular markers or other related studies) through a literature review. It is our hope that this review will help to paint a picture of the extant flax gene pool with molecular features for plant breeding, germplasm conservation, and evolutionary research.

2.2   Flax Gene Pool

Many taxonomic, cytological, and evolutionary studies have been conducted for the genus *Linum* (e.g., see Winkler 1931; Ockendon and Walters 1968; Gill and Yermanos 1967a, b; Harris 1968; Chennaveeraiah and Joshi 1983; McDill et al. 2009; McDill and Simpson 2011; Fu et al. 2016), but no reports have been found specifically on the classification of flax genetic resources into various flax gene pools, following the concept of the crop gene pool developed by Harlan and de Wet (1971) and/or its modification by Gepts and Papa (2003) (Diederichsen 2007). Here, we have made the first attempt, based on the knowledge acquired from literature, to draw a picture of the flax gene pool for further research and flax improvement on the boundaries of the conceptualized gene pools.

2.2.1   Primary Gene Pool

The flax primary gene pool should consist of cultivated flax (*L. usitatissimum*) and its progenitor pale flax (*L. bienne*) (Diederichsen 2007). Both species are diploid, inbreeding species with 2n = 30 chromosomes and homostylous flowers. They can easily hybridize with each other in both directions to yield fertile hybrids (Gill 1966). Cytologically, cultivated flax is differentiated from pale flax by one chromosome translocation (Gill 1966). Previous studies confirmed that pale flax is the progenitor of cultivated flax (Tammes 1928; Diederichsen and Hammer 1995; Fu et al. 2002a; Allaby et al. 2005; Uysal et al. 2010). The domestication of pale flax occurred in the Near East several thousand years ago (Heer 1872; Zohary and Hopf 2000), although there is also evidence of flax use in Neolithic cultures as a source of fiber. Flax fiber has been identified in prehistoric sites in Israel, Syria, and Georgia (van Zeist and Bakker-Heeres 1975; Kvavadze et al. 2009). Flax was grown in Egypt between 4500 and 4000 BC and reached Switzerland around 3000 BC (Helbaek 1959).

Pale flax has several synonyms, including *L. usitatissimum* subsp. *angustifolium* (Huds.) Thell, *L. ambiguum* Jord., *L. hohenhackeri* Boiss., *L. usitatissimum* ssp. *hispanicum* Thell., *L. dehiscens* Vav. et Ell. subsp. *angustifolium* (Huds.) Vav. et Ell., and *L. angustifolium* Huds. (Hammer 2001). It is indigenous to the geographical territory bordering the Mediterranean Sea, Iran, and the Canary Islands (Diederichsen and Hammer 1995). Currently, there are 350–400 accessions of pale flax conserved in several major genebanks. However, the largest original collection of 120 accessions is conserved at Leibniz-Institut für Pflanzengenetik und Kulturpflanzenforschung (IPK), Gatersleben, Germany, including collections from botanical gardens and from natural habitats in Europe. Two recent collections were made: 36 accessions from Turkey (Uysal et al. 2010) and 16 accessions from the Balkan countries (Gutaker 2014). Overall, the existing germplasm collections have large geographical gaps while attempting to represent the species distribution, and efforts are needed to collect germplasm from across the complete natural geographical range of species distribution.

There were more than 40,000 accessions of cultivated flax currently conserved in 81 genebanks around the world (FAO 2010), and possibly only 10,000–15,000 accessions are genetically unique (Diederichsen 2007). These accessions represent primitive flax, flax landraces, released cultivars, and breeding lines that were collected in different countries over time. Specifically, they can be divided into six major groups based on morphological and qualitative traits: (1) fiber flax; (2) oil flax; (3) dual-purpose flax that is an intermediate form between the first two types cultivated for fiber and oil; (4) large seeded flax with a set of specific morphological features and cultivated for oil in the Mediterranean region and North Africa; (5) winter flax cultivated for fiber and oil in the Caucasus, Turkey, Balkans, and some other south regions of Europe; and (6) dehiscent flax, the primitive flax form with dehiscent capsules (Dillman 1953; Kulpa and Danert 1962; Diederichsen 2007; Melnikova et al. 2014). Detailed analyses of ex situ flax germplasm distributions were made by Maggioni et al. (2002) and Diederichsen (2007). Four infraspecific groups, proposed by Kulpa and Danert (1962), have been characterized using phenotypic and molecular characteristics (Diederichsen and Fu 2006).

Large efforts have been made to characterize the germplasm of cultivated flax, as flax breeding has focused largely on either well-adapted materials or genetic diversity within cultivated flax (Vromans 2006; Diederichsen and Fu 2008; Hall et al. 2016). Characterization includes the application of various molecular markers, such as RAPD, AFLP, SSR, and SNP to assess genetic diversity, structure, and relationship among flax germplasm, and the genetic and genomic analyses of breeding materials. The recent sequencing of the flax genome (Wang et al. 2012) provides valuable tools for genomics-assisted breeding of this crop. The potential of pale flax as a source of useful genetic variation for cultivated flax was also explored through genomic analysis of 125 accessions (Soto-Cerda et al. 2014c).

2.2.2 Secondary Gene Pool

This gene pool consists of wild flax species which can be hybridized with cultivated flax to give a partially fertile hybrid. Several studies on the crossability of cultivated flax with wild flax species revealed successful crosses with the following *Linum* species: *L. nervosum, L. pallescens, L. africanum, L. corymbiferum, L. decumbens, L. hirsutum, L. floccosum,* and *L. tenue* (Gill and Yermanos 1967a; Bari and Godward 1970; Seetharam 1972). Specifically, Gill and Yermanos (1967a) found the following successful crosses:

L. usitatissimum x L. africanum
L. corymbiferum x L. usitatissimum
L. usitatissimum x L. decumbens
L. nervosum x L. usitatissimum
L. pallescens x L. usitatissimum,

Seetharam (1972) reported the following successful hybridizations:

L. hirsutum x *L. usitatissimum*
L. floccosum x *L. usitatissimum*
L. tenue x *L. usitatissimum*
L. africanum x *L. usitatissimum*
L. pallescens x *L. usitatissimum,*

and Bari and Godward (1970) performed the following successful crosses:

L. africanum x *L. usitatissimum*
L. pallescens x *L. usitatissimum*

Seetharam (1972) and Gill and Yermanos (1967b) attempted crosses between species having different chromosome numbers, but without any success.

Based on these reports, one could position these eight species as members of the flax secondary gene pool. However, as Diederichsen (2007) stressed, verification is warranted on species crossability, due to some uncertain species determinations. All the assessed species may not have 2n = 30 chromosomes as previously reported (Melnikova et al. 2014). For example, Melnikova et al. (2014) confirmed *L. decumbens* has only 2n = 16 chromosomes. Therefore, if there is a successful cross of cultivated flax with *L. decumbens*, one could reason that the chance of success of a cross with its close relative *L. grandiflorum* Desf with 2n = 16 chromosomes would be high. Interestingly, this gene pool has not yet been exploited for modern flax breeding (Diederichsen and Fu 2008).

2.2.3   Tertiary Gene Pool

Based on the gene pool concept developed by Harlan and de Wet (1971), the tertiary gene pool should include all of the 200 other species of the genus *Linum* which cannot be hybridized with cultivated flax but could be exploited using advanced biotechnological tools. Currently, 54 *Linum* species for 5 botanic sections of the genus are reportedly represented in the world's ex situ collections with a total of about 900 accessions (Diederichsen 2007) and include those flax species from the flax primary and secondary gene pools. However, little exploitation has been made of the wild species, although some taxonomic and phylogenetic analyses were reported (e.g., McDill et al. 2009, 2011; Fu et al. 2016).

Taxonomic distinction of sections and species within the genus has been based on morphological differences (Winkler 1931; Ockendon and Walters 1968). Five sections are largely defined: *Linum, Dasylinum* (Planch.) Juz., *Linastrum* (Planchon), Bentham, *Syllinum* Griseb., and *Cathartolinum* (Reichenb.) Griseb (Ockendon and Walters 1968). The genus is not monophyletic and has two major lineages: a yellow-flowered clade (sections *Linopsis, Syllinum*, and *Cathartolinum*) and a blue-flowered clade (sections *Linum, Dasylinum*, and *Stellerolinum*) (McDill et al. 2009).

There are no reports that wild flax species have been used as a genetic resource in flax breeding. However, several *Linum* species are used or have the potential to be used as ornamental plants: *L. perenne*, *L. narbonense*, *L. flavum*, and *L. grandiflorum* (Hegi 1925; Winkler 1931; Cullis 2011). These species display considerable morphological diversity in plant height, growth habit, flower size, and flower color and thus deserve exploitation toward breeding for ornamental plants (Diederichsen and Fu 2008).

2.2.4 Quaternary Gene Pool

Based on the modification of the original gene pool model by Gepts and Papa (2003) for genetic engineered crops, flax has a quaternary gene pool with the genetically modified (GM) flax CDC Triffid (McHughen et al. 1997). This transgenic sulfonyl-urea herbicide-resistant flax cultivar was developed through biotechnology to acquire the modified acetolactate synthase gene from *Arabidopsis* and was released in 1996 by the University of Saskatchewan's Crop Development Centre. This culti-var provided a broadleaf cropping option to summer fallowing conditions or con-tinuous cropping systems to cereals, in fields previously treated with residual sulfonylurea herbicides. However, it was de-registered in 2001 by the Canadian government, and illegal GM flax contamination was reported in 2009 (European Commission 2009), closing down Canadian flax export markets for a short time (see also the chapter by Ryan and Smyth). The development of testing protocols for the transgene in exported seeds and rigorous testing prior to shipping resulted in the return of export markets.

2.3 Molecular Features of the Primary Gene Pool

Many molecular markers have been developed and applied to characterize diverse flax germplasm, including RAPD, AFLP, SSR, ISSR, and SNP (e.g., see Aldrich and Cullis 1993; Everaert et al. 2001; Vromans 2006; Cloutier et al. 2009; Deng et al. 2010; Uysal et al. 2010; Smýkal et al. 2011; Fu and Peterson 2012; Kumar et al. 2012; Gutaker 2014). For example, RAPD markers were applied to character-ize the flax germplasm collection consisting of 2813 active accessions of cultivated flax from 69 countries and 54 accessions of 26 wild species of the genus (Fu et al. 2002b). Retrotransposon-based markers were applied to characterize 708 flax accessions comprising 143 landraces, 387 varieties, and 178 breeding lines (Smýkal et al. 2011). SSR markers were employed to characterize the flax core collection of 407 accessions (Diederichsen et al. 2013; Soto-Cerda et al. 2013, 2014a, b). These characterizations provide an opportunity to identify molecular features present in the primary gene pool.

2.3.1   P1. Increased Genomic Resources Have Been Developed for Cultivated and Pale Flax

The last decade has seen increased genomic resources developed for cultivated and pale flax. First, a draft genome of flax cultivar "CDC Bethune" was assembled and released (Wang et al. 2012), in which 116,602 contigs (302 Mb) or 88,384 scaffolds (318 Mb) representing ~81% of the flax genome were reported and 43,384 protein-coding genes were predicted. The genome assembly has been improved to delineate complete chromosomes (You et al. 2016c; the chapter by You and Cloutier). Many genomic and transcriptomic resources have been developed for cultivated and pale flax through next-generation sequencing (Fu and Peterson 2012; Fu et al. 2016). The transcriptomic resources include more than 286,000 *L. usitatissimum* ESTs, which are deposited in the National Center for Biotechnology Information (NCBI) database (Cloutier et al. 2009; Sveinsson et al. 2013; Gutaker 2014). Second, linkage and BAC-based physical maps were developed (Spielmeyer et al. 1998; Cloutier et al. 2011, 2012; Ragupathy et al. 2011). Third, many whole-genome-based SNP markers have been developed and used to genotype flax germplasm (e.g., see Fu and Peterson 2012; Kumar et al. 2012; Gutaker 2014). These genomic resources will form a base for further genomic analysis of flax germplasm.

2.3.2   P2. Cultivated Flax Has Substantial Genetic Diversity But Weak Population Structure

A genetic characterization of the Canadian flax core collection comprising 407 accessions using 448 microsatellite markers revealed abundant genetic diversity with 5.32 alleles per locus (Soto-Cerda et al. 2013). Population differentiation was weak between fiber and linseed flax groups (Fst = 0.094). The average genome-wide linkage disequilibrium was 0.036, with a relatively fast decay of 1.5 cM. These findings are compatible with those obtained from a similar analysis of 60 diverse flax accessions (Soto-Cerda et al. 2012). Such genetic features suggest that whole-genome scans through association mapping are feasible to detect loci influencing complex traits. Proof of principle for genomic selection employed in biparental populations of flax for seed yield, oil content, iodine value, linoleic, and linolenic acid content showed some promising genetic gain for linseed breeding (You et al. 2016a).

2.3.3   P3. A Flax Accession Can Be Genetically Homogenous or Heterogeneous

A RAPD analysis of the genetic variability within the flax cultivar CDC Normandy showed extreme genetic uniformity (Fu et al. 2003a). Only 2 (out of 98) RAPD loci derived from 16 informative RAPD primers were polymorphic among 72 assayed plants, indicating the genetic homogeneity in CDC Normandy. This is expected, as it was a somatic variant derived from the flax cultivar McGregor and its genetic background is expected to be close to a pure line (Rowland et al. 2002). An ISSR analysis of 18 Indian flax cultivars revealed a significant intravarietal variation (7.4%) (Rajwade et al. 2010). Similarly, an SSR characterization of six Austrian flax accessions revealed a range of gene diversity values from 0.102 to 0.451 per accession, but some of these accessions were found to be intermixed from the same seed source (Halbauer et al. 2016). These findings indicate flax accessions can be genetically heterogeneous, even with a self-fertilization rate of 95% or higher reported for this species (Robinson 1937). This feature is useful not only for variety identification in flax cultivar development and registration (Everaert et al. 2001) but also for assessing germplasm integrity in genebanks.

2.3.4   P4. A Trend of Genetic Narrowing Was Revealed in North American Linseed Cultivars

A RAPD analysis of 54 linseed cultivars released in North America from 1908 to 1998 showed a trend of genetic narrowing with 2.5% of 108 variable RAPD loci fixed over the last century (Fu et al. 2003b). More RAPD variation was found in old Canadian linseed cultivars (before 1980) than those released after 1980. These findings are consistent with the recent pedigree analysis of 82 flax cultivars registered in Canada since 1910, showing a narrow genetic base of the Canadian cultivars released in the last three decades (You et al. 2016b). Such findings are not unique to flax breeding, as genetic narrowing was also found in Canadian wheat and oat gene pools (Fu and Dong 2015) but help to demonstrate the genetic consequence of long-term flax breeding on the flax primary gene pool. Interestingly, little research has been done to assess the genetic base of the flax breeding gene pools in other countries (Vromans 2006).

2.3.5   P5. Variable Patterns of Genetic Diversity Were Observed Over Different Types of Germplasm

Analyses of RAPD variations in the 2727 flax accessions held in Plant Gene Resources of Canada were made with respect to different types of flax germplasm (Fu et al. 2002b; Fu 2005). It was found that RAPD variations observed in flax were generally low, which is largely expected for flax (a highly selfing species) and the breeding methods commonly used in flax. Measured in the proportion of fixed recessive RAPD loci, cultivated flax had relatively lower RAPD variation than its

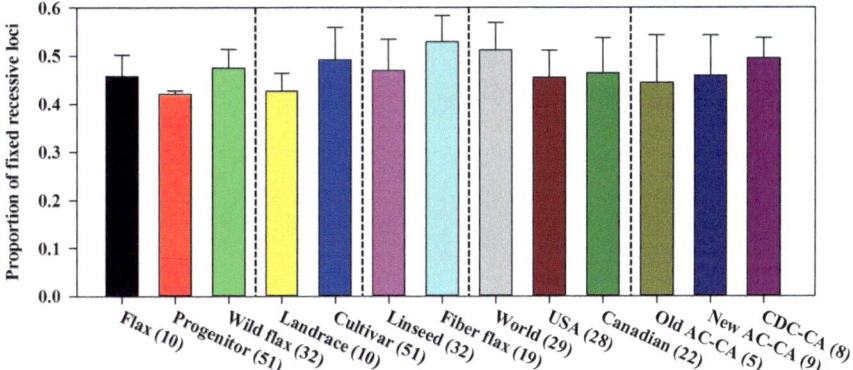

Fig. 2.1 Comparison of genetic diversity among various groups of flax germplasm measured with the proportion of fixed recessive RAPD loci. More fixed loci mean less genetic diversity. The number of accessions for the group is shown in parenthesis. The 22 Canadian cultivars (CA) were further divided into 3 groups reflecting the cultivars developed by Agriculture and Agri-Food Canada (AC) over time and Crop Development Centre (CDC), University of Saskatchewan. (Adapted from Fu et al. 2002c, 2003b)

progenitor pale flax; landrace germplasm displayed higher RAPD variations than cultivars; and linseed cultivars revealed more RAPD variations than fiber flax, as illustrated in Fig. 2.1. These patterns of genetic variation were consistent with those obtained using AFLP (Vromans 2006), ISSR (Rajwade et al. 2010), and retrotransposon-based (Smýkal et al. 2011) markers. For example, the estimates of Nei gene diversity for wild relatives, landraces, cultivars, and breeding lines were 0.46, 0.26, 0.22, and 0.21, respectively (Smýkal et al. 2011). These findings indicate that flax cultivars have a narrow genetic background.

2.3.6   P6. The Geographic Diversity Patterns Matched with Vavilov's Four Centers of Flax Diversity

The patterns of RAPD variation in 2727 flax accession over 12 geographic regions (see Fig. 2.2) were largely in line with the early formulation by Vavilov (1926, 1951) on the four centers of flax diversity (i.e., Abbyssinian, Mediterranean, Central Asia, and Near East) (Fu 2005). Grouping the accessions into 12 major regions explained 8.2% of the RAPD variation. Accessions from East Asia and European regions were most diverse. Accessions from West Asia region were genetically more related to those from the Africa region and less to those from the Indian sub-continent. These diversity patterns are useful not only for germplasm conservation but also for the search for genetic variants by flax breeders.

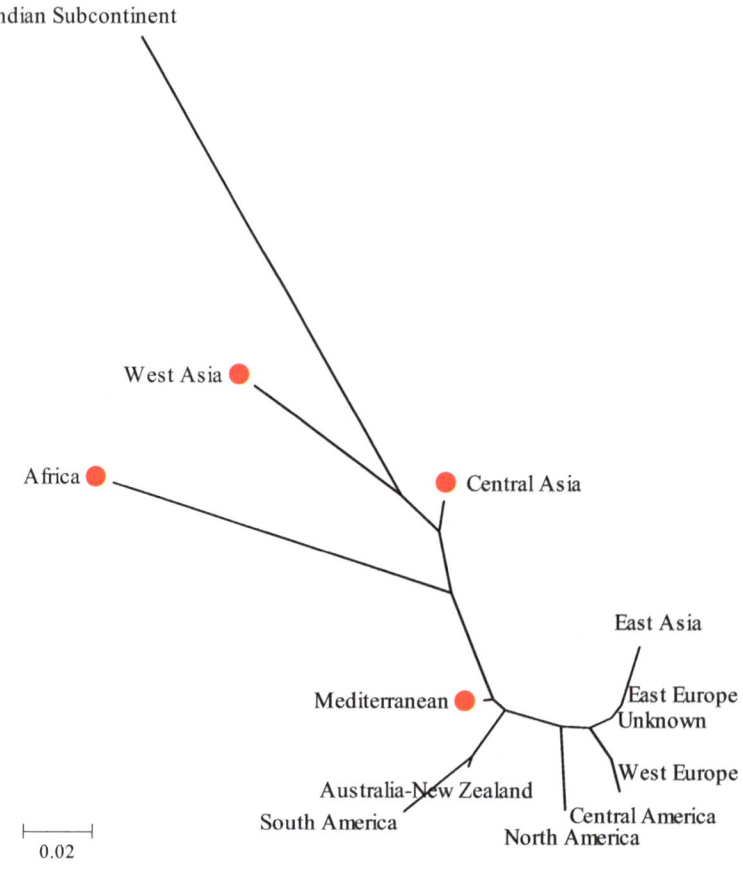

Fig. 2.2 Genetic relationships of 2727 flax accessions representing 12 proposed regions and 1 group of unknown origin, as illustrated in the neighbor-joining tree based on the differences in 149 RAPD markers. The four centers of flax diversity formulated by Vavilov in 1926 are highlighted in the regions with red nodes. (Adapted from Fu 2005)

2.3.7 *P7. Two Distinct Geographic Groups Were Identified in Cultivated Flax Germplasm*

The RAPD-based clustering of 2727 flax accessions from 12 geographic regions also revealed 2 distinct geographical groups of flax accessions from Indian subcontinent and Africa, followed by the accessions from West Asia and Central Asia (Fu 2005). Such distinct grouping (Fig. 2.2) supported the multiple centers of flax variation. Flax germplasm from the Indian Subcontinent and Africa deserves more attention in conservation and breeding. Also, the genetic distinctness present in these regions was useful for the identification of distinct flax germplasm to develop the flax core collection (Fu 2006; Diederichsen et al. 2013) for germplasm utilization.

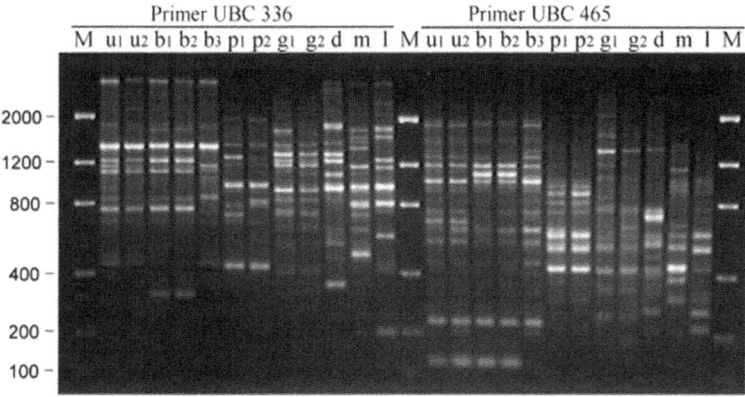

Fig. 2.3 The first RAPD evidence for close genetic relationship between cultivated and pale flax, as illustrated in 12 flax accessions representing 7 flax species with 2 RAPD primers (UBC 336 and UBC 465). M is a ladder DNA. u1 and u2 represented the samples for 2 accessions of *L. usitatissimum*; b1, b2, and b3 for 3 accessions of *L. bienne*; p1 and p2 for 2 accessions of *L. perenne*; g1 and g2 for 2 accessions of *L. grandiflorum*; d for 1 accession of *L. decumbens*; m for 1 accession of *L. mesostylum*; and l for 1 accession of *L. leonii*. (Adapted from Fu et al. 2002a)

2.3.8   P8. Molecular Evidence Was Found for the Wild Progenitor of Cultivated Flax

The wild progenitor of cultivated flax has been long hypothesized to be *L. bienne*, largely based on evidence from several phytogeographic, cytogenetic, and pheno-typic studies (e.g., Tammes 1928; Diederichsen and Hammer 1995). A RAPD anal-ysis of 12 accessions representing 7 flax species (Fu et al. 2002a) revealed that cultivated and pale flax species had a higher RAPD similarity than the other pairs of flax species, as illustrated in Fig. 2.3. Also, these two species were consistently clustered in the same group with all three similarity coefficients used (Fu et al. 2002a). Some cytogenetic analysis further confirmed the close relationship between cultivated and pale flax (Muravenko et al. 2010), and the evidence of higher SSR marker transferability was also found between cultivated and pale flax (Fu and Peterson 2010; Soto-Cerda et al. 2011).

2.3.9   P9. Pale Flax Is Genetically Diverse and Displays Significant Spatial Genetic Autocorrelation

Genetic diversity analysis of pale flax using ISSR markers revealed that pale flax is genetically diverse, having similar levels of genetic variation as cultivated flax (Uysal et al. 2010). It also displayed significant spatial genetic autocorrelation in natural populations. Genetic distances among the pale flax accessions were significantly associated with their geographic distributions and elevational differences among sites, suggesting their strong local adaptation (Uysal et al. 2010). Similar patterns of genetic variation were also found with SSR markers (Soto-Cerda et al. 2014c). A 5sRNA marker analysis of Turkish pale flax accessions revealed two major groups (Christopher Cullis, personal communication), suggesting the existence of different genetic backgrounds in the Turkish pale flax accessions.

2.3.10   P10. Pale Flax Has Potential as a Source of Useful Genetic Variation for Cultivated Flax

Soto-Cerda et al. (2014c) demonstrated the potential of pale flax as a source of useful genetic variation for cultivated flax through molecular diversity and association analyses. They identified favorable alleles with a potentially positive effect to improve yield per se through yield components in pale flax. Pale flax displayed larger variation in vegetative plant parts and growth habit than cultivated flax and more heterogeneity within accessions. Within pale flax, a higher degree of variation was observed in many generative parts such as flower characteristics compared to capsule and seed traits (Uysal et al. 2012). SSR marker transferability was assessed between cultivated and pale flax (Fu and Peterson 2010; Soto-Cerda et al. 2011), and many SNP markers were developed in pale flax (Fu and Peterson 2012).

2.3.11   P11. Flax Was Probably Domesticated First for Oil Use and Later for Fiber Production

Flax is known as one of the founder crops in the Near East, but the detailed history of its domestication remains elusive. Studies using molecular tools provided some evidence for multiple paths of flax domestication for oil-associated traits before selection of other domestication-associated traits of seed dispersal loss and fiber production (Allaby et al. 2005; Fu et al. 2012). However, 30,000-year-old fiber fragments of wild flax were unearthed from the Dzudzuana Cave, Georgia, suggesting their use by prehistoric hunter-gatherers (Kvavadze et al. 2009; Bergfjord et al. 2010). Other genetic evidence for early flax domestication was based on capsular dehiscence (Fu 2011). Based on population-based resequencing, an ancestral winter group of cultivated flax was also revealed having a significant implication for the flax domestication process (Fu 2012). A RNA-Seq analysis showed that pale flax contributed to the adaptation of cultivated flax in the European climate through post-domestication gene flow (Gutaker 2014).

2.4   Molecular Features of the Secondary and Tertiary Gene Pools

The phylogenetic relationships and evolutionary history of diverse species in the *Linum* genus have been investigated using cytogenetic, genetic, and genomic approaches (Gill and Yermanos 1967a, b; Chennaveeraiah and Joshi 1983; Fu et al. 2002a, 2016; Allaby et al. 2005; McDill et al. 2009; Muravenko et al. 2009, 2010; Fu and Allaby 2010; Sveinsson et al. 2013; Melnikova et al. 2014). Four major features were identified for the secondary and tertiary gene pools.

2.4.1   ST1. The Genus Has Two Major Lineages

A phylogenetic analysis of the genus based on chloroplast DNAs showed that the genus has two major lineages: a yellow-flowered clade (sections *Linopsis*, *Syllinum*, and *Cathartolinum*) and a blue-flowered clade (sections *Linum*, *Dasylinum*, and *Stellerolinum*) (McDill et al. 2009). This finding is consistent with those inferences from the flax transcriptomic analysis (Sveinsson et al. 2013) that two ancient polyploidization events occurred on the base of the genus 41–46 MYA (McDill et al. 2009; McDill and Simpson 2011) and between *L. usitatissimum* and *L. bienne* 5–9 MYA (Wang et al. 2012) and from the chloroplast DNA analysis of 16 *Linum* species (Fu and Allaby 2010).

2.4.2   ST2. Congruent Three-Genome Phylogenetic Relationships Among Botanical Sections

Phylogenetic analyses of 18 *Linum* samples representing 16 species within 4 botanical sections of the *Linum* genus based on SNP datasets of chloroplast, mitochondrial, and nuclear genomes showed congruent three-genome phylogenetic signals for 4 botanical sections of the genus: *Linum*, *Dasylinum*, *Linastrum*, and *Syllinum* (Fu et al. 2016). The finding of congruent three-genome phylogenetic signals, as illustrated with Fig. 2.4, implies that these *Linum* species have little hybridizations among them and less deviated lineage sorting. Such finding was interesting, given that these *Linum* genomes have evolved in different inheritance pathways over 20–46 million years (McDill et al. 2009) and large variability in chromosome count has been documented (Cullis 2011).

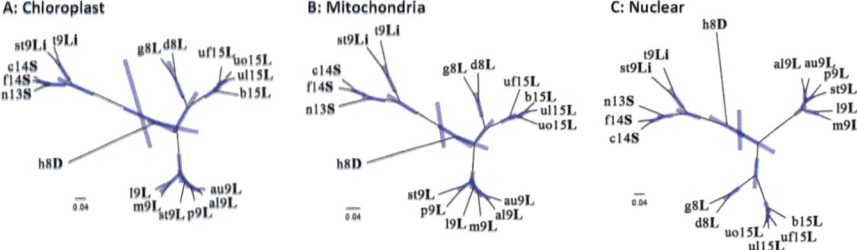

Fig. 2.4 Congruent three-genome phylogenetic signals for four botanical sections of the genus: *Linum*, *Dasylinum*, *Linastrum*, and *Syllinum*, as illustrated with the phylogenetic trees of 18 *Linum* samples representing 16 species from the BEAST inferences. The inferences were separately made (**a**) based on 4603 chloroplast SNPs, (**b**) 1644 mitochondrial SNPs, or (**c**) 1920 nuclear SNPs. Each tree is shown with branch support and labeled with sample code. The number in the sample code is the haploid chromosome count for the sample. The last bold letter(s) in a sample code stands for its taxonomic section (L *Linum*; Li *Linastrum*; D *Dasylinum*; and S *Syllinum*). (Adapted from Fu et al. 2016)

2.4.3 ST3. Members of Section Linum Are Not as Closely Related as Members of Other Sections

Marker-based studies showed that more genetic diversity is present in the section *Linum* than the other sections of the genus and that members of section *Linum* are not as closely related with each other as members of other sections (Fu and Allaby 2010; Melnikova et al. 2014). However, such differences may reflect the uncertainty in the *Linum* species classification. The difficulty in species identification within the *L. perenne* group and the *L. flavum* group has been long recognized (Ockendon and Walters 1968), and regional forms have often been assigned species names (Ockendon 1968, 1971). Diederichsen (2007) has stressed the need for a clarification of species delimitations within the genus to facilitate flax germplasm collection and management.

2.4.4 ST4. Eight Karyotypes Are Present in Linum Species

A karyogenomic analysis of 24 wild species of the genus *Linum* revealed 8 karyotypic groups (Muravenko et al. 2009, 2010). Chromosome numbers were exactly determined in the karyotypes of all studied species, and all individual chromosomes were identified by C/DAPI banding pattern and localization of 26S rDNA and 5S rDNA. These results not only helped to correct the chromosome counts from earlier cytological studies (e.g., Gill and Yermanos 1967a, b; Chennaveeraiah and Joshi 1983) and confirmed the phylogenetic relationships of the assayed species (Fu and Allaby 2010) but also are significant for further studies of the taxonomy of the genus *Linum* and for genetic mapping and resequencing *Linum* plants.

2.5 Implications and Future Research

The molecular features presented here largely reflect our interpretation of the current literature and may not be exhaustive. They may also not be fully unique to the flax gene pool and may represent some features for the gene pools of many other crops. More importantly, however, these molecular features are relevant and have implications in flax breeding, germplasm conservation, and evolutionary research.

2.5.1 Flax Breeding

Flax breeding is facing many challenges in the development of new cultivars with durable resistance to diseases, agronomic fitness, and greater yield gain and stability (Green et al. 2008; Hall et al. 2016). These include the narrow genetic base used for flax breeding (Fu et al. 2002c, 2003b; Cloutier et al. 2009; Smýkal et al. 2011; You et al. 2016b), the few related species to incorporate new variation, the lack of hybrid production systems (Green et al. 2008), and the insufficient genomic tools for molecular breeding (Cloutier et al. 2011, 2012; Fu and Peterson 2012). However, some features revealed here may be useful for flax breeding. First, the primary gene pool has a lot of genetic variability for exploitation. This is encouraging, as the breeding gene pool in North America has shown some genetic narrowing (Fu et al. 2003b; You et al. 2016b). Efforts to exploit the genetic diversity in the primary gene pool are warranted. Second, a core collection of flax germplasm was developed (Diederichsen et al. 2013) and can be useful to improve the access to flax genetic diversity. Inclusion of some flax germplasm originating from East Asia and East Europe regions should be considered, at least for initial trait screening in breeding (Fu 2005). Third, natural genetic variation can be acquired from pale flax, as it has showed the potential as a useful source of genetic variability for flax genetic improvement (Soto-Cerda et al. 2014c). Fourth, increased genomic resources make genomics-assisted breeding possible and flax breeding more efficient by incorporating both marker-assisted selection as well as genomic selection (You et al. 2016a).

In spite of these implications, further research is still needed to facilitate the access by flax breeders to natural genetic variation from the gene pool. Advanced genomic tools and biotechnology can be applied to search for new, useful genetic variants from the flax gene pool (Soto-Cerda et al. 2014c; Brozynska et al. 2016). Genomics-assisted breeding can increase breeding efficiency (Sorrells 2015; You et al. 2016a). Research on species crossability in the genus is also needed to assess the gene pool boundaries and to evaluate the extent of gene flow among species of the gene pool (Jhala et al. 2008). Genetic and genomic analyses of hybrids are also needed to expand the useful gene pool (Kumar et al. 2016), and alien introgressions into flax breeding lines should be explored using a combination of cytology, gene expression analysis, and genome editing (Gill et al. 2011; Quetier 2016).

2.5.2   Germplasm Conservation

The molecular features displayed in the flax gene pool are important for developing a strategy to collect and conserve flax germplasm. The major gap lies in the collection of pale flax and other wild relatives. Even though some efforts had been made to collect pale flax in Turkey and Balkan countries, more efforts are needed to cover the other regions of the species distribution. The geographic patterns of genetic variability in cultivated flax were instrumental in identifying genetically unique flax germplasm (Fu 2006) and provided guidance to enhance the existing flax collections, particularly for allelic richness in Asia and Europe and for allelic distinctness such as in the Indian subcontinent and Africa regions. The patterns of genetic variability in different types of germplasm over the different regions are useful in classifying intraspecific diversity of cultivated flax (Diederichsen and Fu 2006). Also obvious is the need for a systematic treatment of the *Linum* taxon (Diederichsen 2007) to validate the formalized gene pool and to guide the collection of flax wild relatives.

2.5.3   Evolutionary Research

Flax is a dual-purpose crop cultivated for oil and fiber productions (Allaby et al. 2005). It provides a good model to infer crop domestication processes (Fu 2012). More genomic research can be carried out to shed light on the flax domestication processes over the last 10,000 years. Recent finding of congruent three-genome phylogenetic relationships present among four botanical sections of the genus (Fu et al. 2016) provides another excellent model for evolutionary studies of plant genome evolution. Little is known about the generality of the finding on the evolution of the organelle and nuclear genomes in other sections of the genus and other plants. Also, research on the evolution of chromosome number in the genus (Harris 1968; Cullis 2011) will be fruitful and provide some insight into flax genome evolution (McDill and Simpson 2011; Sveinsson et al. 2013).

2.6   Concluding Remarks

We reviewed the literature to formalize the concept of the flax gene pool for further research and to improve the boundaries of the secondary and tertiary gene pools. Clearly, more research is needed on species crossability and taxonomic treatments within the genus. We identified 11 molecular features for the primary gene pool and 4 features for the secondary and tertiary gene pools. These features not only provide a molecular view of the formalized gene pool but also are useful for flax breeding, germplasm conservation, and evolutionary research. Research on flax domestication and genome evolution will be fruitful. Advanced genomic tools and biotechnological methods such as transformation and CRISPR-cas9-based gene editing can be applied to facilitate the accelerated access by flax breeders to natural genetic variation in the flax gene pool.

Acknowledgments We would like to thank Drs. Ken Richards, Axel Diederichsen, Frank You, Petr Smýkal, Braulio Soto-Cerda, and Raja Ragupathy for their helpful comments on the earlier version of the manuscript. This work was supported by an A-Base research project of Agriculture and Agri-Food Canada to YBF.

References

Aldrich J, Cullis CA (1993) RAPD analysis in flax: optimization of yield and reproducibility using Klen *Taq* 1 DNA polymerase, Chelex 100, and gel purification of genomic DNA. Plant Mol Biol Rep 11:128–141

Allaby R, Peterson GW, Merriwether A, Fu YB (2005) Evidence of the domestication history of flax (*Linum usitatissimum*) from genetic diversity of the *sad2* locus. Theor Appl Genet 112:58–65

Bari G, Godward MBE (1970) Interspecific crosses in *Linum*. Euphytica 19:443–446

Bergfjord C, Karg S, Rast-Eicher A, Nosch M-L, Mannering U, Allaby RG, Murphy BM, Holst B (2010) Comment on "30,000-year-old wild flax fibers". Science 328:1634

Brozynska M, Furtado A, Henry RJ (2016) Genomics of crop wild relatives: expanding the gene pool for crop improvement. Plant Biotechnol J 14:1070–1085

Castañeda-Álvarez NP, Khoury CK, Achicanoy HA et al (2016) Global conservation priorities for crop wild relatives. Nat Plants 2:16022

Chennaveeraiah MS, Joshi KK (1983) Karyotypes in cultivated and wild species of *Linum*. Cytologia 48:833–841

Cloutier S, Niu Z, Datla R, Duguid S (2009) Development and analysis of EST-SSRs for flax (*Linum usitatissimum* L.). Theor Appl Genet 119:53–63

Cloutier S, Ragupathy R, Niu Z, Duguid S (2011) SSR-based linkage map of flax (*Linum usitatissimum* L.) and mapping of QTLs underlying fatty acid composition traits. Mol Breed 28:437–451

Cloutier S, Ragupathy R, Miranda E et al (2012) Integrated consensus genetic and physical maps of flax (*Linum usitatissimum* L.). Theor Appl Genet 125:1783–1795

Cullis C (2011) Linum. In: Kole C (ed) Wild crop relatives: genomic and breeding resources oilseeds. Springer, New York, pp 177–189

Deng X, Long S, He D, Li X, Wang Y, Liu J, Chen H (2010) Development and characterization of polymorphic microsatellite markers in *Linum usitatissimum*. J Plant Res 123:119–123

Diederichsen A (2007) *Ex situ* collections of cultivated flax (*Linum usitatissimum* L.) and other species of the genus *Linum* L. Genet Resour Crop Evol 54:661–678

Diederichsen A, Fu YB (2006) Phenotypic and molecular (RAPD) differentiation of four infraspecific groups of cultivated flax (*Linum usitatissimum* L. subp. *usitatissimum*). Genet Resour Crop Evol 53:77–90

Diederichsen A, Fu YB (2008) Flax genetic diversity as the raw material for future success. In: FAO/ESCORENA (eds) Proceedings of the 2008 international conference on flax and other bast plants, Saskatoon, Canada, July 21–23, 2008, pp 270–280

Diederichsen A, Hammer K (1995) Variation of cultivated flax (*Linum usitatissimum* L. subp. *usitatissimum*) and its wild progenitor pale flax (subsp. *angustifolium* (Huds.) Thell.). Genet Resour Crop Evol 42:263–272

Diederichsen A, Kusters PM, Kessler D, Bainas Z, Gugel RK (2013) Assembling a core collection from the flax world collection maintained by Plant Gene Resources of Canada. Genet Resour Crop Evol 60:1479–1485

Dillman AC (1953) Classification of flax varieties, 1946. USDA technical bulletin no. 1054. United States Department of Agriculture, Washington, DC, 56 pp

European Commission (2009) Report on the verification of the performance of a construct-specific assay for the detection of flax CDC Triffid event fp967 using real-time PCR. http://gmo-crl.jrc.ec.europa.eu/flax.htm. accessed on 18 August 2016

Everaert I, Riek JD, Loose MD, Waes JV, Bockstaele EV (2001) Most similar variety grouping for distinctness evaluation of flax and linseed (*Linum usitatissimum* L.) varieties by means of AFLP and morphological data. Plant Var Seeds 14:69–87

FAO (2010) The second report on the state of the world's plant genetic resources for food and agriculture. FAO, Rome

Ford-Lloyd BV, Schmidt M, Armstrong J et al (2011) Crop wild relatives–undervalued, underutilized and under threat? Bioscience 61:559–565

Fu YB (2005) Geographic patterns of RAPD variation in cultivated flax. Crop Sci 45:1084–1091

Fu YB (2006) Redundancy and distinctness in flax germplasm as revealed by RAPD dissimilarity. Plant Genet Resour 4:117–124

Fu YB (2011) Genetic evidence for early flax domestication with capsular dehiscence. Genet Resour Crop Evol 58:1119–1128

Fu YB (2012) Population-based resequencing revealed an ancestral winter group of cultivated flax: implication for flax domestication processes. Ecol Evol 2:622–635

Fu YB, Allaby RG (2010) Phylogenetic network of *Linum* species as revealed by non-coding chloroplast DNA sequences. Genet Resour Crop Evol 57:667–677

Fu YB, Dong Y (2015) Genetic erosion under modern plant breeding: case studies in Canadian crop gene pools. In: Ahuja MR, Jain SM (eds) Genetic diversity and Erosion in plants. Springer International Publishing Switzerland, Chapter 4, pp 89–104

Fu YB, Peterson GW (2010) Characterization of expressed sequence tag-derived simple sequence repeat markers for 17 *Linum* species. Botany 88:537–543

Fu YB, Peterson GW (2012) Developing genomic resources in two *Linum* species via 454 pyrosequencing and genomic reduction. Mol Ecol Resour 12:492–500

Fu YB, Peterson G, Diederichsen A, Richards KW (2002a) RAPD analysis of genetic relationships of seven flax species in genus *Linum* L. Genet Resour Crop Evol 49:253–259

Fu YB, Diederichsen A, Richards KW (2002b) Molecular characterization of 2800 flax accessions at Plant Gene Resources of Canada with RAPD markers. pp. 144–149. In: Proceedings of 59th Flax Institute of the United States, Fargo, ND, USA

Fu YB, Diederichsen A, Richards KW, Peterson G (2002c) Genetic diversity within a range of cultivars and landraces of flax (*Linum usitatissimum* L.) as revealed by RAPDs. Genet Resour Crop Evol 49:167–174

Fu YB, Guerin S, Peterson G, Diederichsen A, Rowland G, Richards KW (2003a) RAPD analysis of genetic variability of regenerated seeds in the Canadian flax cultivar CDC Normandy. Seed Sci Technol 31:207–211

Fu YB, Rowland GG, Duguid SD, Richards KW (2003b) RAPD analysis of 54 North American flax cultivars. Crop Sci 43:1510–1515

Fu YB, Diederichsen A, Allaby RG (2012) Locus-specific view of flax domestication history. Ecol Evol 2:139–152

Fu YB, Dong Y, Yang M-H (2016) Multiplexed shotgun sequencing reveals congruent three-genome phylogenetic signals for four botanical sections of the flax genus *Linum*. Mol Phylogenet Evol 101:122–132

Gepts P (2000) A phylogenetic and genomic analysis of crop germplasm: a necessary condition for its rational conservation and utilization. In: Gustafson J (ed) Proc. Stadler Symposium. Plenum, New York, pp 163–181

Gepts P, Papa R (2003) Possible effects of (trans) gene flow from crops on the genetic diversity from landraces and wild relatives. Environ Biosaf Res 2:89–103

Gill KS (1966) Evolutionary relationship among *Linum* species. Ph.D. diss. University of California, Riverside, CA

Gill KS, Yermanos DM (1967a) Cytogenetic studies on the genus *Linum* I. Hybrids among taxa with 15 as the haploid chromosome number. Crop Sci 7:623–627

Gill KS, Yermanos DM (1967b) Cytogenetic studies on the genus *Linum* I. Hybrids among taxa with 9 as the haploid chromosome number. Crop Sci 7:627–631

Gill BS, Friebe BR, White FF (2011) Alien introgressions represent a rich source of genes for crop improvement. Proc Natl Acad Sci U S A 108:7657–7658

Gladis TH, Hammer K (2002) The relevance of plant genetic resources in plant breeding. FAL Agr Res Special Issue 228:3–13

Godfray C, Beddington JR, Crute IR et al (2010) Food security: the challenge of feeding 9 billion people. Science 327:812–818

Green AG, Chen Y, Singh SP, Dribnenki JCP (2008) Flax. In: Kole C, Hall TC (eds) Compendium of transgenic crop plants: transgenic oilseed crops. Blackwell Publishing Ltd., Oxford, pp 199–226

Gutaker R (2014) The genetic variation of cultivated flax (*Linum usitatissimum* L.) and the role of its wild ancestor (*Linum bienne* Mill.) in its evolution. A Ph.D. thesis, The University of Warwick, p 186

Halbauer E-M, Bohinec V, Wittenberger M, Hansel-Hohl K, Sehr EM (2016) Genetic diversity of Austrian flax accessions – a case study for ex situ germplasm characterization. Abstracts of the 20th EUCARPIA genetic congress, 29 August – 1 September 2016, Zurich, Switzerland. Agroscope, Institute for Sustainability Sciences ISS

Hall LM, Booker H, Siloto RMP, Jhala AJ, Weselake RJ (2016) Flax (*Linum usitatissimum* L.). In: McKeon TA, Hayes DG, Hildebrand DF, and Weselake RJ (eds.) Industrial Oil Crops. Elsevier B.V. Netherlands, Chapter 6, pp 157–194

Hammer K (1998) Genepools – structure, availability and elaboration for breeding (German, English summary). Schriften Gen Res 8:4–14

Hammer K (2001) Linaceae. In: Hanelt P, Institute of Plant Genetics and Crop Plant Research (eds) Mansfeld's encyclopedia of agricultural and horticultural crops (except ornamentals), vol 2. Springer, Berlin Heidelberg, pp 1106–1108

Harlan JR, de Wet JMJ (1971) Towards a rational classification of cultivated plants. Taxon 20:509–517

Harris BD (1968) Chromosome numbers and evolution in North American species of *Linum*. Am J Bot 55:1197–1204

Heer O (1872) Über den Flachs und die Flachskultur im Altertum. Neujahrsblatt der Naturforschenden Gesellschaft Zürich 74:1–26

Hegi G (1925) Illustrierte Flora von Mitteleuropa, vol 5, Part 1. Lehmanns Verlag, Mu"nchen, pp 3–38

Helbaek H (1959) Domestication of food plants in the Old World. Science 130:365–372

Henry RJ (2014) Genomics strategies for germplasm characterization and the development of climate resilient crops. Front Plant Sci 5:68

Henry RJ, Nevo E (2014) Exploring natural selection to guide breeding for agriculture. Plant Biotechnol 12:655–662

Jhala A, Hall LM, Hall JC (2008) Potential hybridization of flax with weedy and wild relatives: an avenue for movement of engineered genes? Crop Sci 48:825–840

Kannenberg LW, Falk DE (1995) Models for activation of plant genetic resources for crop breeding programs. Can J Plant Sci 75:45–53

Kelly JD, Kolkman JM, Schneider K (1998) Breeding for yield in dry bean (*Phaseolus vulgaris* L.). Euphytica 102:343–356

Kulpa W, Danert S (1962) Zur Systematik von *Linum usitatissimum* L. [On the systematics of Linum usitatissimum L.]. Kulturpflanze 3:341–388

Kumar S, You FM, Cloutier S (2012) Genome wide SNP discovery in flax through next generation sequencing of reduced representation libraries. BMC Genomics 13:684

Kumar N, Paul S, Chaudhary HK, Jamwal NS, Singh AD (2016) Wide hybridization and charac-terization of hybrids of *Linum usitatissimum* L. for crossability, agro-morphological traits and rust resistance. SABRAO J Breed Genet 48:136–144

Kvavadze E, Bar-Yosef O, Belfer-Cohen A, Boaretto E, Jakeli N, Matskevich Z, Meshveliani T (2009) 30,000-year-old wild flax fibers. Science 325:1359

Maggioni L, Pavelek M, van Soest LJM, Lipman E (2002) Flax genetic resources in Europe. Ad hoc meeting: 7–8 December 2001, Prague, Czech Republic. International Plant Genetic Resources Institute, Rome, Italy

Maxted N, Hawkes JG, Guarino L, Sawkins M (1997) The selection of taxa for plant genetic con-servation. Genet Resour Crop Evol 7:337–348

Maxted N, Kell S, Ford-Lloyd B, Dulloo E, Toledo Á (2012) Toward the systematic conservation of global crop wild relative diversity. Crop Sci 52:774–785

McDill J, Simpson BB (2011) Molecular phylogenetics of Linaceae with complete generic sam-pling and data from two plastid genes. Bot J Linn Soc 165:64–83

McDill J, Repplinger M, Simpson BB, Kadereit JW (2009) The phylogeny of *Linum* and Linaceae subfamily Linoideae, with implications for their systematics, biogeography, and evolution of heterostyly. Syst Bot 34:386–405

McHughen A, Rowland GG, Holm FA, Bhatty RS, Kenaschuk EO (1997) CDC Triffid transgenic flax. Can J Plant Sci 77:641–643

Melnikova NV, Kudryavtseva AV, Zelenin AV et al (2014) Retrotransposon-based molecular mark-ers for analysis of genetic diversity within the genus *Linum*. Biomed Res Int 2014:1. https://doi.org/10.1155/2014/231589

Muir A, Wescott ND (2003) Flax, the genus *Linum*. Taylor & Francis, London, UK

Muravenko OV, Yurkevich OY, Bolsheva NL et al (2009) Comparison of genomes of eight spe-cies of sections *Linum* and *Adenolinum* from the genus *Linum* based on chromosome banding, molecular markers and RAPD analysis. Genetica 135:245–255

Muravenko OV, Bol'sheva NL, Yurkevich OY et al (2010) Karyogenomics of species of the genus *Linum* L. Russ J Genet 46:1339–1342

Ockendon DJ (1968) Biosystematic studies in the *Linum perenne* group. New Phytol 67:787–813

Ockendon DJ (1971) Taxonomy of the *Linum perenne* group in Europe. Watsonia 8:205–235

Ockendon DJ, Walters SM (1968) *Linum* L. In: Tutin TG, Heywood VH (eds) Flora Europaea, vol 2. Cambridge University Press, Cambridge, pp 206–211

Quetier F (2016) The CRISPR-Cas9 technology: closer to the ultimate toolkit for targeted genome editing. Plant Sci 242:65–76

Ragupathy R, Rathinavelu R, Cloutier S (2011) Physical mapping and BAC-end sequence analysis provide initial insights into the flax (*Linum usitatissimum* L.) genome. BMC Genomics 12:217

Rajwade AV, Arora RS, Kadoo NY, Harsulkar AM, Ghorpade PB, Gupta VS (2010) Relatedness of Indian flax genotypes (*Linum usitatissimum* L.): an inter-simple sequence repeat (ISSR) primer assay. Mol Biotechnol 45:161–170

Robinson BB (1937) Natural cross-pollination studies in fiber flax. J Amer Soc Agron 29:644–649

Rowland GG, McHughen AG, Hormis YA, Rashid KY (2002) CDC Normandy flax. Can J Plant Sci 82:425–426

Seetharam A (1972) Interspecific hybridization in *Linum*. Euphytica 21:489–495

Sharma S, Upadhyaya HD, Varshney RK, Gowda CL (2013) Pre-breeding for diversification of primary gene pool and genetic enhancement of grain legumes. Front Plant Sci 4:309

Smartt J (1984) Gene pools in grain legumes. Econ Bot 38:24–35

Smýkal P, Bačová-Kerteszová N, Kalendar R, Corander J, Schulman AH, Pavelek M (2011) Genetic diversity of cultivated flax (*Linum usitatissimum* L.) germplasm assessed by retrotrasnsposon-based markers. Theor Appl Genet 122:1385–1397

Sorrells ME (2015) Genomic selection in plants: empirical results and implications for wheat breeding. In: Ogihara Y, Takumi S, Handa H (eds) Advances in wheat genetics: from genome to field. Springer Tokyo, pp 401–409

Soto-Cerda BJ, Urbina Saavedra H, Navarro Navarro C, Mora Ortega P (2011) Characterization of novel genic SSR markers in *Linum usitatissimum* (L.) and their transferability across eleven *Linum* species. Electron J Biotechnol 14:6

Soto-Cerda BJ, Maureira-Butler I, Muñoz G, Rupayan A, Cloutier S (2012) SSR-based population structure, molecular diversity and linkage disequilibrium analysis of a collection of flax (*Linum usitatissimum* L.) varying for mucilage seed-coat content. Mol Breed 30:875–888

Soto-Cerda BJ, Diederichsen A, Ragupathy R, Cloutier S (2013) Genetic characterization of a core collection of flax (*Linum usitatissimum* L.) suitable for association mapping studies and evidence of divergent selection between fiber and linseed types. BMC Plant Biol 13:78

Soto-Cerda BJ, Duguid S, Booker H, Rowland G, Cloutier S (2014a) Genomic regions underlying agronomic traits in linseed (*Linum usitatissimum* L.) as revealed by association mapping. J Integr Plant Biol 56:75–87

Soto-Cerda BJ, Duguid S, Booker H, Rowland G, Diederichsen A, Cloutier S (2014b) Association mapping of seed quality traits using the Canadian flax (*Linum usitatissimum* L.) core collection. Theor Appl Genet 127:881–896

Soto-Cerda BJ, Diederichsen A, Duguid S et al (2014c) The potential of pale flax as a source of useful genetic variation for cultivated flax revealed through molecular diversity and association analyses. Mol Breed 34:2091–2107

Spielmeyer W, Green AG, Bittisnich D, Mendham N, Lagudah ES (1998) Identification of quantitative trait loci contributing to Fusarium wilt resistance on an AFLP linkage map of flax (*Linum usitatissimum*). Theor Appl Genet 97:633–641

Spillane C, Gepts P (2001) Evolutionary and genetic perspectives on the dynamics of crop gene pools. In: Cooper HD, Spillane C, Hodgkin T (eds) Broadening the genetic base of crop production. CABI, Wallingford, Oxon, UK, pp 25–70

Sveinsson S, McDill J, Wong GKS, Li J, Li X, Deyholos MK, Cronk QCB (2013) Phylogenetic pinpointing of a paleopolyploidy event within the flax genus (*Linum*) using transcriptomics. Ann Bot 113:753–761

Tammes T (1928) The genetics of the genus *Linum*. Bibliogr Genet 4:1–36

Uysal H, Fu YB, Kurt O, Peterson GW, Diederichsen A, Kusters P (2010) Genetic diversity of cultivated flax (*Linum usitatissimum* L.) and its wild progenitor pale flax (*Linum bienne* Mill.) as revealed by ISSR markers. Genet Resour Crop Evol 57:1109–1119

Uysal H, Kurt O, Fu YB, Diederichsen A, Kusters P (2012) Variation in phenotypic characters of pale flax (*Linum bienne* Mill.) from Turkey. Genet Resour Crop Evol 59:19–30

van Zeist W, Bakker-Heeres JAH (1975) Evidence for linseed cultivation before 6000 BC. J Archaeol Sci 2:215–219

Vavilov NI (1926) Studies on the origin of cultivated plants. Inst. Bot. Appl. et d' Amelior. des Plantes. Leningrad, State Press. (In Russian and English)

Vavilov NI (1951) The origin, variation, immunity and breeding of cultivated plants. Chron Bot 13:1–366

Vromans J (2006) Molecular genetic studies in flax (*Linum usitatissimum* L.), Wageningen University

Wang Z, Hobson N, Galindo L et al (2012) The genome of flax (*Linum usitatissimum*) assembled *de novo* from short shotgun sequence reads. Plant J 72:461–473

Winkler H (1931) Linaceae, Trib. I. 3. Linoideae-Eulineae. In: Engler A (ed) The natural plant families with their genera and most important species, in particular of used plants, 2nd edn. W. Engelmann, Leipzig, pp 111–120

You FM, Booker HM, Duguid SD, Jia G, Cloutier S (2016a) Accuracy of genomic selection in biparental populations of flax (*Linum usitatissimum* L.). Crop J 4:290–303

You FM, Duguid SD, Lam I, Cloutier S, Rashid KY, Booker H (2016b) Pedigrees and genetic base of flax cultivars registered in Canada. Can J Plant Sci 96:837–852

You FM, Li P, Ragupathy R, et al (2016c). The draft flax genome pseudomolecules. In the 66th flax Institute of the United States (Fargo, North Dakota, USA), pp. 17–24

Zohary D, Hopf M (2000) Domestication of plants in the Old World, 3rd edn. Oxford University Press, Oxford, pp 125–132

Chapter 3
Flax Breeding and Cultivar Registration in Canada

Helen Mary Booker

3.1 Cultivar Development in Canada

Canada has produced and exported more flax than any other country since 1994. Europe was the major export market prior to 2009, but transgenic flax discovered in a shipment of Canadian flax halted such exports, initially resulting in a steep decline in flax planted across the Prairies. The entire flax value chain worked together to institute the farm stewardship program, and the Crop Development Centre (CDC, University of Saskatchewan) reconstituted its commercial flax cultivars to alleviate concerns about transgene contamination. As a result, huge changes in flax production and export dynamics occurred over the past 3–5 years; Canada now accounts for ~30% of world production and Canadian exports for ~50% of global flax trade (Booker et al. 2017).

Three major breeding programs have contributed to cultivar development in Canada: the Agriculture and Agri-Food Canada (AAFC) program located at the Research and Development Centre in Morden, Manitoba; the Crop Development Centre (CDC) program located at the University of Saskatchewan in Saskatoon, Saskatchewan; and the Saskatoon R&D facility of Crop Production Services Canada Inc. (CPS). Another breeding program targeting flax for crop diversification in Québec is active at CÉROM (Centre de recherche sur les grains), located in Saint Mathieu-de-Beloeil. Some seed companies are also introducing cultivars from other countries. Disease resistance to rust and wilt has been emphasized by all of these programs to keep such problems under control. Thus, all registered flax varieties are immune to endemic races of flax rust and must have moderate resistance to fusarium wilt. Since 1973, when the last outbreak of rust occurred, the resistance to flax rust

H. M. Booker (✉)
Flax Breeder, Crop Development Centre, Department of Plant Sciences, College of Agriculture and Bioresources, University of Saskatchewan, Saskatoon, SK, Canada
e-mail: helen.booker@usask.ca

© Springer Nature Switzerland AG 2019
C. A. Cullis (ed.), *Genetics and Genomics of Linum*, Plant Genetics and Genomics: Crops and Models 23, https://doi.org/10.1007/978-3-030-23964-0_3

has continued to hold. The first occurrence of powdery mildew disease in the Prairies was reported in the 1990s, and, as of 2017, moderate resistance to powdery mildew is a requirement for cultivars registered for production in Canada. Finally, Western Canada has gone from three flax breeding programs to one in the last couple of years, leaving the University of Saskatchewan's CDC Flax Breeding Program as the sole developer of improved flax cultivars for the Canadian Prairies.

3.1.1   Agriculture and Agri-Food Canada Program

Since the early 1900s, Agriculture and Agri-Food Canada and its predecessors were developing flax varieties for Canada and, in particular, for the Canadian Prairies. The initial program at the Central Experimental Farm in Ottawa produced varieties such as Diadem, Ottawa 770B, Ottawa 829C, and Novelty. During the 1950s, this program was particularly active, releasing varieties such as Linott, Raja, and Rocket. The 1950s and 1960s also marked the beginning of an evolution and transition in flax breeding in Canada. A new program was initiated at the Indian Head Experimental Farm and the Winnipeg Cereal Breeding Laboratory, which led to the development of the variety Cree. In parallel, a breeding program conducted at the Fort Vermillion Experimental Farm and Beaverlodge Research Station, Alberta, in the 1960s produced the variety Noralta, the predominant variety grown in Northern Alberta and Saskatchewan. The breeding programs were eventually consolidated and moved to Winnipeg in 1960 and then later to Morden. Past varieties from the AAFC program and released by the Morden Research Centre include Dufferin, McGregor, NorLin, NorMan, AC Linora, AC McDuff, AC Emerson, AC Carnduff, AC Lightning, Hanley, and Macbeth. More recent varieties released from the AAFC breeding program include AAC Bravo, AAC Bright, AAC Marvelous, AAC Prairie Sunshine, Prairie Blue, Prairie Grande, Prairie Sapphire, Prairie Thunder, and Shape. The current AAFC flax program at Morden has shifted from breeding to agronomy research.

3.1.2   Crop Development Centre Program, University of Saskatchewan

The CDC flax breeding program focuses on the development of cultivars for Western Canada. A modest breeding program was carried out at the University of Saskatchewan from the 1920s through the 1960s, which produced the varieties Royal and Redwood 65. The program was enlarged in 1974 when the CDC initiated a flax breeding program that delivered commercially available cultivars such as Vimy, CDC Bethune, CDC Sorrel, and CDC Sanctuary; more recently, CDC Glas, CDC Neela, CDC Plava, CDC Buryu, CDC Rowland, and specialty yellow seed coat CDC Melyn and CDC Dorado were developed for the emerging human health and animal nutrition market. The CDC flax breeding efforts focus on developing improved flax cultivars adapted to the Canadian Prairies. Reductions in flax breeding activities by other organizations, including AAFC and CPS Canada, have made the CDC flax breeding efforts critical for the broader flax industry.

3.1.3 CÉROM

In 2000, a small breeding program was initiated in Québec with the intent of providing oilseed flax as an additional crop to Québec growers. Though flax was first introduced in Canada by French settlers, this crop is no longer a major crop grown in Québec. However, flax would be an interesting crop rotation alternative, especially in more northern regions such as the lower Gaspé Peninsula. The initial germplasm for the program came from the Crop Development Centre in Saskatoon but has been enriched by several sources of exotic germplasm from various countries. No varieties have yet been released, but this program has now reached maturity, and several cultivars should be released in the near future. Both brown and yellow flax varieties are being developed. The focus is on adaptation to the Eastern Canadian environment, high yield, lodging resistance, and enhanced quality.

3.1.4 Crop Production Services Program

In 1987, a flax breeding program was initiated by Biotechnica Canada in cooperation with Australia's Commonwealth Scientific and Industrial Research Organisation (CSIRO) to develop low ALA flax, subsequently known as Solin or Linola™. In 1990, United Grain Growers Ltd. (UGG) purchased Biotechnica's interest in the program and moved the program from Calgary to Manitoba where the facilities at the AAFC Morden Research Centre and the UGG Rosebank Research Farm were utilized. The Solin breeding program produced the following cultivars: Linola¹ᴹ 947, 989, 1084, 2047, 2090, 2126, and 2149.

Following the acquisition of Agricore United by UGG, the program was relocated to the Alberta Research Council/Alberta Innovates Technology Futures (ARC/AITF) research centre in Vegreville in 2006. Shortly thereafter, the Linola™ breeding program was cancelled, and a NuLin™ (high ALA) program was established and then later transferred to Viterra when the Saskatchewan Wheat Pool merged with Agricore United in 2007. After a few years, the program also started developing brown and yellow seeded flax varieties with traditional levels of ALA. Another move occurred in 2011 when the program relocated to the Viterra R&D facility in Saskatoon, Saskatchewan. CPS acquired the Viterra breeding program in late 2013. Varieties released from the Viterra program include NuLin™ 50 (VT50), WestLin 70, WestLin 71, and WestLin 72. Glencore purchased Viterra in

2012 with assets going to Agrium and shortly thereafter came a cessation of flax cultivar development by the company.

Note: In the past, yellow seed coat cultivars were Solin varieties; breeders used the yellow seed coat to identify cultivars with a low ALA content (developed using mutation breeding) so those cultivars would not get confused with commodity flax, which has a high ALA content (wild type). However, the market for low ALA flax oil never significantly developed. So, developers asked the Canadian Food Inspection Agency to decouple the yellow seed coat trait from the low ALA trait. They agreed, and Solin varieties were deregistered.

The characteristics of current oilseed flax cultivars registered for production in Canada are described in Table 3.1, whilst cultivar yield performance is presented in Table 3.2.

3.2 Breeding Methodology

Breeding in an autogamous (self-pollinating) crop plant such as flax involves generating variability, selection of superior recombinants, and fixing genotypes through inbreeding, resulting in homozygous pure lines using a method referred to as pedigree breeding (You et al. 2016). The first step involves the selection of parental material, which depends on the trait(s) of interest, purpose of the cross, and relative importance of characters other than yield. Pedigree breeding begins with hybridization of parental purelines followed by selfing of F1 hybrid (or crossed) plants to produce an F2 (bulk) population from which single plants are grown as progeny rows in the F3 generation. More recently, more complex crosses are being done such as multi-parent advanced generation crosses (MAGIC) to diversify the genetic base of the flax breeding working collection. Selection and inbreeding are done together in the pedigree system (commonly from the F3 through F6 generations). Selection at each generation is primarily for visible characters such as vigour of stand and maturity. Oil/linolenic acid content is either determined on the bulk seed from each selected progeny row or on a single plant selection (within a progeny row), and these data are used to further scale down the populations and remove undesirable phenotypes. Early generation F3 or F4 testing and a bulk method may also be used to advance material. Additionally, field trials of bulk populations can be used to identify superior families for advancement, whereby single plants are selected from increase plots and grown as single plant progeny rows or hillplots. Higher priority populations may also be fast-tracked by using single seed descent in the growth chamber and/or a contra-season (winter) nursery. Bulk seed from (normally F5 plant generation) is used to establish yield plots, and sub-lines of these are grown in the F6 generation. The bulk seed from an F_6 row may then be used to establish a small increase plot for seed development.

Table 3.1 Characteristics of oilseed flax cultivars

Cultivar[a] (Year of registration)	Seed coat colour[a]	Maturity[b]	Resistance to lodging[d]	Seed size[e]	Oil content[f]	Oil quality[g] Iodine value[h]	ALA content[i]	Resistance to Wilt	Powdery mildew
AAC Bravo (2012)	B	L	G	L	44.6	194.0	60.9	MR	MR
AAC Bright (2017)	Y	L	G	M	48.8	192.1	56.2	MR	–
AAC Marvelous (2017)	B	L	G	M	47.1	192.1	55.8	MR	–
AAC Prairie Sunshine (2016)	B	L	VG	M	47.7	192.5	59.3	MR	–
CDC Bethune (1998)	B	L	G	M	45.6	187.6	54.2	MR	MR
CDC Buryu (2016)	B	L	G	M	46.1	192.6	56.0	MR	–
CDC Dorado (2017)	Y	M	F	M	45.2	204.2	64.0	MR	MR
CDC Glas (2012)	B	L	VG	M	47.4	196.0	58.4	MR	MR
CDC Melyn (2016)	Y	L	G	S	46.6	199.4	61.0	MR	–
CDC Neela (2013)	B	L	G	M	46.5	198.3	60.7	MR	MR
CDC Plava (2015)	B	M	G	M	47.0	195.8	57.8	MR	–
CDC Rowland (2018)	B	L	VG	L	44.6	194.5	59.3	MR	MR
CDC Sanctuary (2009)	B	L	F	M	45.3	188.2	56.2	MR	–
CDC Sorrel (2005)	B	L	G	L				MR	–
NuLin™ 50 (2009)	Y	L	G	M	48.8	210.3	67.8	MR	–
Prairie Blue (2003)	B	L	VG	S	45.9	191	56.8	MR	MR
Prairie Grande (2007)	B	M	VG	M	45.2	194.9	58.4	MR	MR

(continued)

Table 3.1 (continued)

Cultivar[a] (Year of registration)	Seed coat colour[a]	Maturity[b]	Resistance to lodging[d]	Seed size[e]	Oil content[f]	Oil quality[g]		Resistance to	
						Iodine value[h]	ALA content[i]	Wilt	Powdery mildew
Prairie Sapphire (2009)	B	L	G	M	49.0	194.3	57.8	MR	MR
Prairie Thunder (2006)	B	M	VG	M	44.4	196.5	58.9	MR	R
Omega (interim 2015–17)	Y	L	F	M	44.3	184.7	51.8	MR	MR
Shape (2008)	B	L	G	L	50.2	196.8	59.1	MR	MR
WestLin 60 (2016)	B	M	G	M	46.6	198.2	60.5	MR	–
WestLin 70 (2013)	B	L	G	M	46.8	198.4	60.8	MR	MR
WestLin 71 (2013)	B	L	VG	M	47.4	199.2	61.2	MR	MS
WestLin 72 (2014)	B	L	G	M	47.0	192.4	56.8	MR	MR

Based on data from pre-registration trials in the Prairie Provinces
[a]*B* Brown, *Y* Yellow
[b]*E* Early, *M* Medium, *L* Late
[c]*VG* Very good, *G* Good, *F* Fair, *P* Poor
[d]*S* Small, *M* Medium, *L* Large
[e]Oil content: results are reported as percent, calculated on a moisture-free basis
[f]Oil quality of flax is based on the amount of linolenic acid measure in the seed or as measured by iodine value calculated from the fatty acid composition of the seed. A higher iodine value and/or higher ALA content indicates a higher overall oil quality in the seed
[g]Iodine: iodine number is calculated from fatty acid composition
[h]ALA (alpha-linolenic acid): percent of total fatty acid composition
[i]*S* Susceptible, *MS* Moderately susceptible, *MR* Moderately resistant, *R* Resistant

3.2.1 Adaptation Testing

Canada's Prairies are representative of a diversity of agro-ecological zones based on distinct soil type and fertility, length of growing season, rainfall patterns, etc. Multiple location testing (MET) of experimental lines is required to identify lines adapted to Western Canada production zones. To start with the F6 generation, superior breeding lines are identified in replicated yield plots and the best pure lines are advanced to MET where experimental breeding lines are evaluated in advanced yield trials in the Prairie Provinces at sites representative of the brown soil and black and grey soil (shorter versus longer growing season) zones. Selection of elite breeding lines for advancement to pre-registration (cooperative) testing is made based on 6 location years of agronomic data in Western Canada.

Table 3.2 Yield performance of oilseed flax cultivars

Variety[a]	Soil zone[b]		
	Black and grey (long growing season)	Brown and dark brown	Black and grey (shorter growing season)
CDC Bethune (1998)	100	100	100
AAC Bravo (2012)[c]	105 (12)	96 (12)	99 (12)
AAC Marvelous (2017)	108 (5)	101 (7)	103 (5)
AAC Prairie Sunshine (2016)	105 (5)	104 (6)	105 (4)
CDC Buryu (2016)	108 (8)	108 (11)	101 (10)
CDC Glas (2012)[d]	102 (7)	98 (6)	103 (8)
CDC Neela (2013)[e]	109 (6)	105 (7)	105 (7)
CDC Plava (2015)	–	–	106 (12)
CDC Sanctuary (2009)	98 (14)	105 (12)	93 (11)
CDC Rowland (2018)	117 (6)	116 (6)	105 (7)
NuLin™ 50 (2009)	102 (13)	101 (7)	–
Omega (interim 2015–17)	83 (6)	96 (4)	83 (1)
Prairie Grande (2007)	93 (13)	93 (10)	97 (12)
Prairie Sapphire (2009)[f]	97 (14)	95 (11)	93 (11)
Prairie Thunder (2006)	102 (13)	98 (10)	98 (9)
Shape (2008)[g]	96 (12)	91 (10)	93 (12)
WestLin 60 (2016)	–	–	96 (7)
WestLin 70 (2013)	97 (6)	94 (7)	99 (7)
WestLin 71 (2013)	94 (4)	100 (8)	95 (5)
WestLin 72 (2014)	113 (5)	106 (6)	103 (4)

Based on data from pre-registration trials in the Prairie Provinces

[a]Number in brackets is the site-years in cooperative trials in the Prairie Provinces available for comparison with the check variety, CDC Bethune. The more site-years used, the more dependable the result. The check CDC Bethune is present in all site-years. Yield of varieties is expressed relative to CDC Bethune in the same number of site-years. Because the number of site-years varies with varieties, yield performance of a variety can only be compared to CDC Bethune and not to other varieties

[b]See Fig. 3.1 soil zones of Western Canada

[c]Duguid et al. (2014a)

[d]Booker et al. (2014b)

[e]Booker et al. (2014a)

[f]Duguid and Rashid (2013)

[g]Duguid et al. (2014b)

3.2.2   Pre-registration (Cooperative) Trials

Candidate lines are lines evaluated in pre-registration (cooperative) trials for another 2–3 years at test sites representative of the different agro-ecological zones of the northern (Canadian) Prairies. Pre-registration testing in flax is carried out under the auspices of the Prairie Recommending Committee on Oilseeds (PRCO). The purpose of the trial is to collect agronomic performance data for consideration in the Committee's decision regarding their support for registration of a candidate line as a cultivar for production in Canada. The PRCO's flax workers group agrees by consensus on candidate lines to test based on provisional datasets presented by the developer and the disposition of those lines in pre-registration testing. This group also makes decisions with regard to the standards (check cultivars) and protocols to be followed for the registration trials, reporting back to the PRCO. Currently, the Flax Breeder at the CDC (University of Saskatchewan) provides oversight for the randomization of the test at each site, field books, data collated for grain yield and other agronomic traits, seed quality testing, data analysis, and preparation of the linseed cooperative report. Candidate lines are currently tested for 2 years (minimum 10 station years), and those exhibiting merit are brought forward to the PRCO for the recommendation for registration in Canada. In general, cultivar development takes 7 years using a contra-season (or winter) nursery and up to 10 years without.

3.2.3   Post-registration (Provincial Regional) Trial

Once a candidate line receives support for registration in Canada, it may be entered by the developer or agent (e.g., a seed company or distributor) in one of the Prairie provincial trials, normally for 5 years of adaptation testing. The relative performance (to CDC Bethune, the yield standard) under different agro-ecological zones of each province is presented in the Provincial Seed Guides, which are updated and published on an annual basis and widely available to producers. Seed quality and disease resistance may also be included in the tables, with this information obtained from the pre-registration test data for the candidate line (new cultivar). At the recommendation of the CDC Flax Breeder, the pre-registration and regional testing for flax in Saskatchewan were consolidated. Consequently, a 26-entry 3-rep test was conducted in 2018, including check cultivars, experimental (candidate) lines, and (new flax cultivars) lines supported for registration in Canada grown at provincial test sites representative of the different soil types and length of growing season across the Prairies (Fig. 3.1).

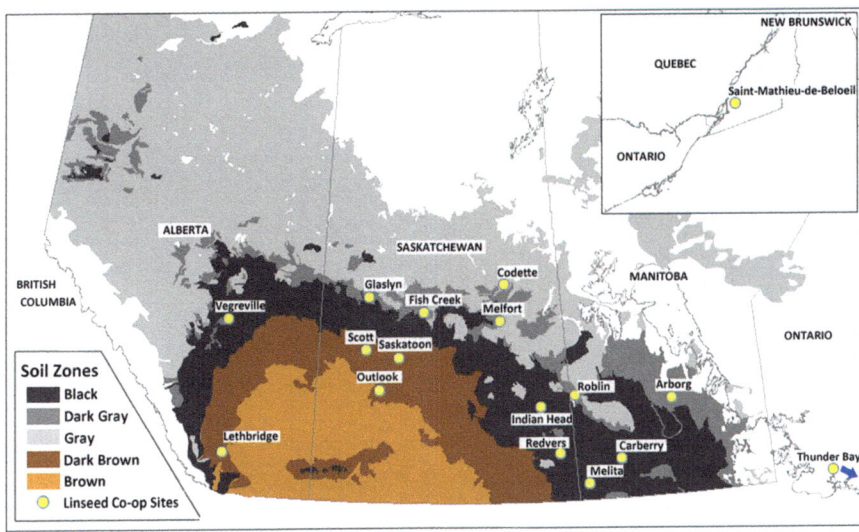

Fig. 3.1 Soil zones of Western Canada and cooperative test sites (2018 Linseed Test)

3.3 Northern Adaptation Traits

Canadian flax production is largely limited to the southern parts of the Prairie Provinces due to the significant risk in the northern part of the grain belt of a killing frost before crop maturity. However, benefits of growing crops in the northern climate include higher yield potential due to lower chances of drought and heat blast (inhibition of seed formation by high temperatures during flowering) (Cross et al. 2003) and better oil quality (higher alpha-linolenic acid content) due to cooler temperatures (pers. comm. P. Dribnenki, Senior Plant Breeder Consultant, Dribnenki Consulting) (Sosulski and Gore 1964). Thus, there is potential to expand flax acres in non-traditional growing areas, such as the northern grain belt, whilst improving the quality of the crop. Traits that increase the suitability of flax to more northern latitudes include earlier flowering, day-length neutrality (for flower induction), early maturation, frost tolerance, high yield, and suitable oil and protein content (see, e.g., Miller et al. 2001). Reduced time to flowering and earlier maturation are also major goals for other crops, including Ethiopian mustard (*Brassica carinata*; Getinet et al. 1996) and chickpea (*Cicer arietinum*; Anbessa et al. 2006). In the Canadian Prairies, indeterminate crops such as chickpea, lentil, and canola (*Brassica juncea* L.) have been improved by breeding for earlier flowering and a longer reproductive duration (lentil and canola; Bueckert and Clarke 2013).

Day-length sensitivity is a key factor controlling flowering time in flax. In most flax cultivars or accessions, flowering is accelerated under long days; the flower induction period ranges from 22 to 34 days after emergence for early and late flowering flax genotypes (Sizov 1955; Zhang 2013) (Table 3.3). Research involves

Table 3.3 Mean height, plant branching, days to flowering (DTF), growing degree days to flowering (GDDF), days to maturity, growing degree days to maturity (GDDM), and seed yield for cultivars and accessions grown at Kernen Crop Research Farm in 2011

Name	Flowering type	Convariety	Height (cm)[a]	Plant branching (1–6)[b]	DTF[c]	GDDF	Days to maturity[d]	GDDM	Seed yield (kg/ha)
CDC Bethune		usitatissimum	60	4	49	777	92	1540	2852
Flanders		usitatissimum	57	4	47	746	101	1694	2707
CN98807	Early		39	3	46	729	102	1705	1146
CN97530	Early	elongatum	72	5	46	729	90	1504	1204
CN98440	Early	usitatissimum	41	3	49	777	104	1728	519
CN98974	Early	mediterraneum	30	2	49	777	115	1905	334
CN97529	Early	usitatissimum	59	4	45	704	87	1450	1511
CN98237	Early	usitatissimum	40	3	48	753	110	1862	851
CN98973	Early	mediterraneum	42	4	47	737	95	1594	2347
CN98370	Early	mediterraneum	38	4	48	753	101	1694	1436
CN97350	Early	usitatissimum	38	4	43	672	88	1463	1302
CN98794	Early		36	3	48	753	94	1574	1579
CN98397	Early	usitatissimum	31	3	49	777	114	1890	637
CN98683	Early	elongatum	71	4	47	746	90	1504	1638
CN98014	Early	usitatissimum	39	3	45	704	88	1463	1583
CN98398	Early	usitatissimum	28	3	49	777	114	1890	379
CN98468	Early	usitatissimum	38	3	48	753	100	1673	858
CN98135	Early	usitatissimum	37	2	49	769	110	1862	693
CN98286	Early	elongatum	63	4	46	729	88	1471	1429
CN97180	Late	elongatum	80	5	50	798	106	1779	1999
CN96992	Late	usitatissimum	66	4	50	787	104	1728	2489
CN97004	Late	usitatissimum	36	1	49	769	94	1574	1156
CN98150	Late	elongatum	75	6	49	769	92	1540	1357

CN100837	Late	crepitans	41	3	48	753	112	1879	1521
CN101421	Late	elongatum	79	6	48	753	90	1504	1183
CN100828	Late	usitatissimum	52	3	49	769	106	1779	2439
CN101052	Late	elongatum	104	6	50	787	91	1522	927
CN101416	Late	elongatum	102	6	50	798	97	1634	1509
CN101419	Late	elongatum	102	6	50	798	97	1634	1477
CN97129	Late	usitatissimum	57	4	53	854	106	1779	2060
CN101559	Late	usitatissimum	74	5	51	817	98	1650	1986
CN101572	Late	usitatissimum	78	5	49	777	103	1719	2195
CN96991	Late	usitatissimum	59	3	62	1024	116	1920	1385
CN40081	Late	usitatissimum	78	5	49	777	95	1594	1871
CN101610	Late	usitatissimum	57	4	49	769	101	1694	2174
CN97584	Late	usitatissimum	60	5	49	777	92	1540	1984

Source: Adapted from Zhang (2013)

[a]Measured from the uppermost plant part to ground at capsule maturity

[b]1 = 1/1, 2 = 1/2, 3 = 1/3, 4 = 1/4, 5 = 1/5, and 6 = 1/6 of total stem length branched, as described in Diederichsen and Richards (2003)

[c]From sowing to 5% flowering of the plot

[d]From sowing to 80% capsule maturity, i.e., brown bolls of the plot

identifying day-length insensitive lines to breed flax that flowers earlier in the short season zone of the Canadian Prairies, as these plants will use other environmental or internal (genetic or developmental) cues to initiate flowering (Sun et al. 2018; Vasudevan 2019). Furthermore, late flowering diverts energy and photosynthates away from seed development and maturity, reducing seed yield and oil content. Late flowering also tends to cause stems to remain green, which delays and complicates harvest. Thus, earlier flowering genotypes with a determinate flowering habit that produce a set number of flowers (or less indeterminant) are more desirable for northern adaptation. Selection for northern adaptation in flax involves scoring breeding lines in replicated yield plots for characteristics such as early season vigour (on a 1–9 scale from 1 = extremely weak to 9 = extremely vigorous), determinate growth habit (on a 1–9 scale from 1 = >40% of plants in plot with flowers or unopened buds to 9 = no plants in plot with flowers or unopened buds), and stem/ straw dry down (on a 1–9 scale from 1 = all stems grass green to 9 = all stems brown) (personal communication, Paul Dribnenki, Senior Plant Breeder, Dribnenki consultants).

3.3.1 Pathology: Flax Rust

Flax rust, caused by *Melampsora lini* (Ehrenb.) Lev., occurs worldwide and can cause severe losses in seed yield and reduced fibre quality. In Canada, flax rust is currently controlled by single or multiple race-specific resistant genes in all registered flax cultivars. At least 34 locus/allele combinations confer resistance to rust and occur in 7 groups (K, L, M, N, P, D, and Q); each group segregates together as closely linked alleles/genes. Rust resistance has been a requirement for Canadian flax varieties for decades, with a possible seven alleles of the K, L, and M groups conferring immunity to flax rust (race 371) present in all flax cultivars registered for production in Canada (Rashid and Kenaschuk 1994). However, the alleles for Canadian rust resistance (race 371) are not effective everywhere; for example, Canadian flax varieties were completely obliterated by rust in Ethiopia (a centre of diversity for the crop) (J.S. (Pat) Heslop-Harrison, University of Leicester, pers. comm). This is problematic as flax breeders work to increase the diversity of Canadian flax by introgressing germplasm from other countries. To maintain the current high level of rust resistance in Canadian varieties, we must maintain the Canadian alleles in future lines containing germplasm from other countries. The loci currently providing rust resistance in Canada are K, L, and M (Table 3.4). Restriction fragment length polymorphism (RFLP) and polymerase chain reaction (PCR)/restriction enzyme markers are available to detect alleles of the L loci; however, these markers have the drawbacks of being dominant-only markers with a lower specificity and low throughput. The sequences of primers to detect the resistant L^2, L^6, and L^{11} alleles have been published (Hausner et al. 1999) as has the resistance-providing M^3 gene sequence (Lawrence et al. 2010). These known sequences were used to identify the exact physical loci of these genes using the

Table 3.4 Canadian flax varieties with different rust resistance alleles

	L6	L9 (ineffective vs. North American rust)	Unidentified L	L11 (possible replacement for L6)	L2	M	M3	K1
Birio[a]	+							
AC Linora[a]	+					+		
CDC Normandy[b]	+							
Andro[b]	+					+		
Flanders[b]	+							
Linola947[c]	+							
Linola989[c]	+							
Linott[a]	+					+		
McGregor[a]	+							
NorMan[a]	+							
NorLin[a]	+							
Somme[b]	+					+		
Vimy[b]	+					+		
AC Emerson[a]	−	+					+	
AC McDuff[a]	−	+						
AC Watson[a]	−	+						
Dufferin[a]	−	+						
Raja[a]	−		+					
Noralta[a]	−		+					
Redwing[a]	−		+			+		
Stewart[a]					+			
Cass[a]		+					+	
Linore[a]			+					

[a]Agriculture Agri-Food Canada germplasm
[b]Crop Development Centre, University of Saskatchewan germplasm
[c]CPS Canada germplasm

CDC Bethune pseudomolecule reference genome (Ravichandran et al. 2017). Current research involves updating the published information on the older markers to modern competitive allele-specific PCR (KASP), TaqMan, high-resolution melt (HRM), and most recently STARP markers. The new markers have the advantage of being allele-specific, high throughput, and able to distinguish between homozygotes and heterozygotes (meaning fewer lines and generations will need to be used to reach homozygosity). Currently, the CDC flax breeding program routinely uses molecular marker-assisted selection (MMAS; HRM assay that detects L gene variants) to identify recombinants with the L6 allele, which confers immunity to rust race 371. Moreover, indoor rust screening (with race 317) is routinely conducted to ensure immunity to rust in new flax cultivars. Rust inoculum (which can be maintained frozen for long periods of time) is warmed and suspended in oil, and an atomizer is used to disperse the inoculum onto the leaf surface of plants, which are maintained under high humidity for a period of time and then rated after 2 weeks (immune, hypersensitive, resistant, or susceptible). Breeding for immunity to rust race 371 will continue to use the indoor screening protocol for parental lines (of exotic origin) and elite (and candidate) lines in the MET as immunity to flax rust (race 371) is a requirement for registration of a new flax cultivar for production in Canada.

3.3.2 Pathology: Fusarium Wilt

Fusarium oxysporum f. sp. *lini* (*Fol*) causes fusarium wilt in flax. Fusarium wilt can cause plant mortality in susceptible cultivars, resulting in severe yield losses. It is widespread in the Canadian Prairies, and control of fusarium using fungicides is not possible due to long-lived soil-borne spores (Rashid 2003a, b). As a consequence, the development of varieties resistant to fusarium wilt is imperative. Several challenges are associated with breeding fusarium wilt resistance. A number of different pathotypes are present within Western Canada, each resulting in different disease severities (Mpofu and Rashid 2001; Edirisinghe 2016). Plants with resistance to one particular pathotype may be susceptible to others. Another challenge when breeding for fusarium wilt resistance is the possible existence of multiple major resistance loci (Spielmeyer et al. 1997, 1998; Edirisinghe 2016). This can make identifying the loci and combining the major loci in a single variety challenging. Furthermore, fusarium resistance genes identified in other plants have hypervariable regions that are different between plant varieties. For example, there are three classes of leucine-rich repeat (LRR) genes, all of which have multiple members that play a role in disease resistance. The differences between family members mainly occur in the LRR domain of the protein and are thought to be responsible for the specificity of the protein to specific pathotypes in the gene-for-gene model of disease resistance (Ellis et al. 2000; Diederichsen et al. 2013). As a consequence, resistant plants with different genetic backgrounds are likely to have different subsets of resistance genes/alleles. Robust genetic markers for wilt resistance would allow flax breeders to perform early and efficient selection of segregants resistant to wilt. Less obviously, but of equal importance, these markers would also benefit the improvement of traits that are not disease related. This is because Canadian flax breeders are required to include North American germplasm in their crosses as sources of wilt resistance (e.g., wilt-resistant Bison somewhere in the pedigree; You et al. 2016) and are therefore limited in the diversity of germplasm they can use to generate new populations. Thus, a better understanding of wilt resistance would also positively impact agronomic traits such as grain yield and/or adaptation to wider climatic zones. To maintain an adequate level of resistance to fusarium in Canadian cultivars, elite breeding lines in performance testing are screened annually for reaction to fusarium wilt resistance (scored on a scale of 0 to 9, (0 = healthy and vigorous, 9 = severely wilted or dead plants) at established wilt nurseries at Indian Head and Saskatoon, Saskatchewan, and Morden, Manitoba. The wilt nursery (at Preston, Saskatoon) was moved from the University of Saskatchewan Investigation land in the late 1970s as there had been a wilt nursery on the Investigation plots as late as the 1950s. The area was seeded with wilt susceptible Novelty flax and, lo and behold, wilt was found 20 years after flax would have been grown on the land. The soil was moved to the Preston area and began building up wilt in the nursery. The original soil (inoculum) for the Investigation nursery was from Indian Head, which also supplied the soil to the Morden Agriculture Agri-Food Canada station. The origin of the Indian Head nursery can be traced to Fargo, North Dakota, United States (personal communication G.G. Rowland, retired Flax Breeder, Crop Development Centre, University of Saskatchewan).

3.3.3 Pathology: Pasmo

Pasmo disease caused by *S. linicola* (Speg.) Garassini is prevalent and widespread in flax in Western Canada, and Canadian flax cultivars are susceptible. Genetic resistance has not been identified in Canadian germplasm, and this is why breeding for disease resistance has made minimal progress. Flax grown in areas with high humidity and temperature exhibit rapid disease development, as was evident in 2010 and 2014. Seed yield of susceptible varieties infected during flowering can be reduced by as much as 75% (Sackston 1959; Gillis et al. 2008; Rashid 2010; Rashid and Duguid 2010), and this disease continues to be a problem in flax production areas of Canada. In terms of seed yield loss, time of inoculation appears more important than the severity of symptoms at maturity (Islam 2018). Pasmo infection also reduces quantity and quality of straw from linseed flax, impacting fibre yield. The *S. linicola* fungus survives from one season to the next on infected stems left in the field. Field sanitation methods, such as tillage, are not advisable due to conflict with soil conservation (no till) practices. The most practical method of controlling this disease is the production and use of resistant varieties, combined with effective crop rotations. Collaborative research (AAFC & University of Saskatchewan) characterizing the flax germplasm (Canadian world collection housed at Plant Gene Resources Canada) for resistance to pasmo will allow breeders to include resistant lines in crossing, ultimately leading to new cultivars that farmers can use to reduce the potential impact of this disease on their crop. To determine the level of resistance to pasmo, elite (and candidate) breeding lines are evaluated under epiphytotics using the 0–9 scale (0 = no sign of infection, 9 ≥ 90% leaf area affected) in the Prairies.

3.3.4 Pathology: Powdery Mildew

Powdery mildew (PM), caused by the obligate biotrophic ascomycete *Oidium lini* Skoric, is a common, widespread, and easily recognized foliar disease of flax present in most growing areas worldwide (Gill 1987; Aly et al. 2012). Canadian flax varieties registered for production in Canada as of 2017 must have PM resistance. Previous work using simple sequence repeats (SSRs) and single nucleotide polymorphisms (SNPs) identified three QTL responsible for PM resistance in a recombinant inbred line (RIL) population derived from a cv. Norman (PM susceptible) × cv. Linda (PM resistant) cross (Asgarinia et al. 2013). More recently, four loci that putatively provide resistance to Canadian strains of powdery mildew were identified in a segregation population derived from a cross of French fibre flax cultivars (Young et al. 2017). Current research involves identifying underlying candidate genes conferring resistance to PM and development of DNA markers for these same loci for PM resistance present in both segregating populations, which is likely as 97% of the variability is reported to be associated with these regions. Until such time, breeding for resistance to PM (scored on a scale of 0–9: 0 = no sign of infection, 9 ≥ 90% leaf area affected) to remove susceptible genotypes relies on epiphytotics, which are not common across the Prairies.

3.4   Flaxseed Quality Evaluation

Flaxseed quality traits such as high oil content, desirable fatty acid composition, and high protein content are important quality factors under genetic control (Kumar et al. 2015; You et al. 2018). There is sufficient diversity in the cultivated germplasm collection to ensure continued improvement of most quality traits (Soto-Cerda et al. 2014; You et al. 2017). To increase efficiency and enhance wet chemistry evaluations, near-infrared (NIR) technology is extensively used to evaluate flaxseed quality traits such as oil content and iodine value (a measure of unsaturated linoleic and linolenic fatty acids), which estimates the drying potential of the oil (important for industrial purposes).

3.4.1   Oil Content

Oil content of current Canadian cultivars ranges between 45 and 50% (Table 3.5). Oil content is the most important quality parameter for oilseed flax (linseed). Accordingly, high oil content is selected for in early generations. Oil content is determined by near-infrared measurements calibrated against nuclear magnetic resonance (NMR) spectroscopy, which is calibrated against extraction reference method ISO659-2009. Results are reported as percent, calculated on a moisture-free basis. Oil content can be influenced by the cultivar and environment, and grain oil content can vary by as much as 15% from farm to farm (Duguid 2010). During handling, different grain lots are mixed, and the oil content of Canadian flax is averaged out as reported in Table 3.5.

Table 3.5 Oil content and fatty acid profile summary of the variety composite data from 2010 to 2014 (Source: Oilseed Unit, Canadian Grain Commission). Oil content is determined by near-infrared measurement calibrated against the FOSFA extraction reference method. Results are reported as percent calculated on a moisture-free basis. Iodine value is calculated from fatty acid composition according to AOCS Cd 1 (Daun and Marek 1983). Linolenic acid composition is determined by gas-liquid chromatography of the fatty acid esters according to the AOCS Ce-91 method with esters prepared by the AOCS 2-66((3) method. Protein content is determined by near-infrared measurement calibrated against the Combustion Nitrogen Analysis Reference Method and expressed on an N × 6.25, dry whole seed, and dry defatted meal basis

Variety[a]	Province	Year(s)	No. of samples	Oil content[b] %	Prot. content[c] %	Oil-free protein %	FFA[d] %	Fatty acid composition[e]						
								C16:0 %	C18:0 %	C18:1 %	C18:2 %	C18:3 %	Sats[f] %	Iodine value[g]
AC Watson	SK	2011	5	44.9	21.2	38.6	0.17	4.8	3.6	16.1	15.2	59.4	8.7	196
CDC arras	SK	2012, 2014	11	44.7	21.4	38.7	0.18	5.9	3.5	16.4	15.3	58.2	9.8	193
CDC Bethune	MB	2010–2014	87	44.8	23.1	41.8	0.26	4.8	3.4	19.9	15.2	55.6	8.7	189
CDC Bethune	SK	2010–2014	311	45.6	21.2	39.0	0.14	4.9	3.4	18.9	15.7	56.2	8.6	191
CDC Glas	AB	2014	5	45.7	23.1	42.6	0.11	5.4	3.3	18.9	15.3	56.1	9.1	190
CDC Sorrel	AB	2011, 2014	12	46.2	21.6	40.2	0.10	5.1	3.2	18.0	14.1	58.8	8.5	194
CDC Sorrel	MB	2010–2014	62	44.8	22.3	40.4	0.29	5.0	3.4	18.9	13.6	58.1	8.8	192
CDC Sorrel	SK	2010–2014	210	45.5	20.4	37.5	0.16	5.0	3.1	17.7	14.0	59.3	8.5	194
Flanders	AB	2011, 2013	9	45.6	23.1	42.5	0.11	4.8	4.3	18.3	14.5	57.1	9.5	191
Hanley	AB	2014	6	43.9	23.8	42.4	0.22	5.5	2.5	15.3	17.0	58.7	8.4	196
Hanley	MB	2010–2012	19	43.6	23.3	41.4	0.24	5.4	2.7	14.7	16.5	59.9	8.4	198
Lightning	MB	2011, 2012, 2014	19	46.2	24.0	44.6	0.22	4.7	3.8	18.2	15.8	56.4	8.9	191
Prairie sapphire	AB	2014	10	47.0	23.0	43.3	0.18	4.8	3.6	19.1	15.7	55.9	8.7	190
Prairie sapphire	SK	2014	5	47.0	21.7	40.9	0.15	4.8	3.5	17.1	16.1	57.6	8.7	193

(continued)

Table 3.5 (continued)

Variety[a]	Year(s)	Province	No. of samples	Oil content[b] %	Prot. content[c] %	Oil-free protein %	FFA[d] %	Fatty acid composition[e]					Sats[f] %	Iodine value[g]
								C16:0 %	C18:0 %	C18:1 %	C18:2 %	C18:3 %		
Taurus	2012	MB	6	45.1	23.9	43.5	0.16	5.1	3.0	20.5	14.2	56.1	8.5	189
Taurus	2010, 2012, 2013	SK	25	45.5	21.7	39.9	0.14	5.2	2.7	19.2	14.3	57.6	8.3	192
Vimy	2010–2014	SK	86	45.5	21.4	39.4	0.12	5.4	3.2	16.8	14.8	59.0	9.0	194
MEAN				45.4	22.4	41.0								

[a] As indicated on submitted harvest survey envelopes
[b] Moisture-free basis
[c] %N × 6.25; moisture-free basis
[d] % free fatty acids in oil, expressed as percentage oleic acid
[e] % of fatty acids including palmitic (C16:0), stearic (C18:0), oleic (C18:1), linoleic (C18:2), and linolenic (C18:3)
[f] Saturated fatty acids: defined as the sum of C16:0, C18:0, C20:0, C22:0, and C24:0
[g] Calculated from fatty acid composition

3.4.1.1 Fatty Acid Composition

Linseed oil is composed of five main fatty acids: palmitic (~5%), stearic (~3%), oleic (~18%), linoleic (~14%), and alpha-linolenic (>50%) (Table 3.5). Fatty acid composition is determined by gas-liquid chromatography of the fatty acid esters according to AOCS method Ce-91 with esters prepared by AOCS method Ce 2–66(93). ALA is likewise determined. Iodine number is calculated from fatty acid composition according to AOCS method Cd 1c-85 or NIR. Flax oil is one of the most unsaturated common plant oils, and its iodine value is greater than 185 Wijs units (Duguid 2010). The level of unsaturation varies with both cultivar and environment. Environmental studies have shown that flax grown under cool conditions has higher levels of linolenic acid and overall iodine value (pers. comm. P. Dribnenki, Senior Plant Breeder Consultant, Dribnenki Consulting) (Sosulski and Gore 1964).

3.4.2 Protein Composition

After crushing, the defatted linseed meal can be fed to livestock. Oil-free protein content is determined by NIR on the seed, with NIR calibrated against the Combustion Nitrogen Analysis Reference Method and expressed on an N × 6.25, whole seed dry basis, and on an oil-free meal basis. Canadian flaxseed has been found to contain about 22% crude protein, which translates into a meal protein content of 41% (Table 3.5). Protein and oil content of flaxseed tend to be inversely related, but flax lines have been developed with high oil and protein content (Duguid 2010; You et al. 2017).

The range of Canadian exports of flaxseed by cultivar is summarized in Table 3.5.

3.5 Seed Coat Colour

Seed coat colour in flax can vary and includes yellow, olive, light, and dark brown seeds as well as mottled seeds (Diederichesen and Raney 2008). The brown seed coat is the wild type and yellow (golden), or mottled seed coat colour in flax is conditioned by four loci (Y, d, g, and b^1b^{1vg}), with the proanthocyandin (PA)-accumulating layer (endothelium) thinner and unpigmented in yellow seed coat mutants (Mittapalli and Rowland 2003; Sudarshan et al. 2017). Recessive alleles conditioning the yellow seed coat are associated with a more fragile seed coat prone to cracking, whereas genotypes possessing the dominant Y allele possess a more robust seed coat (personal communication, Paul Dribnenki, Senior Plant Breeder, Dribnenki Consultants). Primers for Y and d markers developed under the Genome Canada TUFGEN project (2009–2014) were kindly provided to the CDC by G. Selvaraj (Research Scientist, National Research Council, Saskatoon) for use in the flax breeding program. The Y marker (KASP assay) has been used to identify breeding lines that are mixed or contain heterozygous plants. As Y is dominant, heterozygous seeds are yellow in colour; however, they would not breed true as brown-seeded individuals would appear in subsequent generations. Thus, the Y KASP assay is currently being used to ensure genetic purity of breeding lines in the working collection to develop lines with visual phenotypes associated with different quality traits for animal and human consumption markets. The use of MMAS for seed coat colour genes increases the efficiency of the development of yellow seed coat lines; determination of Y or d introgression requires an extra generation, as expression of the genes occurs in maternal tissue, and thus the true breeding genotype/phenotype is masked for a generation. Current research at the CDC involves identifying SNP markers as KASP assays for the g and $b1^{vg}$ alleles conditioning the yellow and mottled seed coat traits, respectively.

3.6 Flaxseed Anti-nutritionals

Whole and milled flaxseeds possess GRAS (generally recognized as safe) status and are used as ingredients in conventional foods (GRAS notice No. 000280). Flaxseed meal, a co-product of oil extraction, is utilized in various foods. The meal contains some residual oil and most of the health-promoting components, including dietary fibre (mucilage), protein, lignin, and other minor chemicals of the whole seed. Saskatchewan-based Bioriginal Food Science Corporation and CanMar Foods market several flaxseed and meal products for different bakery products, cereal and energy bars, and health supplements for domestic and international markets. Flax recently (2014) received a Health Canada endorsement stating that consumption of 40 g of ground flaxseed per day significantly reduces blood cholesterol levels. Further increases in the market value of flaxseed and flaxseed products are based on the health-promoting constituents and their availability; any anti-nutrients present may have a negative effect with respect to the healthy food qualities of flaxseed products.

3.6.1 Cyanogenic Glycosides

Flax naturally produces cyanogenic glycosides (CGs; especially linustatin and neolinustatin) in both its seeds and vegetative tissues and releases cyanide in response to herbivory or pathogenesis. CG levels in flaxseed are not considered toxic but may bar export of flaxseed to certain destinations. For example, enforcement of Japanese food sanitation laws has resulted in five violations occurring between 2000 and 2008 affecting products from Canada and the USA. Increasingly stringent regulations that will limit the amount of these compounds in flax are likely to restrict markets; conversely, markets for flaxseed might expand if CG content is lowered.

Information on CG content and hydrogen cyanide (HCN) production of Canadian flaxseed varieties and products is very limited. Canadian flax-importing countries such as Japan require testing of HCN and mandate levels to be <10 mg/kg in seed. The same standard level established by the FAO/WHO joint committee on food standards as safe for cassava is applied to flaxseed. In Europe, the standard for CG is 250 mg/kg for flaxseed and 350 mg/kg for flax meal for animal feed use (Directive 2002/32/EC). Current breeding efforts focus on incorporating wet chemistry methods (gas or high-performance chromatography and NMR spectroscopy) of methanol extract from whole seed to detect and quantify CGs and identify recombinant genotypes with lower CG content.

3.6.2 Cadmium (Cd)

Flax bio-accumulates Cd leading to measurable levels in the seeds (range 0.232–0.716 ppm; Table 3.6). Cadmium content in flaxseed is influenced by geography, soil type, and fertilizer usage (Oomah et al. 2007; Garrett and Thorleifson 1999; Grant and Lafond 1997). There are no maximum limits for Cd in flaxseed in Canada, for important markets such as the EU, as there are for cereals and pulses. However, buyers may set their own specifications on the concentration of Cd in flaxseed. A general awareness of healthy food choices and health-related benefits is increasing worldwide, and these are a focus for flax, especially because flaxseed (linseed) is considered a health food. Future research will focus on identification of Cd accumulation in a diverse panel (of flax accessions) grown under low- and high-Cd field conditions to determine the underlying genetic control of this trait and identify low-Cd accumulating lines to enhance breeding for reduced Cd accumulation in flax.

Table 3.6 Summary of variety composite cadmium (Cd) data for 2014

	Year	Province[a]	N samples in composite	Cd (mg/kg)
CDC arras	2014	SK	5	0.411
CDC Bethune	2014	MB	12	0.282
CDC Bethune	2014	SK	66	0.374
CDC Glas	2014	AB	5	0.669
CDC Sorrel	2014	AB	6	0.646
CDC Sorrel	2014	MB	11	0.303
CDC Sorrel	2014	SK	60	0.436
Hanley	2014	AB	6	0.475
Lightning	2014	MB	8	0.232
Prairie sapphire	2014	AB	10	0.716
Prairie sapphire	2014	SK	5	0.233
Vimy	2014	SK	15	0.556

Source: S. Tittlemier, Trace Organic Trace Element Analysis, Canadian Grain Commission, confidential comm.
[a]Prairie provinces of Canada: Alberta (AB), Saskatchewan (SK), Manitoba (MB)

3.7 Conclusion

An overview is presented on flax breeding and cultivar development in Canada, the largest producer and exporter of flax globally. Crop characteristics of economic importance are described including current knowledge and research to enhance breeding. Much progress has been made with regard to cultivar improvement considering the long breeding history of the crop in Canada. The introduction of genomics-assisted breeding has begun to bring efficiencies to flax breeding, which will be critical going forward as there is now one breeding program in Western Canada to serve the needs of producers and consumers of this valuable product. Importantly, Canada is the first country to support a health claim for flax, an oilseed crop that provides the highest plant-based source of unsaturated fatty acids. New value-added markets for flax for human health and animal nutrition will be important to revive this important oilseed crop in Canada and her export markets.

References

Aly AA, Mansour MTM, Mohamen HI, Abd-Elsalam KA (2012) Examination of correlations between several biochemical components and powdery mildew resistance of flax cultivars. Plant Pathol J 28(2):149–155

Anbessa Y, Warkentin T, Bueckert R, Vandenberg A (2006) Short internode, double podding and early flowering effects on maturity and other agronomic characters in chickpea. Field Crops Res 102:43–50

Asgarinia P, Cloutier S, Duguid SD, Rashid KY, Mirlohi AF, Banik M, Saeidi G (2013) Mapping QTL for powdery mildew resistance in flax (Linum usitatissimum L.). Crop Sci 53(6):2462–2472. https://doi.org/10.2135/cropsci2013.05.0298

Booker HM, Rowland GG, Kutcher HR, Rashid KY (2014a) CDC Neela oilseed flax. Can J Plant Sci 94:1313–1314. doi:CJPS-2014-174

Booker HM, Rowland GG, Rashid KY (2014b) CDC Glas oilseed flax. Can J Plant Sci 94:451–452. https://doi.org/10.4141/CJPS2013-158

Booker HM, Lamb EG, Smyth SJ (2017) Ex-post assessment of genetically modified, low level presence in Canadian flax. Transgenic Res 26(3):399–409

Bueckert RA, Clarke JM (2013) Review: annual crop adaptation to abiotic stress on the Canadian prairies: six case studies. Can J Plant Sci 93(3):375–385

Cross RH, McKay SAB, McHughen AG, Bonham-Smith PC (2003) Heat-stress effects on reproduction and seed set in *Linum usitatissimum* L. (flax). Plant Cell Environ 26:1013–1020

Daun JK, Marek CJ (1983) Use of gas liquid chromatography for monitoring the fatty acid composition of Canadian rapeseed. J Amer Chem Soc 60:1751–1754

Diederichesen A, Raney JP (2008) Pure-lining of flax (Linum usitatissimum L.) genebank accessions for efficiently exploiting and assessing seed character diversity. Euphytica 164(1):255–273

Diederichsen A, Richards KW (2003) Cultivated flax and the genus Linum L. – taxonomy and germplasm conservation. In: Muir A, Westcott N (eds) Flax, the genus Linum. Taylor & Francis, London, UK, pp 22–54

Diederichsen A, Kusters PM, Kessler D, Bainas Z, Gugel RK (2013) Assembling a core collection from the flax world collection maintained by Plant Gene Resources of Canada. Genet Resour Crop Evol 60:1479–1485. https://doi.org/10.1007/s10722-012-9936-1

Duguid SD (2010) Flax. In: Vollmann J, Rajcan I (eds) Oil Crops. Springer, New York, pp 233–255
Duguid SD, Rashid KY (2013) Prairie Sapphire flax. Can J Plant Sci 93:1271–1275
Duguid SD, Rashid KY, Busch H, Schaupp H (2014a) AAC Bravo flax. Can J Plant Sci 94:153–156
Duguid SD, Rashid KY, Kenaschuk EO (2014b) Shape flax. Can J Plant Sci 94:157–160
Edirisinghe P (2016) Characterization of Flax Germplasm for Resistance to Fusarium Wilt Caused
   by Fusarium oxysporum f. sp. lini. M.Sc. Thesis. Department of Plant Sciences, College of
   Agriculture and Bioresources, University of Saskatchewan, Saskatoon, Saskatchewan, 120 p
Ellis J, Dodds P, Prior T (2000) The generation of plant disease resistance gene specificities.
   Trends Plant Sci 5:373–379
Garrett RG, Thorleifson LH (1999) The provenance of Prairie tills and its importance in mineral
   exploration, p. 155–162. In: Ashton KE, Harper CT (eds) Advances in Saskatchewan geology
   and mineral exploration, Saskatchewan Geological Society, Regina, Spec. Publ. 14
Getinet A, Rakow G, Downey RK (1996) Agronomic performance and seed quality of Ethiopian
   mustard in Saskatchewan. Can J Plant Sci 76:387–392
Gill KS (1987) Linseed. Publications and information division. Indian Council of Agricultural
   Research, New Delhi, p 386
Gillis EK et al (2008) The pasmo pathogen of flax, investigation potential resistance and
   characteristics of infection. Proceedings of the 62nd flax Institute of the United States, March
   26-28, Fargo, ND, pp 78–81
Grant CA, Lafond GP (1997) The effect of fertilizer, environment and location on cadmium
   accumulation in flax. Final Report 18 pages, available at https://saskflax.com//quadrant/media/
   Research%20Reports/1997_Cadmium_in_Flaxseed_Final_Report.pdf
Hausner G, Rashid KY, Kenaschuk EO, Procunier JD (1999) The development of codominant
   PCR/RFLP based markers for the flax rust-resistance alleles at the L locus. Genome 42:1–8
Islam T (2018) Fungicide Management of Pasmo of Flax and Fungicide Sensitivity of *Septoria
   linicola*. M.Sc. Thesis. Department of Plant Sciences, College of Agriculture and Bioresources,
   University of Saskatchewan, Saskatoon, Saskatchewan, 128 p
Kumar S, You FM, Duguid S, Booker H, Rowland G, Cloutier S (2015) QTL for fatty acid
   composition and yield in linseed (*Linum usitatissimum* L.). Theor Appl Genet 128:965–984.
   https://doi.org/10.1007/s00122-015-2483-3
Lawrence GJ, Anderson PA, Dodds PN, Ellis JG (2010) Relationships between rust resistance
   genes at the M locus in flax. Molec Plant Pathol 11:19–32
Miller PR, McDonald CL, Derksen DA, Waddington J (2001) The adaptation of seven broadleaf
   crops to the dry semiarid prairie. Can J Plant Sci 81:29–43
Mittapalli O, Rowland G (2003) Inheritance of seed color in flax. Crop Sci 43:1945–1951
Mpofu SI, Rashid KY (2001) Vegetative compatibility groups within *Fusarium oxysporum* f.sp.
   lini from *Linum usitatissimum* (flax) wilt nurseries in western Canada. Can J Bot 79:836–843
Oomah B, Berekoff B, Li-Chan E, Mazza G, Kenaschuk E, Duguid S (2007) Cadmium-binding
   protein components of flaxseed: influence of cultivar and location. Food Chem 100(1):318–325
Rashid KY (2003a) Principal diseases of flax. In: Muir A, Westcott N (eds) Flax, the genus Linum.
   Taylor & Francis, London, pp 92–124
Rashid KY (2003b) Diseases of flax. In: Bailey KL, Gossen BD, Gugel RK, Morrall RAA (eds)
   Diseases of field crops in Canada. 3rd ed. Saskatoon: the Canadian Plant Phytopathological
   Society, pp. 147–154
Rashid KY (2010) Efficacy of fungicides in reducing pasmo and yield loss in flax. Proceedings of
   the 63rd flax Institute of the United States, March 25–26, 2010, Fargo, ND, pp 73–77
Rashid KY, Duguid S (2010) Promising resistance to pasmo in flax. Proceedings of the 63rd flax
   Institute of the United States, March 25–26, 2010, Fargo, ND, pp 78–81
Rashid KY, Kenaschuk EO (1994) Genetics of resistance to flax rust in six Canadian flax cultivars.
   Can J Plant Pathol 16:266–272
Ravichandran S, You FM, Rashid KY, Young L, Booker HM, Cloutier S (2017) Structural
   organization and haplotypes of rust resistance genes in flax. Joint Meeting Canadian
   Phytopathological Society and Canadian Society of Agronomy, Winnipeg, June 18–22, P3

Sackston WE (1959) Pasmo – past, present and future. Proc. 29th Flax Institute of the United States. Fargo, ND, pp 3–5

Sizov IA (1955) Flax. Selhozgiz, Moscow, pp 97–101

Sosulski FW, Gore RF (1964) The effect of photoperiod and temperature on the characteristics of flaxseed oil. Can J Plant Sci 44(4):381–382

Soto-Cerda BJ, Duguid S, Booker H, Rowland G, Diederichsen A, Cloutier S (2014) Association mapping of seed quality traits using the flax (*Linum usitatissimum* L.) core collection. Theor Appl Genet 127(4):881–896. https://doi.org/10.1007/s00122-014-2264-4

Spielmeyer W, Lagudah ES, Mendham N, Green AG (1997) Inheritance of resistance to flax wilt (*Fusarium oxysporum* f.sp. lini Schlecht) in a doubled haploid population of *Linum usitatissimum L*. Euphytica 101:287–291

Spielmeyer W, Green AG, Bittisnich D (1998) Identification of quantitative trait loci contributing to fusarium wilt resistance on an AFLP linkage map of flax (*Linum usitatissimum*). Theor Appl Genet 97:633–641

Sudarshan GP, Kulkarni M, Akhov L, Ashe P, Shaterian H, Cloutier S, Rowland G, Wei Y, Selvaraj G (2017) QTL mapping and molecular characterization of the classical *D* locus controlling seed and flower color in *Linum usitatissimum* (flax). Sci Rep 7:15751

Sun J, Young LW, Daba K, Booker HM (2018) Photoperiod sensitivity of Canadian flax cultivars and 5-azacytidine treated early flowering derivative lines, BMC Plant Biol (9)

Vasudevan A (2019), Mapping of Genomic Regions Underlying Early Flowering Trait in 'RE2', a Mutant Derived from Flax (*Linum utitatiissimum* L. Cultivar 'Royal'. M.Sc. Thesis. Department of Plant Sciences, College of Agriculture & Bioresources, University of Saskatchewan

You FM, Duguid SD, Lam I, Cloutier S, Rashid KY, Booker HM (2016) Pedigrees and genetic base of flax cultivars registered in Canada. Can J Plant Sc 96(5):837–852. https://doi.org/10.1139/cjps-2015-0337

You FM, Jia G, Xiao J, Duguid SD, Rashid KY, Booker HM, Cloutier S (2017) Genetic variability of 27 traits in a core collection of flax (*Linum usitatissimum* L.) reveal divergent selection between fibre and linseed types. Front Plant Sci 8. https://doi.org/10.3389/fpls.2017.01636

You FM, Xiao J, Li P, Yao Z, Jia G, He L, Kumar S, Soto-Cerda B, Duguid SD, Booker HM, Rashid KY, Cloutier S (2018) Genome-wide association study and selection signatures detect genomic regions associated with seed yield and oil quality in flax. Int J Mol Sci 19(8). https://doi.org/10.3390/ijms19082303

Young LW, Trouve JP, Speck A, You F, Robinson S, Rashid K, Booker H (2017) Using QTLseq to identify powdery mildew resistance loci in flax. 2nd International Symposium on Innovations in Plant and Food Science, University of Saskatchewan, Saskatoon, SK

Zhang T (2013) Characterization of the flax Core collection for earliness and canopy traits. MSc thesis, Department of Plant Sciences, College of Agriculture & Bioresources, University of Saskatchewan, 81 p

Chapter 4
The First Flax Genome Assembly

Michael K. Deyholos

4.1 Origins of the Flax Genome Sequencing Project

The original flax genome sequencing project (Wang et al. 2012) was motivated by both economic and scientific considerations. Although flax had been used to produce either oil (from linseed types) or bast fiber (from fiber types) for millennia, in the last half of the twentieth century, flax cultivation fell out of favor as flax products were replaced by cotton and synthetic alternatives (Oplinger et al. 1989; Zohary and Hopf 2000). By the end of the twentieth century, flax had therefore received comparatively little attention from molecular geneticists. Canada was at this time a leading producer of linseed (700,000 T in 2000; www.fao.org), and so various governments and organizations in Canada expressed interest in expanding flax production, particularly as an alternative to wheat and canola that had come to dominate rotations on farms in the Canadian prairies. In this context, efforts to obtain funding for genome sequencing began around 2004, with the first success coming from an unusual source, the province of Alberta's Department of Energy. This was leveraged to obtain further funding from Genome Canada, as part of the TUFGEN (Total Utilization of Flax Genomics) project. TUFGEN's objective was to develop genomic resources to accelerate improvement in the oil, fiber, and nutraceutical qualities of flax, eventually facilitating the development of a dual-purpose crop. Given the state of the art in genome assembly when the project was funded, it was not expected that the available funding would be sufficient to deliver anything other than a low-coverage, low-quality assembly, but as explained below, rapid advances in sequencing technology and assembly software allowed for a relatively complete genome sequence to be produced.

M. K. Deyholos (✉)
University of British Columbia, Okanagan, Kelowna, BC, Canada
e-mail: michael.deyholos@ubc.ca

© Springer Nature Switzerland AG 2019
C. A. Cullis (ed.), *Genetics and Genomics of Linum*, Plant Genetics and Genomics: Crops and Models 23, https://doi.org/10.1007/978-3-030-23964-0_4

From a scientific perspective, several observations were used to justify the selection of flax as a target of genome analysis. Rapid changes in flax nuclear genome content have been well-characterized in some varieties of flax (see Cullis and Cullis, this volume), and the characterization of the mechanisms underlying these changes will be broadly informative of genome biology across species (Cullis 1973; Chen et al. 2005). Furthermore, although flax is the only agriculturally relevant species within its genus, flax has been selected over millennia for two distinct traits (seed, stem fiber), providing interesting opportunities to study divergent selection (McDill et al. 2009). The genus *Linum* has also attracted the attention of eminent biologists such as Katherine Esau, Harold Flor, and Charles Darwin, who used it to study shoot apical meristems, gene-for-gene resistance, and heterostyly (Darwin 1863; Esau 1942; Flor 1955), and the genus remains a useful system for studying this least-understood pollination control mechanism (Ushijima et al. 2012). *Linum*, and indeed the entire Linaceae family, is a remarkably diverse family, in terms of both its morphology and ecology, with representatives found in tropical, subtropical, and temperate regions, and diverse forms ranging from lianas to canopy trees, and small herbaceous annuals and perennials (McDill et al. 2009; McDill and Simpson 2011). The Linaceae is placed in the Malpighiales order, along with poplar (*Populus trichocarpa*), which was the first tree species sequenced (Tuskan et al. 2006), and it was expected that the flax genome might provide an interesting comparison to poplar.

4.2   Challenges of Plant Genome Sequencing

All eukaryotic genomes contain repetitive sequences in clusters that may extend from dozens to millions of bases in length. Indeed, most of the DNA in a typical eukaryote is derived from various types of repetitive sequences, including transposable elements. This fact was established by analyses of DNA reassociation kinetics, which were conducted decades prior to the first genome sequencing projects (Waring and Britten 1966). Such analyses in flax showed that 44% of the genome was a single-copy sequence (Cullis 1981).

Whole genome sequencing projects depend on the ability to rebuild (ideally) chromosome-scale strings of the original nuclear DNA, by joining overlapping regions of the much shorter sequence reads produced by a sequencing instrument. However, if the length of a sequence read is shorter than the length of a repeat unit, it is impossible to accurately assemble regions that are rich in repeats. This is a major problem, because genomes such as flax contain thousands of regions of repetitive DNA interspersed between and within genes. More precisely, repetitive DNA limits contiguity, i.e., the length of fragments that can be successfully assembled as a continuous string. Contiguity is most often reported by a descriptive statistic called the N_{50} value, which is the length of the smallest fragment in the assembly for which that fragment and all larger fragments together contain 50% of the total assembly. Larger N_{50} values therefore generally indicate more successful assemblies.

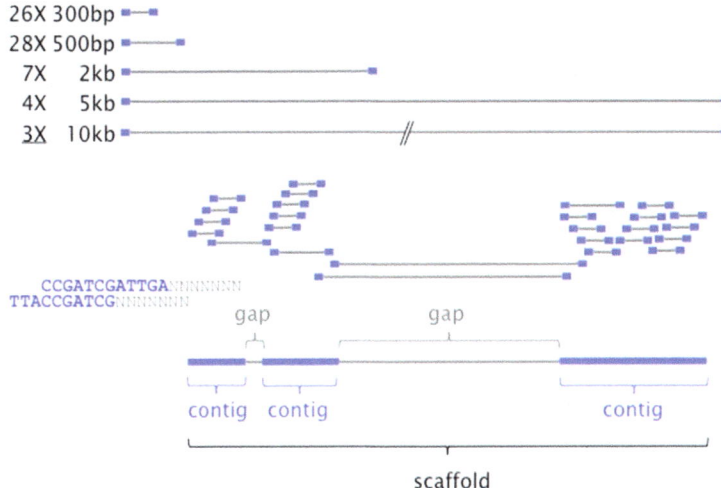

Fig. 4.1 Sequence strategy used for assembly of the flax genome. Fragments of 300 bp, 500 bp, 2 kb, 5 kb, and 10 kb were isolated and sequenced as paired ends at coverage ranging from three-fold to 28-fold. Thicker lines represent known sequence. Thin lines represent DNA of unknown sequence, but known length. Regions of known sequence can be assembled into contigs and contigs can be joined by paired-end reads into scaffolds

One way to overcome the problem of repetitive DNA is to essentially bridge repetitive or otherwise problematic regions using paired-end reads, including mate-pair technology (Fig. 4.1). As implied by the name, paired-end reads contain sequence from both ends of a DNA fragment that is longer than the reads themselves (Korbel et al. 2007). Importantly, the DNA fragment itself can be very long, and genomic DNA can be randomly sheared and size selected so that only fragments of a known length are used for as paired-end sequencing templates. Thus, paired-end technology allows fragments of a precise length to be sequenced, but only for a short distance from each end. The known sequence at each end is sufficient to allow the paired-end fragment to be incorporated into the assembly, thereby joining clusters of reads that would otherwise not be able to be connected to each other because of long repeats or other sources of ambiguity. To restate this using the terminology of genome sequencing, paired-end reads join contigs to create scaffolds. Thus, scaffolds contain regions of unknown sequence, but of known length, whereas all of the sequence within a contig is known.

4.3   Flax at a Turning Point in Genome Sequencing

Flax was approximately the thirtieth plant species for which a whole genome assembly was published, yet flax was at the forefront of a major advance in genome sequencing strategies. Twelve years before the flax genome was published, the first

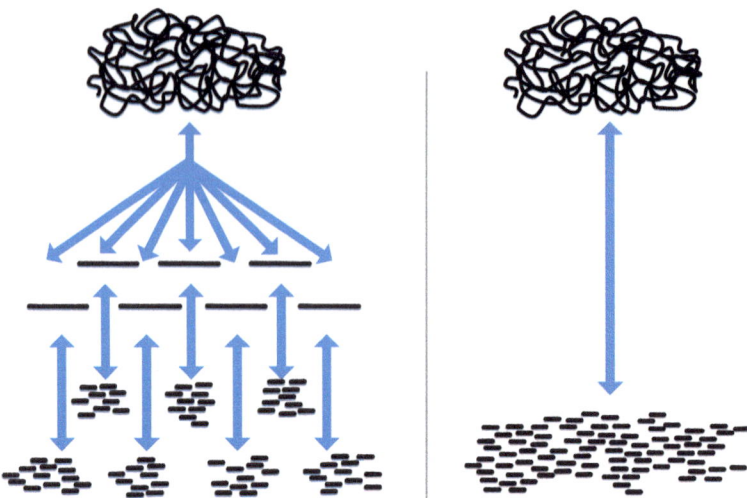

Fig. 4.2 Two strategies for genome sequencing. Clone-by-clone sequencing (on left) requires the production of a set of ordered sub-clones (e.g., BACs), which are sequenced and assembled individually. In contrast, whole genome shotgun sequencing (on right) attempts to reassemble chromosome-scale fragments directly from individual sequencing reads

whole genome sequence reported for any plant was completed, using a physical map made from fragments of the chromosomes of Arabidopsis (Arabidopsis Genome Initiative 2000, Fig. 4.2). These fragments were cloned in BACs (bacterial artificial chromosomes), and through a laborious process, a minimal set of overlapping BACs were chosen that together contained essentially all of the DNA in the Arabidopsis genome. Using Sanger technology, which produces read lengths of up to 700 bp, each BAC was sequenced independently. The BAC sequences were then assembled with each other to constitute the whole genome. In total, the mapping and sequencing of the 135 Mbp Arabidopsis genome took 10 years, at a cost of $100 million. Two separate technological advances increased the efficiency of subsequent plant genome projects (Goff et al. 2014). First, the map-based or clone-by-clone sequencing approach used in Arabidopsis was gradually replaced by whole genome shotgun (WGS) strategies in which the entire genome, rather than individual sub-fragments, was used as a template for the sequencing reaction. Second, the use of next-generation sequencing (NGS) technologies (initially Roche/454, then almost exclusively Illumina) reduced the per-base cost of sequencing by several orders of magnitude, as compared to the Sanger technique. NGS reads at the time were significantly shorter than those that could be obtained through Sanger sequencing, but this limitation could be partially overcome through the use of paired-end or mate-pair reads, as described above. Almost all of the plant genomes that preceded flax used either WGS or NGS, but not both, and most also incorporated some additional information (e.g., low-coverage physical maps, optical maps, or Sanger sequence reads). However, the flax project was the first plant genome assembly to show that WGS and NGS could be used together, exclusive of any other information, and that this combination alone was sufficient to produce assemblies with high contiguity (e.g., scaffold N_{50} values >500 kb) (Wang et al. 2012). Indeed, during the planning of the flax sequencing project, most North American experts were very doubtful that Illumina reads alone would be sufficient for assembly into anything with contiguity of more than a few kb. However, BGI pioneered this approach and demonstrated its feasibility in higher organisms, first in the sequencing of the giant panda (Li et al. 2010). The approach of using exclusively Illumina reads in a WGS assembly is now by far the most common approach used to sequence plant genomes, although additional information is still required to assemble sequences into chromosome-scale fragments.

4.4   Generating the First Flax Genome Sequence

The biological material chosen for whole genome sequencing was the variety CDC Bethune, which is a spring linseed type bred by Gordon Rowland. This was the most widely grown variety in Canada at the onset of the project and has the benefits of high yield and good lodging resistance (Rowland et al. 2002). This variety was chosen for sequencing to ensure that the outcome of the flax project would have maximum relevance to breeders and linseed producers in Canada. Furthermore, seeds with well-defined pedigrees of the variety could be obtained, specifically a line that had been self-fertilized over 19 generations, the last 8 of which were single-seed descent and could therefore be assumed to have very low heterozygosity.

The template used as input for the flax genome sequencing project was simple: over 50 mg of relatively crude DNA was extracted from surface-sterilized seeds that were germinated on sterile media, and this was shipped in Tris-EDTA at ambient temperature to BGI's sequencing facility in Shenzen, China. There, five different types of libraries were prepared from size-selected inserts of 300, 500, 2 kb, 5 kb, and 10 kb (Fig. 4.1, Wang et al. 2012). These were sequenced as paired-ends (or mate-pairs), at coverage ranging from 28× (for the 500 bp library) to just 3.1 X for the 10 kb library and using an Illumina Genome Analyzer II with read lengths of 44, 75, or 100 bp. After removing adapter sequences and low-quality reads, the final sequencing output was 35 Gbp, or over 350 million reads.

The 35 Gbp of clean sequence generated by the sequencing instruments was used as input to the SOAPdenovo assembler (Li et al. 2009), which is a software program written at BGI and critical to the success of projects like the flax genome. SOAPdenovo was one of the first programs to solve the problems created by next-generation sequencing: the short length of the reads and the large number of reads made it otherwise computationally impractical to assemble a large genome. SOAPdenovo overcame this issue by modeling the assembly problem as a de Bruijn graph. This mathematical framework is a particularly efficient way of finding overlaps between short sequences and eliminating redundancies in sequencing reads. Although previous assembly programs had also used de Bruijn graphs, SOAPdenovo was the first implementation that could produce contigs and scaffolds of high contiguity, through a four-phase process: error correction, contig assembly, scaffolding, and gap closure.

The published flax genome assembly consisted of 318 Mbp (in scaffolds) or 302 Mbp (in contigs only) (Wang et al. 2012). This assembly therefore contained approximately 85% of the actual genome, as measured by flow cytometry. The remaining 15% of the genome not captured in the assembly was presumed to consist mostly of highly repetitive sequences, such as that found around the centromeres and telomeres. These are generally ignored in genome projects. The assembly was divided between over 88, 384 scaffolds—this was considerably more than the 15 chromosomes that would result from a theoretical perfect assembly. However, most of these scaffolds were very small and did not contribute to the useful part of the assembly: only 1458 scaffolds were over 2 kb. Because shotgun assemblies often contain many small fragments that are of little practical relevance, a metric called N_{50} is commonly used to describe the size of assembly units (contigs and scaffolds). The N_{50} value for the assembly was 693 kb (scaffold), meaning that 50% of the total assembly (i.e., 159 Mbp) could be found in scaffolds 693 kb or larger and the N_{90} value (82 kb) indicated that 90% of the assembly was found in scaffolds 82 kb or larger. Because the average gene length in a plant such as Arabidopsis is under 3 kb, it is expected that an assembly with these characteristics will contain enough contiguous genes to be useful for most purposes (e.g., describing gene families, identifying regulatory elements, mapping polymorphisms, investigating evolutionary relationships, conducting reverse genetics experiments), even if it still contains thousands of unconnected scaffolds.

To validate the assembly, the scaffolds were compared to several independent sources of genomic sequence information (Wang et al. 2012). ESTs (expressed sequence tags) are Sanger-derived, partial sequences of mRNA transcripts. More than 239,000 ESTs of at least 400 bp had been reported for flax, and 93% of these aligned at high stringency (BLASTN e-value $<10^{-20}$) to the flax genome assembly, indicating that assembly was nearly complete in terms of its representation of the regions that contained genes. Separately, a few BACs and fosmids were created from an aliquot of the same DNA that had been used for Illumina sequencing. These large-insert clones were sequenced and assembled using Sanger technology and then were aligned to the WGS scaffolds. Wherever large blocks of scaffolds and BACs aligned, the nucleotides were more than 99% identical, showing that the Illumina sequencing and subsequent filtering and error correction were very accurate at the base-pair level. However, three of the seven BACs had segments that aligned best to non-contiguous regions of different scaffolds. These inconsistencies in the arrangement of large blocks of sequence indicated that the accuracy of the long-range continuity of the WGS assembly, which depends on the large insertion (e.g., 10 kb) mate-pairs, was limited. Furthermore, sequences from the ends of BACs had previously been obtained from DNA of the CDC Bethune variety, with a mean insert size of 135–150 kb; 9688 of these sequences could be aligned to a single scaffold concordantly (meaning both ends of a given BAC aligned to the same scaffold, with high confidence). For 15% of these aligned BACs, the distance on the scaffolds between the positions where the BAC end sequences aligned was much higher or lower than the expected insert size, which was further evidence that the scaffolds contained many mis-assemblies, and that further information would be required to generate an assembly with higher long-range accuracy. Nevertheless, the flax genome assembly proved to be sufficient as a resource for many subsequent studies (>200 citations to date). Considering that the cost of sequencing the flax genome was approximately $100,000 (1000× less than Arabidopsis) and the sequencing and assembly were completed in 6 weeks, the trade-off between accuracy and efficiency was justified.

4.5 Structure and Composition of the Flax Genome

At a high level, the flax genome was found to be generally similar to other sequenced eudicots (Goff et al. 2014). Using a combination of computer predictions and empirical data (e.g., ESTs), the flax genome was annotated with a predicted 43,384 genes, a number within the range of other eudicots (Wang et al. 2012). Furthermore, the distribution of exon and intron lengths in flax was very similar to five other representative eudicots tested. There was also evidence of many insertions of chloroplast DNA into the flax nuclear genome: 1356 insertions, ranging from 23 to 5899 bp in length, were identified. The frequency and distribution of these NUPT (nuclear plastid DNA) insertions is also consistent with what has been reported in other species (Richly and Leister 2004). The G + C content of the flax assembly, at 40%, was the highest of any genome reported at the time and was found also in an analysis of the BAC end sequences (Ragupathy et al. 2011). The significance of this observation is still unknown but could perhaps be related to the abundance of particular types of repetitive elements. Transposable elements constituted 23% of the flax WGS assembly; this is probably an underestimate of the content in the actual genome, since the sequencing and assembly process used is biased against transposable elements and other repetitive sequence. One-third of a eudicot genome is generally expected to contain identifiable transposable elements (Gonzalez and Deyholos 2012). The flax WGS was, however, somewhat distinct from most other genomes in the types of transposable it contained: Copia-type LTR transposons comprised 38% of all of the transposable elements (9.8% of the assembly), but Gypsy-type LTRs made up only 31% of the transposable elements (8.3% of the assembly); in most other eudicots, Gypsy elements tend to be more abundant than Copia-type.

   The predicted proteins of flax included 12 PFAM domains that were significantly overrepresented in flax as compared to 15 other plant genomes that were surveyed (Wang et al. 2012). Among the overrepresented domains were agglutinins, basic secretory proteins, alginate lyases, self-incompatibility, and GRAS family transcription factor. The genome has also been used to produce detailed characterizations of over one dozen protein families, including ABC transporters, agglutinins, beta-galactosidases, cellulose synthases, chitinases, cinnamyl alcohol dehydrogenase, orbitides, tubulins, and pectin methylesterases (Pinzón-Latorre and Deyholos 2013; Mokshina et al. 2014; Chantreau et al. 2015; Faruque et al. 2015; Burnett et al. 2016; Lane et al. 2016; Gavazzi et al. 2017; Yurkevich et al. 2017; Preisner et al. 2018). The success of these gene family characterization projects is evidence of the utility of the flax genome assembly.

References

Burnett PGG, Olivia CM, Okinyo-Owiti DP, Reaney MJT (2016) Orbitide composition of the Flax Core Collection (FCC). J Agric Food Chem 64(25):5197–5206

Chantreau M, Chabbert B, Billiard S, Hawkins S, Neutelings G (2015) Functional analyses of cellulose synthase genes in flax (Linum usitatissimum) by virus-induced gene silencing. Plant Biotechnol J 13(9):1312–1324

Chen YM, Schneeberger RG, Cullis CA (2005) A site-specific insertion sequence in flax genotrophs induced by environment. New Phytol 167(1):171–180

Cullis C (1973) DNA differences between flax genotrophs. Nature 243:515–516

Cullis CA (1981) DNA-sequence organization in the flax genome. Biochim Biophys Acta 652(1):1–15

Darwin C (1863) On the existence of two forms, and on their reciprocal sexual relations, in several species of the genus Linum. J Linn Soci Bot 7:69–83

Esau K (1942) Vascular differentiation in the vegetative shoot of Linum. I. The procambium. Am J Bot 29(9):738–747

Faruque K, Begam R, Deyholos MK (2015) The Amaranthin-Like Lectin (LuALL) genes of flax: a unique gene family with members inducible by Defence hormones. Plant Mol Biol Report 33(3):731–741

Flor HH (1955) Host-parasite interaction in flax rust—its genetics and other implications. Phytopathology 45:680–685

Gavazzi F, Pigna G, Braglia L, Giani S, Breviario D, Morello L (2017) Evolutionary characterization and transcript profiling of beta-tubulin genes in flax (Linum usitatissimum L.) during plant development. BMC Plant Biol 17:237

Goff SA, Schnable JC, Feldmann KA (2014) The evolution of plant gene and genome sequencing. Advances in Botanical Research 69:47–90

Gonzalez LG, Deyholos MK (2012) Identification, characterization and distribution of transposable elements in the flax (Linum usitatissimum L.) genome. BMC Genomics 13:644

Initiative AG (2000) Analysis of the genome sequence of the flowering plant Arabidopsis thaliana. Nature 408:796–815

Korbel JO, Urban AE, Affourtit JP, Godwin B, Grubert F, Simons JF, Kim PM, Palejev D, Carriero NJ, Du L, Taillon BE, Chen ZT, Tanzer A, Saunders ACE, Chi JX, Yang FT, Carter NP, Hurles ME, Weissman SM, Harkins TT, Gerstein MB, Egholm M, Snyder M (2007) Paired-end mapping reveals extensive structural variation in the human genome. Science 318(5849):420–426

Lane TS, Rempe CS, Davitt J, Staton ME, Peng YH, Soltis DE, Melkonian M, Deyholos M, Leebens-Mack JH, Chase M, Rothfels CJ, Stevenson D, Graham SW, Yu J, Liu T, Pires JC, Edger PP, Zhang Y, Xie YL, Zhu Y, Carpenter E, Wong GKS, Stewart CN (2016) Diversity of ABC transporter genes across the plant kingdom and their potential utility in biotechnology. BMC Biotechnol 16:47

Li R, Yu C, Li Y, Lam T-W, Yiu S-M, Kristiansen K, Wang J (2009) SOAP2: an improved ultrafast tool for short read alignment. Bioinformatics 25(15):1966–1967

Li RQ, Fan W, Tian G, Zhu HM, He L, Cai J, Huang QF, Cai QL, Li B, Bai YQ, Zhang ZH, Zhang YP, Wang W, Li J, Wei FW, Li H, Jian M, Li JW, Zhang ZL, Nielsen R, Li DW, Gu WJ, Yang ZT, Xuan ZL, Ryder OA, Leung FCC, Zhou Y, Cao JJ, Sun X, Fu YG, Fang XD, Guo XS, Wang B, Hou R, Shen FJ, Mu B, Ni PX, Lin RM, Qian WB, Wang GD, Yu C, Nie WH, Wang JH, Wu ZG, Liang HQ, Min JM, Wu Q, Cheng SF, Ruan J, Wang MW, Shi ZB, Wen M, Liu BH, Ren XL, Zheng HS, Dong D, Cook K, Shan G, Zhang H, Kosiol C, Xie XY, Lu ZH, Zheng HC, Li YR, Steiner CC, Lam TTY, Lin SY, Zhang QH, Li GQ, Tian J, Gong TM, Liu HD, Zhang DJ, Fang L, Ye C, Zhang JB, Hu WB, Xu AL, Ren YY, Zhang GJ, Bruford MW, Li QB, Ma LJ, Guo YR, An N, Hu YJ, Zheng Y, Shi YY, Li ZQ, Liu Q, Chen YL, Zhao J, Qu N, Zhao SC, Tian F, Wang XL, Wang HY, Xu LZ, Liu X, Vinar T, Wang YJ, Lam TW, Yiu SM, Liu SP, Zhang HM, Li DS, Huang Y, Wang X, Yang GH, Jiang Z, Wang JY, Qin N, Li L, Li JX, Bolund L, Kristiansen K, Wong GKS, Olson M, Zhang XQ, Li SG, Yang HM, Wang J, Wang J (2010) The sequence and de novo assembly of the giant panda genome. Nature 463(7279):311–317

McDill J, Repplinger M, Simpson BB, Kadereit JW (2009) The phylogeny of Linum and Linaceae subfamily Linoideae, with implications for their systematics, biogeography, and evolution of heterostyly. Syst Bot 34(2):386–405

McDill JR, Simpson BB (2011) Molecular phylogenetics of Linaceae with complete generic sampling and data from two plastid genes. Bot J Linn Soc 165(1):64–83

Mokshina N, Gorshkova T, Deyholos MK (2014) Chitinase-Like (CTL) and Cellulose Synthase (CESA) gene expression in gelatinous-type cellulosic walls of flax (Linum usitatissimum L.) Bast fibers. PLoS One 9(6):e97949

Oplinger ES, Oelke EA, Doll JD, Bundy LG, Schuler RT (1989) Flax. Alternative field crops manual. University of Minnesota, St. Paul, MN

Pinzón-Latorre D, Deyholos MK (2013) Characterization and expression of the pectin methylesterase (PME) and pectin methylesterase inhibitor (PMEI) gene families in flax (Linum usitatissimum). BMC Genomics 14:742

Preisner M, Wojtasik W, Kostyn K, Boba A, Czuj T, Szopa J, Kulma A (2018) The cinnamyl alcohol dehydrogenase family in flax: differentiation during plant growth and under stress conditions. J Plant Physiol 221:132–143

Ragupathy R, Rathinavelu R, Cloutier S (2011) Physical mapping and BAC-end sequence analysis provide initial insights into the flax (*Linum usitatissimum* L.) genome. BMC Genomics 12(1):217

Richly E, Leister D (2004) NUPTs in sequenced eukaryotes and their genomic organization in relation to NUMTs. Mol Biol Evol 21(10):1972–1980

Rowland GG, Hormis YA, Rashid KY (2002) CDC Bethune flax. Can J Plant Sci 82(1):101–102

Tuskan GA, DiFazio S, Jansson S, Bohlmann J, Grigoriev I, Hellsten U, Putnam N, Ralph S, Rombauts S, Salamov A, Schein J, Sterck L, Aerts A, Bhalerao RR, Bhalerao RP, Blaudez D, Boerjan W, Brun A, Brunner A, Busov V, Campbell M, Carlson J, Chalot M, Chapman J, Chen GL, Cooper D, Coutinho PM, Couturier J, Covert S, Cronk Q, Cunningham R, Davis J, Degroeve S, Dejardin A, Depamphilis C, Detter J, Dirks B, Dubchak I, Duplessis S, Ehlting J, Ellis B, Gendler K, Goodstein D, Gribskov M, Grimwood J, Groover A, Gunter L, Hamberger B, Heinze B, Helariutta Y, Henrissat B, Holligan D, Holt R, Huang W, Islam-Faridi N, Jones S, Jones-Rhoades M, Jorgensen R, Joshi C, Kangasjarvi J, Karlsson J, Kelleher C, Kirkpatrick R, Kirst M, Kohler A, Kalluri U, Larimer F, Leebens-Mack J, Leple JC, Locascio P, Lou Y, Lucas S, Martin F, Montanini B, Napoli C, Nelson DR, Nelson C, Nieminen K, Nilsson O, Pereda V, Peter G, Philippe R, Pilate G, Poliakov A, Razumovskaya J, Richardson P, Rinaldi C, Ritland K, Rouze P, Ryaboy D, Schmutz J, Schrader J, Segerman B, Shin H, Siddiqui A, Sterky F, Terry A, Tsai CJ, Uberbacher E, Unneberg P, Vahala J, Wall K, Wessler S, Yang G, Yin T, Douglas C, Marra M, Sandberg G, Van de Peer Y, Rokhsar D (2006) The genome of black cottonwood, Populus trichocarpa (Torr. & Gray). Science 313(5793):1596–1604

Ushijima K, Nakano R, Bando M, Shigezane Y, Ikeda K, Namba Y, Kume S, Kitabata T, Mori H, Kubo Y (2012) Isolation of the floral morph-related genes in heterostylous flax (Linum grandiflorum): the genetic polymorphism and the transcriptional and post-transcriptional regulations of the S locus. Plant J 69(2):317–331

Wang Z, Hobson N, Galindo L, Zhu S, Shi D, McDill J, Yang L, Hawkins S, Neutelings G, Datla R, Lambert G, Galbraith DW, Grassa CJ, Geraldes A, Cronk QC, Cullis C, Dash PK, Kumar PA, Cloutier S, Sharpe AG, Wong GKS, Wang J, Deyholos MK (2012) The genome of flax (Linum usitatissimum) assembled de novo from short shotgun sequence reads. Plant J 72(3):461–473

Waring M, Britten RJ (1966) Nucleotide sequence repetition – a rapidly REASSOCIATING fraction of mouse DNA. Science 154(3750):791

Yurkevich OY, Kirov IV, Bolsheva NL, Rachinskaya OA, Grushetskaya ZE, Zoschuk SA, Samatadze TE, Bogdanova MV, Lemesh VA, Amosova AV, Muravenko OV (2017) Integration of physical, genetic, and cytogenetic mapping data for Cellulose Synthase (CesA) genes in flax (Linum usitatissimum L.). Front Plant Sci 8

Zohary D, Hopf M (2000) Domestication of plants in the Old World: the origin and spread of cultivated plants in West Asia, Europe and the Nile Valley. Oxford University Press, Oxford, UK

Chapter 5
Assembly of the Flax Genome into Chromosomes

Frank M. You and Sylvie Cloutier

Flax (*Linum usitatissimum* L., $2n = 2x = 30$) has two main products: its seed, rich in nutritional omega-3 fatty acids, and its strong and long-lasting high tensile strength fibre extracted from the straw. Identification and utilization of the genes affecting the yield and quality of these bioproducts are expected to contribute to the improvement of flax and other oil- and fibre-producing species. Genomic resources, especially a complete flax genome reference sequence, promise to facilitate this process. The flax genome was sequenced, and the first draft reference sequence was released in 2012 (Wang et al. 2012). Also, a large number of flax genomic resources have been produced, such as bacterial artificial chromosome (BAC) libraries, a BAC-based physical map, BAC-end sequences (BES) (Ragupathy et al. 2011), simple sequence repeat (SSR) marker-based consensus genetic map (Cloutier et al. 2012), high-density single nucleotide polymorphism (SNP) marker-based genetic map and a BioNano genome optical map (You et al. 2018). Integration of these genomic resources assisted in the validation and ordering of the draft reference sequence into chromosomes, resulting in a higher-quality flax reference sequence, useful for genome evolution studies, comparative genomic analyses, genetic research and molecular breeding (You et al. 2018).

F. M. You (✉) · S. Cloutier
Ottawa Research and Development Centre, Agriculture and Agri-Food Canada,
Ottawa, ON, Canada
e-mail: frank.you@canada.ca

© Springer Nature Switzerland AG 2019
C. A. Cullis (ed.), *Genetics and Genomics of Linum*, Plant Genetics and
Genomics: Crops and Models 23, https://doi.org/10.1007/978-3-030-23964-0_5

5.1 *Linum* Genomes

5.1.1 *Chromosome Numbers*

The genus *Linum* L. belongs to the Linaceae family which comprises approximately 200 diploid species, 75 of which were surveyed for their chromosome numbers that varied greatly with $n = 7, 8, 9, 10, 12, 13, 14, 15, 16, 18, 19, 21, 30, 32, 36, 42$ and 43 (Table 5.1) (Goldblatt 2007; Rice et al. 2014). The numbers $n = 9$ and 15 predominate. In addition, B chromosomes (or supernumerary or accessory chromosomes) were discovered in the lineages *L. capitatum*, *L. flavum* and *L. tauricum* (Nosova 2005; Nosova et al. 2005). The cultivated (*L. usitatissimum* L.) and wild flax (*L. bienne*) possess the same haploid chromosome number of $n = 15$.

Table 5.1 Chromosome numbers of species of the genus *Linum* of the Linaceae family

	Species	Haploid (n)	Diploid ($2n$)		Species	Haploid (n)	Diploid ($2n$)
1	*Linum acuticarpum*	15	30	39	*Linum hologynum*	21	42
2	*Linum africanum*	15	30	40	*Linum julicum*	9	18
3	*Linum album*	14	28	41	*Linum komarovii*	9	18
4	*Linum alexeenkoanum*		27,36	42	*Linum leonii*	9,10	18,20
5	*Linum alpinum*	9	18	43	*Linum lewisii*	9	18
6	*Linum altaicum*	9	18	44	*Linum lundellii*	15	30
7	*Linum amurense*	9	18	45	*Linum maritimum*	10	20
8	*Linum angustifolium*	15	30	46	*Linum mesostylum*	9	18
9	*Linum arboreum*	14	28	47	*Linum monogynum* var. *chathamicum*	42,43	84,86
10	*Linum aristatum*	15	30	48	*Linum mucronatum*	14	28
11	*Linum aroanium*	18	36	49	*Linum mysorense*	30	60
12	*Linum arounicum*	18	36	50	*Linum narbonense*	7,9,14	18,20,28,30
13	*Linum austriacum*	9	18	51	*Linum nervosum*	9	18
14	*Linum austriacum* subsp. *austriacum*	9	18	52	*Linum nodiflorum*	12,13	24,26
15	*Linum austriacum* subsp. *collinum*	9	18	53	*Linum pallescens*	9	18
16	*Linum austriacum* subsp. *euxinum*	9	18	54	*Linum perenne*	9	18

(continued)

Table 5.1 (continued)

	Species	Haploid (n)	Diploid (2n)		Species	Haploid (n)	Diploid (2n)
17	Linum austriacum var. mauritanicum	9	18	55	Linum perenne subsp. alpinum	9	18
18	Linum baicalense	9	18	56	Linum perenne subsp. extraaxillare	9	18
19	Linum bienne	15	30	57	Linum phitosianum	14	28
20	Linum caespitosum	14	28	58	Linum puberulum	16	32
21	Linum capitatum		28 + 1~3B	59	Linum pubescens	8,9	18,16
22	Linum catharticum	8	16	60	Linum pycnophyllum	9	18
23	Linum catharticum subsp. catharticum	8	16	61	Linum salsoloides	9	18
24	Linum comptonii	15	30	62	Linum setaceum	9	18
25	Linum corymbulosum	9	18	63	Linum stelleroides	9,10	18,20
26	Linum crepitans	15	30	64	Linum strictum	9,16	18,32
27	Linum decumbens	8,16	18,32	65	Linum strictum subsp. strictum	18	36
28	Linum elegans	14,15	30,28	66	Linum suffruticosum	18,36	36,72
29	Linum flavum		28 + 1~3B	67	Linum tauricum		28 + 1~3B
30	Linum flavum subsp. flavum	14	28	68	Linum tenue	10	20
31	Linum gracile	15	30	69	Linum tenuifolium	9,18	18,36
32	Linum grandiflorum	8	16	70	Linum trigynum	10	20
33	Linum gyaricum		30 + 1B	71	Linum usitatissimum	15	30
34	Linum gyaricum subsp. icaricum	14	28	72	Linum usitatissimum var. linifolia	15	30
35	Linum hellenicum	14	28	73	Linum usitatissimum var. neelum	15	30
36	Linum heterostylum	15	30	74	Linum vernale	15	30
37	Linum hirsutum	8	16	75	Linum viscosum	8	16
38	Linum hirsutum subsp. hirsutum	8	16				

Data sources: http://www.tropicos.org/Project/IPCN (Goldblatt 2007) and http://ccdb.tau.ac.il (Rice et al. 2014)

5.1.2   Genome Size

The reported size of the flax genome varies with different methods. The haploid genome was estimated to be 685 Mb by calculating DNA content (0.7 pg/1C nucleus) through direct staining of the nuclei (Evans et al. 1972), 350 Mb by reassociation kinetics analysis (Cullis 1981) and 536 Mb by cytometric DNA analysis (Marie and Brown 1993), respectively. Recently, the genome size was estimated to be approximately 373 Mb by flow cytometry analysis (Wang et al. 2012) which is similar to the estimate of 368 Mb obtained based on the BAC-based physical map (Ragupathy et al. 2011). These more recent reports may be more reliable because they are based on methods that were successfully applied to estimate the genome size of other plant species (Chen et al. 2002; Dolezel et al. 2007). A more recent estimation of the size of the flax genome is 317 Mb, a value based on its optical BioNano genome map (You et al. 2018).

5.1.3   Genome Evolution

Polyploidy, a heritable condition of possessing more than two homologous sets of chromosomes, is an evolutionary force that leads to the diversification of plants and shapes their karyotypes (Otto and Whitton 2000). The chromosome number diversity in the *Linum* genus implies chromosomal mutation events in the speciation of this taxon, e.g. polyploidizaion by whole genome duplications, chromosome rearrangements and possible instances of aneuploid reductions or increases (Ray 1944; Rogers 1982).

Evidence in the lineage of cultivated flax and its close relatives showed that it may have undergone two polyploidization events (Figs. 5.1 and 5.2). The first ancient paleoploidization event may have occurred 20–40 million years ago (MYA). Cultivated flax C-banding karyotypes exhibited similarities to the wild relatives *L. austriacum* L. ($n = 9$) and *L. grandiflorum* Desf. ($n = 8$), indicating that it may have originated from an ancient polyploidization event involving both or either species followed by subsequent chromosome losses and/or rearrangements (Muravenko et al. 2001; Muravenko et al. 2003). Next-generation sequencing (NGS) of transcriptomes of 11 *Linum* species was performed using the Illumina platform (Sveinsson et al. 2014). The short reads were de novo assembled to find signatures of polyploidization events from the divergence age distribution of gene paralogs. The results of such analysis revealed a paleopolyploidization event dating 20–40 MYA, which was specific to the lineage of cultivated flax (Fig. 5.1) (Sveinsson et al. 2014). Divergence rates between duplicate gene pairs as measured by Ks, which is calculated as nucleotide substitutions per synonymous site per year, confirmed this estimate in cultivated flax. The small peak observed at approximately 23 MYA is indicative of an ancient whole genome duplication (WGD) (Fig. 5.2) (You et al. 2018).

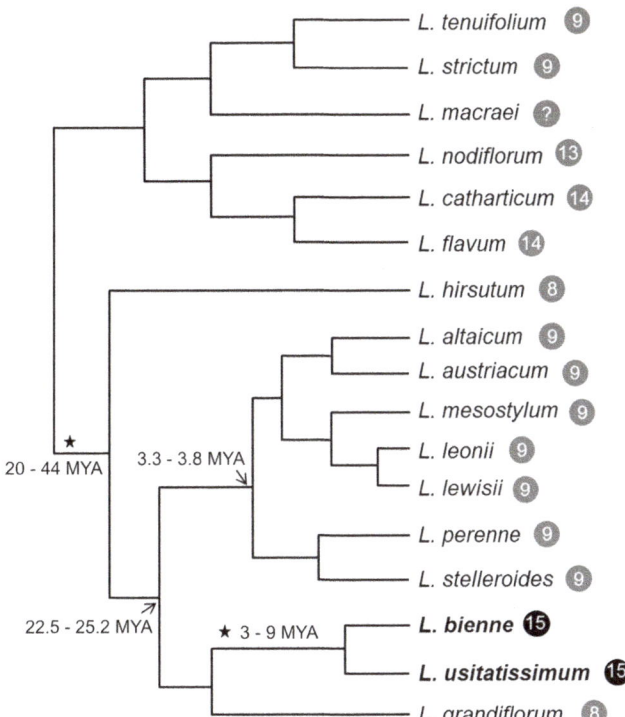

Fig. 5.1 Cladogram of flax (*L. usitatissimum*) and other *Linum* species showing haploid chromosome numbers inside grey or black circles. Stars indicate the approximate placement of the whole-genome duplication that occurred in the common ancestor of cultivated (*L. usitatissimum*) and wild (*L. bienne*) flax. (The tree was drawn based on the integration of data from McDill et al. (2009), Wang et al. (2012), Sveinsson et al. (2014), Fu et al. (2016) and You et al. (2018). Source: You et al. (2018))

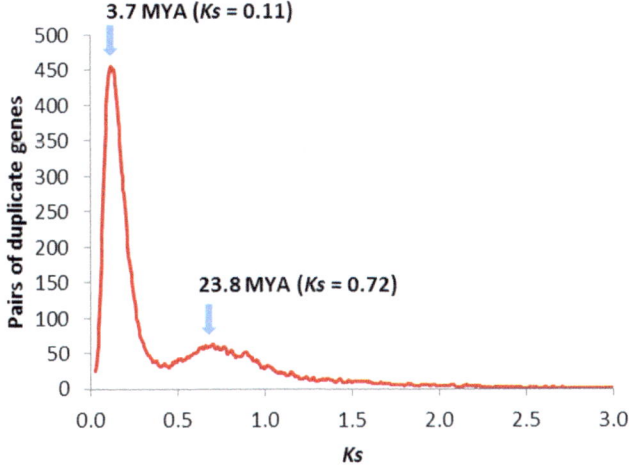

Fig. 5.2 Rates of substitution per synonymous site (*Ks*) within duplicated gene pairs showing one most recent whole genome duplication (WGD) event at approximately 3.7 million year ago (MYA) and one ancient WGD event at approximately 23.8 MYA (You et al. 2018)

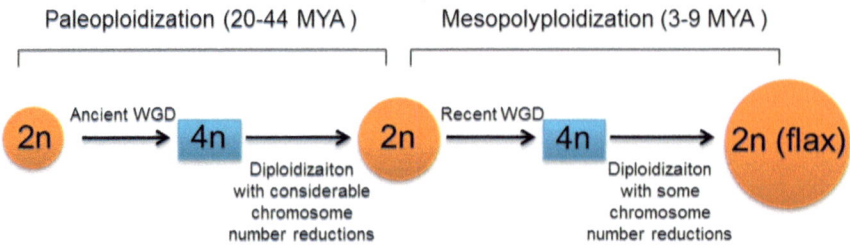

Fig. 5.3 Deduced evolutionary history of flax. MYA: million years ago. (Source: You et al. (2018))

A more recent WGD event has been proposed to have occurred around 5–9 MYA based on data from the whole genome shotgun (WGS) sequence of the flax genome (Wang et al. 2012) or 3.7 MYA based on the Ks of gene pairs in the sorted flax genome (Fig. 5.2) (You et al. 2018). The results from the sorted chromosome-based pseudomolecules suggest that flax has undergone palaeopolyploidization and meso-polyploidization events, followed by rearrangements and deletions or a fusion of chromosome arms from its ancient progenitor having a haploid chromosome number of 8 (Figs. 5.1 and 5.3) (You et al. 2018).

5.2 The First Version of the Flax Genome Sequence

The first version of the flax genome reference sequence has been generated for flax cultivar CDC Bethune (Rowland et al. 2002), using a WGS sequencing strategy based on the Illumina sequencing platform (Wang et al. 2012). Briefly, 25.88 Gb short reads, equivalent to 69× genome coverage, were obtained from seven paired-end and mate-pair libraries ranging in length from 300 bp to 10 Kb. These reads were assembled into 116,602 contigs (302 Mb, 81%) which were then scaffolded into 88,384 super contigs (a number of them representing scaffolds) totalling 318 Mb and representing ~85% of the flax genome, calculated using the 373 Mb estimate obtained by flow cytometry. The N50 sizes (minimum length of contigs and/or scaffolds representing 50% of the assembly) of contigs and scaffolds were 20.1 and 693.5 Kb, respectively. The longest contig was 151.8 Kb and the longest scaffold was 3.09 Mb. Details of the flax WGS assembly are summarized in Table 5.2.

The genome sequence was annotated for protein-coding genes and other elements, including transposable elements (TEs) and functional non protein-coding RNAs (ncRNAs such as tRNAs, rRNAs and miRNAs). A total of 43,384 protein-coding genes were predicted from the assembly, and up to 93% of the published flax ESTs and 86% of *A. thaliana* genes could be aligned to these predicted genes. Therefore, the WGS assembly could be qualified as providing excellent coverage of the gene space. The TEs occupied a total of 73.8 Mb (24.4% of WGS contig assembly), with LTR-type retrotransposons representing the highest proportion (18.4%)

Table 5.2 Statistics of the WGS assembly of the reference genome of flax

	Scaffold length (bp)	No. of scaffolds
N_{90}	82,101	601
N_{50}	693,492	132
Longest	3,087,368	1
≥ 1 kb	300,552,207	3852
With genes	293,519,352	1028
With genes (≥ 1 kb)	293,484,382	980
Total (≥ 100 bp)	318,247,816	88,384

Source: Wang et al. (2012) with some modifications

followed by DNA transposons (3.8%) and long interspersed elements (LINEs; 2.2%). For ncRNA, 297 putative miRNA precursor loci with similarity to known miRNAs, more than 1000 copies of both tRNAs and 5S rRNAs and an unexpectedly low number of 45S rRNAs were identified.

5.3 Improvement of the Flax Draft Reference Sequence

While the flax WGS assembly has achieved a landmark for initial insights into its whole genome, the sequence is still under the "draft" moniker because this sequence information remains highly fragmented. Moreover, comparisons with independently sequenced BACs and fosmids showed some mis-assemblies on a genome-wide scale (Wang et al. 2012). Therefore, ongoing efforts are being made to improve upon the draft reference genome to obtain a higher-quality assembly by taking advantage of available genomic resources as well as new technologies.

5.3.1 Genomic Resources for Improving the Assembly of the Reference Sequence

5.3.1.1 Optical BioNano Genome Map

Optical mapping, now acknowledged as the next-generation mapping method, has been used to improve de novo plant genome assemblies of maize, *Medicago*, *Amborella* and wheat (Zhou et al. 2009; Young et al. 2011; Chamala et al. 2013; Stankova et al. 2016). BioNano Genomics (San Diego, CA) has developed the Irys system, a commercial platform, to construct genome-wide optical maps for a wide range of organisms.

An optical BioNano genome (BNG) map for the flax cultivar CDC Bethune genome was constructed with the aim of improving the flax draft sequence (Table 5.3) (You et al. 2018). Briefly, high molecular weight DNA, isolated from

Table 5.3 Statistics of raw data and contig assembly for BioNano genome map

	No. molecules/contig	Total size	Genome coverage (×)	N50 (Mb)	Longest molecule/contig (Mb)
Raw molecules (≥180 kb)	330,035	82 Gb	200	0.25	2.35
De novo assembly	251	317 Mb	0.85	2.15	7.02
Mapped by scaffolds	211	298.6 Mb	0.80	2.38	7.02
Scaffolded by scaffold sequences	148	298.6 Mb	0.80	3.36	11.47
Scaffolded by BAC contigs	94	298.6 Mb	0.80	6.64	14.83

Source: You et al. (2018)

young leaves of CDC Bethune, was nicked by the endonuclease Nt.*Bsp*QI and labelled at specific sequence motifs. The labelled DNA molecules were then stained, loaded onto the nanochannel array of an IrysChip and imaged automatically by the Irys system. A spectrum of 82 Gb (~200× genome equivalent) raw DNA fragments were collected reflecting a "fingerprint" or "barcode" for each molecule. The image files were converted into BNX files by the AutoDetect software to obtain the basic labelling and DNA length information. The raw DNA molecules in BNX format were aligned, clustered and assembled into contigs using the BioNano Genomics assembly pipeline, and a consensus optical map was derived. The initial optical map was assessed for potential chimera and was refined before the construction of the final BNG map. The map consists of 251 contigs totalling 317 Mb (86%) of the flax genome, with a greatly improved N50 of 2.15 Mb (Table 5.3) compared to the 693.5 Kb of the scaffolds of the draft sequence (Wang et al. 2012). Therefore, the flax BNG map provides a relatively long backbone to guide the arrangement of the flax draft sequence scaffolds.

5.3.1.2 BAC-Based Physical Map and BAC-End Sequences (BESs)

A physical map describes an ordered set of the large genomic DNA inserts, such as BACs. BESs refer to the bidirectional end sequences of the DNA inserts which have been used as sequence tag connectors for the assembly of the whole genome shot-gun sequence of the human genome (Venter et al. 1996). Two BAC libraries (digested by restriction enzymes *Hind*III and *Bam*HI, respectively) were constructed for flax cultivar CDC Bethune which consisted of 92,160 BAC clones. A total of 43,776 of these BAC clones were fingerprinted and assembled into overlapping clones, called BAC contigs. This BAC-based physical map consists of 416 BAC contigs and includes 29,027 BACs, estimating a genome size of approximately 368 Mb (Ragupathy et al. 2011). In addition, all fingerprinted BAC clones were end-sequenced to generate 81,582 high-quality BESs averaging 679 bp (Ragupathy et al. 2011). BESs combined with marker information of the consensus genetic map

(Cloutier et al. 2012) were used to anchor the BAC contigs of the physical map onto the consensus genetic map. The physical map and its BESs are informative for ordering and refining the flax draft reference sequence (Wang et al. 2012).

5.3.1.3 Genetic Maps

A genetic map, also called linkage map, describes the relative position of gene loci or markers on chromosomes or linkage groups as estimated by recombination frequencies. Construction of a genetic map requires a bi-parental mapping population and markers such as SSRs and SNPs. Integration of genetic and physical maps assists the ordering of sequence scaffolds and contigs. As detailed in Chap. 7, an SSR-based consensus genetic map consisting of 770 markers integrated from three independent genetic maps has been developed (Cloutier et al. 2012). In addition, a high-density genetic map based on SNP markers has also been constructed (You et al. 2016) by taking advantage of a genotyping by sequencing strategy (GBS) using the Illumina sequencing platform. A combined flax consensus map based on the integration of common SNPs between the three mapping populations, population-specific SNPs and SSRs of the consensus map (Cloutier et al. 2012) has been constructed. These genetic maps combined with the SSR-based consensus map can be used to align the BNG and BAC contigs and scaffold sequences and assign them to chromosomes.

5.3.2 Strategy for Validation and Arrangement of the Draft Sequence

Two main processes are involved in sorting draft scaffold sequences. The first consists of identifying and refining potential assembly errors in scaffolds, while the second addresses the anchoring of the refined scaffolds to individual chromosomes to generate pseudomolecules. The procedure is separated into seven steps (Fig. 5.4).

5.3.2.1 Validation and Correction

The flax BNG map was used to validate and correct the scaffolds, assisted by the BAC-based physical map for further confirmation. First, the scaffold sequences were in silico digested with Nt.*Bsp*QI, the restriction enzyme used in BioNano genome mapping, generating an in silico map of the scaffolds. Then, the in silico scaffold map was aligned against the BNG map using the IrysView software (BioNano Genomics). The BNG map is believed to be quite accurate because its robustness has been demonstrated in other genomes (Shearer et al. 2014; Stankova et al. 2016). If a scaffold matches completely within, or partly overlaps a unique BNG contig, the scaffold or its matching part is deemed validated, and its

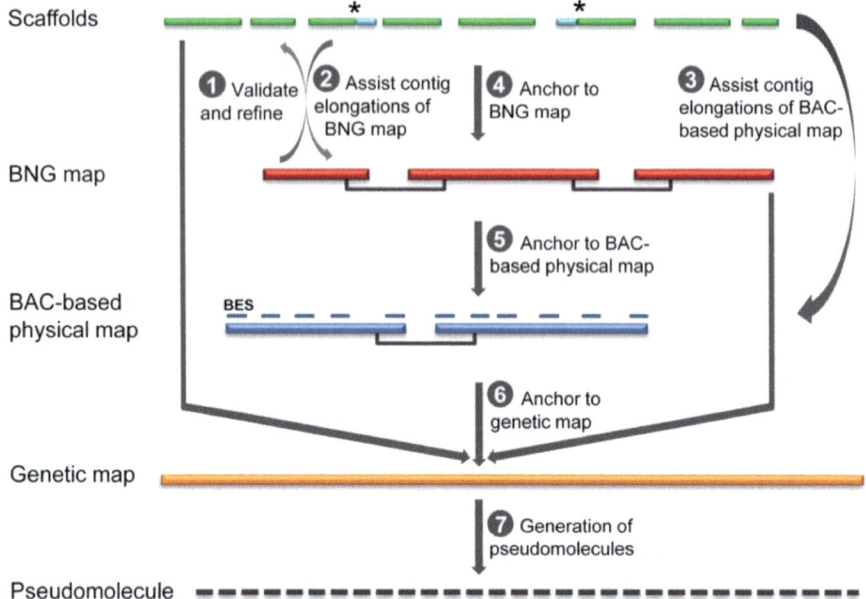

Fig. 5.4 Strategy for anchoring flax scaffolds to chromosomes and for the generation of pseudo-molecules. BNG BioNano genome, BAC bacterial artificial chromosome, BES BAC-end sequence; *indicates mis-assembly sites. (Source: You et al. (2018))

orientation could also be determined (Fig. 5.5a). Conflicts between maps are likely indicative of mis-assemblies in scaffolds, hence needing refinement (Fig. 5.5c). For example, approximately one third (~80Kb) of scaffold422 (243.3 Kb) significantly matched with BNG28 (from 2.18~2.26 Mb) (Fig. 5.6a), while the remaining (~140Kb) matched with BNG189 (from 113.6~248.7 Kb) (Fig. 5.6b), indicating that scaffold422 was mis-assembled and contained two discontiguous fragments that should be reassigned. The BAC-based physical map also confirmed that scaffold422 belonged to two separate BAC contigs with one segment corresponding to contig545 and the other to contig121. A total of 187 (138.2 Mb) out of the 622 scaffolds (282.6 Mb) of the flax WGS assembly having matches to the BNG map displayed such mis-assembly errors and, they were separated into new refined scaffolds using the BioNanoMapper pipeline (You et al. 2018).

5.3.2.2 Prolongation of BioNano Contigs Assisted by Scaffolds in Silico Maps

The scaffold in silico map can be used to extend BioNano contigs if a validated scaffold spans two BioNano contigs (Fig. 5.6b). Here, only the *in silico* maps of the refined scaffolds can be used. For example, scaffold280 aligned completely to five BAC contigs and two BNG contigs, confirming that it was correctly assembled

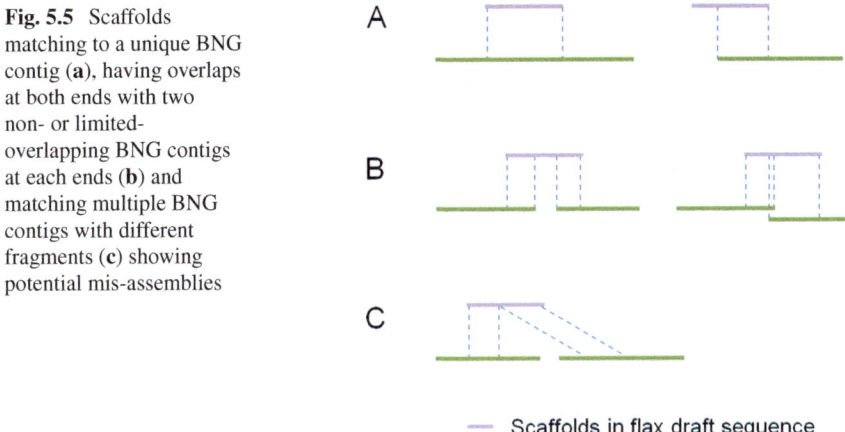

Fig. 5.5 Scaffolds matching to a unique BNG contig (**a**), having overlaps at both ends with two non- or limited-overlapping BNG contigs at each ends (**b**) and matching multiple BNG contigs with different fragments (**c**) showing potential mis-assemblies

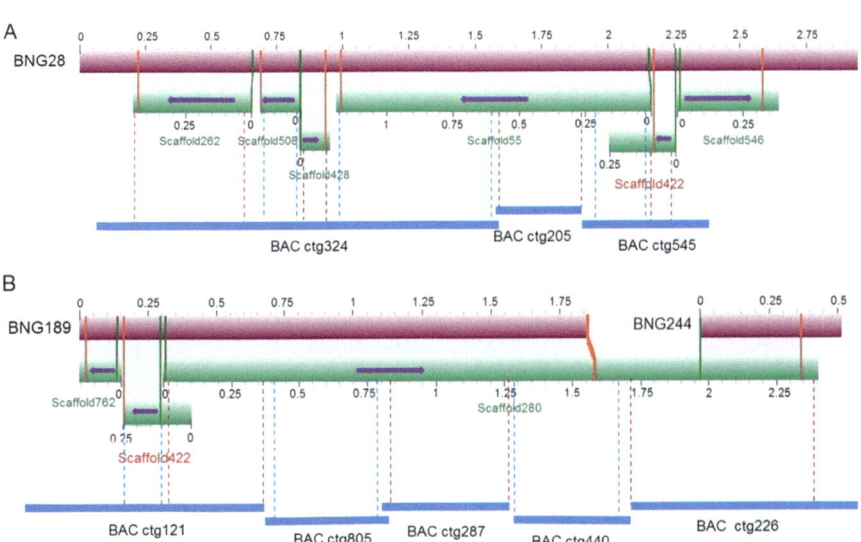

Fig. 5.6 Validation and ordering of the flax scaffolds using the BNG and the BAC-based physical maps. Five scaffolds (scaffold262, 508, 428, 55 and 546) match to the BNG contig BNG28 with significant confidence, while only one third of scaffold422 aligned to BNG28, implying a mis-assembly. BAC contigs ctg324, ctg205 and ctg545 confirmed the structural organization of BNG28 (**a**). The remaining two thirds of scaffold422 corresponded to BNG189. BNG189 and BNG244 were spanned by scaffold280 and merged as a super contig. BAC contigs ctg121, ctg805, ctg287, ctg440 and ctg226 aligned together to confirm the super BAC contig (**b**). (Source: You et al. (2018))

(Fig. 5.6b). Scaffold280 overlapped at both ends with the ends of two BNG contigs, namely, BNG189 and BNG244, thus permitting their merging and filling the gap between them. As a result, 211 BNG contigs were reduced to 148 super BNG contigs with an enlarged N50 of 3.36 Mb and with the longest contig being 11.47 Mb (Table 5.3). The elongated BNG contig set represented a higher-quality optical map that could be exploited to improve the overall genome assembly.

5.3.2.3   Ordering and Orientation of Scaffolds with the BioNano Map

The advanced BNG map can be used to arrange the refined scaffolds for building Mb-long scaffolds with precisely estimated gap sizes between adjacent scaffolds. For instance, six scaffolds totalling 2.38 Mb matched with BNG28 with significant confidence and were assembled into a super scaffold (Fig. 5.6a). The scaffolds covered 80.1% of BNG28. Four scaffolds, namely, scaffold262, scaffold508, scaffold55 and refined scaffold422, were thus oriented and gaps were estimated at 0.57 Mb (Fig. 5.6a). Overall, 831 original and refined scaffolds were ordered and oriented on 148 BNG contigs (Table 5.3).

5.3.2.4   The BAC-Based Physical Map and Genetic Maps Assists Generation of Pseudomolecules

The contigs of the BNG map can also be anchored to the BAC-based physical map and the genetic maps, which in turn facilitates the assembling of scaffolds into pseudomolecules. When a BAC contig spans wholly or partially the last scaffold of a BNG contig and the first scaffold of another with confidence, then the two BNG contigs are declared adjacent as illustrated in Fig. 5.7 where one BNG contig (BNG200) and four super BNG contigs (BNG40 + BNG57, BNG26 + BNG92, BNG76 + BNG86 and BNG247 + BNG234) were assembled. The five BNG contigs can be ordered according to the scaffolds at the proximal or distal ends of two neighbouring BNG contigs

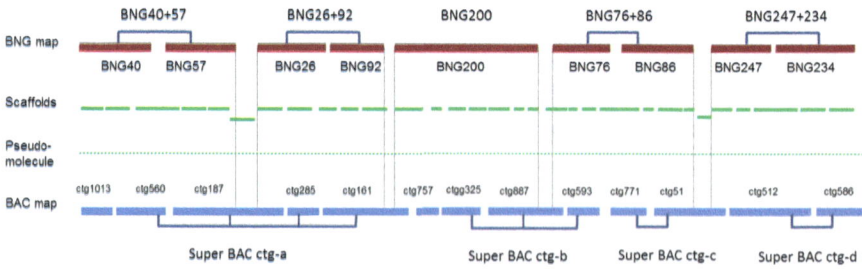

Fig. 5.7 BAC-based physical map assists the ordering of BioNano contigs for pseudomolecule generation. (Source: You et al. (2018))

on the four BAC contigs (contig187, contig161, contig887 and contig51). In this case, BNG contigs as well as scaffolds can be further extended. Using this method, 145 of 148 BNG contigs were anchored to the BAC-based physical map and the 148 BNG contigs were further grouped into 94 super BNG contigs with an average of only six disconnected BNG contigs per chromosome (Table 5.3). Finally, a genetic map of low to medium marker density can be used to sort the super BNG contigs and generate pseudomolecules.

5.3.3 The Second Version of the Flax Reference Sequence

With the flax BNG optical map, the BAC-physical map and consensus genetic map, a total of 609 original scaffolds totalling 284.5 Mb sequences were refined and ordered into 15 individual pseudomolecules representing each chromosome. The pseudomolecules contained ~316 Mb (including gaps) with chromosomes ranging from 15.6 to 29.4 Mb, covering 97% of all annotated genes (Table 5.4). The new pseudomolecule-based reference sequence represents a considerable improvement over the draft flax genome reference sequence (Wang et al. 2012) which will benefit genome-wide SNP discovery, haplotype map construction, QTL identification, association studies and comparative genome analysis.

Table 5.4 The second version of the flax reference genome sequence

Chr.	No. of BNG contigs	BNG size (Mb)	No. of refined scaffolds	Scaffold length (Mb)	Total length (Mb)	No. of genes
1	20	26.9	66	25.2	29.4	4071
2	15	25.4	61	21.7	25.7	2552
3	23	26.5	67	24.1	26.6	3476
4	12	18.2	43	18.3	19.9	2936
5	9	16.4	42	16.5	17.7	2436
6	10	20.7	42	17	18.1	2732
7	9	17.9	53	16.6	18.3	2315
8	17	20.5	73	21.1	23.8	3181
9	10	20.5	70	19.4	22.1	2527
10	8	17.2	32	17.4	18.2	2192
11	9	18.4	38	18.3	19.9	3035
12	15	20.7	41	18.2	20.9	2885
13	8	19.5	48	19.3	20.5	2714
14	20	20.2	55	17.5	19.4	2938
15	20	15.8	41	13.9	15.6	2287
Total	205	304.8	772	284.5	316.2	42,277

Source: You et al. (2018)
Chr. chromosome, *No.* number

References

Chamala S, Chanderbali AS, Der JP, Lan T, Walts B, Albert VA, dePamphilis CW, Leebens-Mack J, Rounsley S, Schuster SC, Wing RA, Xiao N, Moore R, Soltis PS, Soltis DE, Barbazuk WB (2013) Assembly and validation of the genome of the nonmodel basal angiosperm Amborella. Science 342:1516–1517

Chen M, Presting G, Barbazuk WB, Goicoechea JL, Blackmon B, Fang G, Kim H, Frisch D, Yu Y, Sun S, Higingbottom S, Phimphilai J, Phimphilai D, Thurmond S, Gaudette B, Li P, Liu J, Hatfield J, Main D, Farrar K, Henderson C, Barnett L, Costa R, Williams B, Walser S, Atkins M, Hall C, Budiman MA, Tomkins JP, Luo M, Bancroft I, Salse J, Regad F, Mohapatra T, Singh NK, Tyagi AK, Soderlund C, Dean RA, Wing RA (2002) An integrated physical and genetic map of the rice genome. Plant Cell 14:537–545

Cloutier S, Ragupathy R, Miranda E, Radovanovic N, Reimer E, Walichnowski A, Ward K, Rowland G, Duguid S, Banik M (2012) Integrated consensus genetic and physical maps of flax (*Linum usitatissimum* L.). Theor Appl Genet 125:1783–1795

Cullis CA (1981) DNA sequence organization in the flax genome. Biochim Biophys Acta 652:1–15

Dolezel J, Greilhuber J, Suda J (2007) Estimation of nuclear DNA content in plants using flow cytometry. Nat Protoc 2:2233–2244

Evans GM, Rees H, Snell CL, Sun S (1972) The relationship between nuclear DNA amount and the duration of the mitotic cycle. Chromosomes Today 3:24–31

Fu YB, Dong Y, Yang MH (2016) Multiplexed shotgun sequencing reveals congruent three-genome phylogenetic signals for four botanical sections of the flax genus Linum. Mol Phylogenet Evol 101:122–132

Goldblatt P (2007) The index to plant chromosome numbers: past and future. Taxon 56:984–986

Marie D, Brown SC (1993) A cytometric exercise in plant DNA histograms, with 2C values for 70 species. Biol Cell 78:41–51

McDill J, Repplinger M, Simpson BB, Kadereit JW (2009) The phylogeny of Linum and Linaceae subfamily Linoideae, with implications for their systematics, biogeography, and evolution of heterostyly. Syst Bot 34:386–405

Muravenko OV, Samatadze TE, Popov KV, Amosova AV, Zelenin AV (2001) Comparative genome analysis in two flax species by C-banding patterns. Russ J Genet 37:253–256

Muravenko OV, Lemesh VA, Samatadze TE, Amosova AV, Grushetskaia ZE, Popov KV, Semenova O, Khotyleva LV, Zelenin AV (2003) Genome comparisons of three closely related flax species and their hybrids with chromosomal and molecular markers. Genetika 39:510–518

Nosova IV (2005) A study of the Linum species, Sect. Syllinum and Dasylinum. In: Karyology, karyosystematics and molecular phylogeny, pp 73–74, St. Petersburg, Russia.

Nosova IV, Yu Semenova O, Samatadze TE, Amosova AV, Bolsheva NL, Zelenin AV, Muravenko OV (2005) Investigation of karyotype structure and mapping of ribosomal genes on chromosomes of wild Linum species by FISH. Biol Membr 22:244–248

Otto SP, Whitton J (2000) Polyploid incidence and evolution. Annu Rev Genet 34:401–437

Ragupathy R, Rathinavelu R, Cloutier S (2011) Physical mapping and BAC-end sequence analysis provide initial insights into the flax (*Linum usitatissimum* L.) genome. BMC Genomics 12:217

Ray C (1944) Cytological studies on the flax genus, Linum. Am J Bot 31:241–248

Rice A, Glick L, Abadi S, Einhorn M, Kopelman NM, Salman-Minkov A, Mayzel J, Chay O, Mayrose I (2014) The chromosome counts database (CCDB) – a community resource of plant chromosome numbers. New Phytol 206:19–26

Rogers CM (1982) The systematics of Linum sect. Linopsis (Linaceae). Plant Syst Evol 140:225–234

Rowland GG, Hormis YA, Rashid KY (2002) CDC Bethune flax. Can J Plant Sci 82:101–102

Shearer LA, Anderson LK, de Jong H, Smit S, Goicoechea JL, Roe BA, Hua A, Giovannoni JJ, Stack SM (2014) Fluorescence *in situ* hybridization and optical mapping to correct scaffold arrangement in the tomato genome. G3 (Bethesda) 4:1395–1405

Stankova H, Hastie AR, Chan S, Vrana J, Tulpova Z, Kubalakova M, Visendi P, Hayashi S, Luo M, Batley J, Edwards D, Dolezel J, Simkova H (2016) BioNano genome mapping of individual chromosomes supports physical mapping and sequence assembly in complex plant genomes. Plant Biotechnol J 14:1523–1531

Sveinsson S, McDill J, Wong GKS, Li J, Li X, Deyholos MK, Cronk QCB (2014) Phylogenetic pinpointing of a paleopolyploidy event within the flax genus (Linum) using transcriptomics. Ann Bot 113:753–761

Venter JC, Smith HO, Hood L (1996) A new strategy for genome sequencing. Nature 381:364–366

Wang Z, Hobson N, Galindo L, Zhu S, Shi D, McDill J, Yang L, Hawkins S, Neutelings G, Datla R, Lambert G, Galbraith DW, Grassa CJ, Geraldes A, Cronk QC, Cullis C, Dash PK, Kumar PA, Cloutier S, Sharpe AG, Wong GK, Wang J, Deyholos MK (2012) The genome of flax (*Linum usitatissimum*) assembled *de novo* from short shotgun sequence reads. Plant J 72:461–473

You FM, Li P, Ragupathy R, Kumar S, Zhu T, Luo M-C, Duguid SD, Rashid KY, Booker HM, Deyholos MK, Fu YB, Sharpe AG, Cloutier S (2016) The draft flax genome pseudomolecules. In: The 66th Flax Institute of the United States, Fargo, North Dakota, USA, pp 17–24

You FM, Xiao J, Li P, Yao Z, Jia G, He L, Zhu T, Luo M-C, Wang X, Deyholos MK, Cloutier S (2018) Chromosome-scale pseudomolecules refined by optical, physical, and genetic maps in flax. Plant J 95:371–384. https://doi.org/10.1111/tpj.13944

Young ND, Debelle F, Oldroyd GE, Geurts R, Cannon SB, Udvardi MK, Benedito VA, Mayer KF, Gouzy J, Schoof H, Van de Peer Y, Proost S, Cook DR, Meyers BC, Spannagl M, Cheung F, De Mita S, Krishnakumar V, Gundlach H, Zhou S, Mudge J, Bharti AK, Murray JD, Naoumkina MA, Rosen B, Silverstein KA, Tang H, Rombauts S, Zhao PX, Zhou P, Barbe V, Bardou P, Bechner M, Bellec A, Berger A, Berges H, Bidwell S, Bisseling T, Choisne N, Couloux A, Denny R, Deshpande S, Dai X, Doyle JJ, Dudez AM, Farmer AD, Fouteau S, Franken C, Gibelin C, Gish J, Goldstein S, Gonzalez AJ, Green PJ, Hallab A, Hartog M, Hua A, Humphray SJ, Jeong DH, Jing Y, Jocker A, Kenton SM, Kim DJ, Klee K, Lai H, Lang C, Lin S, Macmil SL, Magdelenat G, Matthews L, McCorrison J, Monaghan EL, Mun JH, Najar FZ, Nicholson C, Noirot C, O'Bleness M, Paule CR, Poulain J, Prion F, Qin B, Qu C, Retzel EF, Riddle C, Sallet E, Samain S, Samson N, Sanders I, Saurat O, Scarpelli C, Schiex T, Segurens B, Severin AJ, Sherrier DJ, Shi R, Sims S, Singer SR, Sinharoy S, Sterck L, Viollet A, Wang BB, Wang K, Wang M, Wang X, Warfsmann J, Weissenbach J, White DD, White JD, Wiley GB, Wincker P, Xing Y, Yang L, Yao Z, Ying F, Zhai J, Zhou L, Zuber A, Denarie J, Dixon RA, May GD, Schwartz DC, Rogers J, Quetier F, Town CD, Roe BA (2011) The Medicago genome provides insight into the evolution of rhizobial symbioses. Nature 480:520–524

Zhou S, Wei F, Nguyen J, Bechner M, Potamousis K, Goldstein S, Pape L, Mehan MR, Churas C, Pasternak S, Forrest DK, Wise R, Ware D, Wing RA, Waterman MS, Livny M, Schwartz DC (2009) A single molecule scaffold for the maize genome. PLoS Genet 5:e1000711

Chapter 6
Comparison Between the Genomes of a Fiber and an Oil-Seed Variety of Flax

Christopher A. Cullis and Margaret A. Cullis

6.1 Introduction

The flax crop has been bred for two distinct phenotypes. The fiber varieties' primary attribute is a tall unbranched stem in order to have the most desirable fiber quality. The oil-seed varieties have been developed to have the maximum seed yield and therefore have a more highly branched phenotype. The flax genome reference genome has been developed using the Canadian oil-seed variety Bethune, which is described extensively elsewhere in this volume (Deyholos 2019; Cloutier et al. 2019; You and Cloutier 2019). The fiber flax variety Stormont cirus is of interest because of the variability of the genome when grown under stress environments (Durrant 1958, 1962a, b, 1965, 1971, 1974, 1981; Durrant and Tyson 1960; Evans et al. 1966; Cullis 1977, 1983). A whole genome comparison between fiber and oil-seed flax varieties will shed some light on the regions of the genome that have been highly conserved during the selection of these two crops. This information will also be important when making comparisons between closely related lines derived from a common progenitor to determine if the genomic regions that have been differentially selected in domestication are also included in the subset that may be part of these unusual reorganization events. In addition, this initial characterization will also be useful in the further determination of the variation between Stormont cirus and the derived genotrophs to identify the labile regions of the genome modified in the stress-related genome reorganization.

C. A. Cullis (✉) · M. A. Cullis
Department of Biology, Case Western Reserve University, Cleveland, OH, USA
e-mail: cac5@case.edu

© Springer Nature Switzerland AG 2019
C. A. Cullis (ed.), *Genetics and Genomics of Linum*, Plant Genetics and
Genomics: Crops and Models 23, https://doi.org/10.1007/978-3-030-23964-0_6

6.2   Bethune and Stormont Cirus Genomes

As noted elsewhere in this volume (Deyholos 2019), the 1C genome size of Bethune is 360 Mbp. The size of the Stormont cirus genome has previously been determined by Feulgen cytophotometry (Evans et al. 1972) at 685 Mbp, that is, nearly twice the size of the Bethune genome. This difference is not unexpected considering that considerable DNA variation among *Linum usitatissimum* accessions has been observed (Table 6.1). Therefore, the Bethune and Stormont cirus vary substantially in their total nuclear genome content.

The Bethune genome has been sequenced and assembled into chromosomes (Wang et al. 2012; You et al. 2018; You and Cloutier 2019) and has been used as a reference to compare the sequences present in Stormont cirus that are common in the two varieties. DNA isolated from Stormont cirus was subjected to Illumina next-generation sequencing at approximately 100× coverage. The coverage obtained from the depth of sequencing was consistent with the genome size previously determined, confirming the difference in size between the nuclear DNA contents of the two varieties. The sequence reads were aligned with the Bethune reference genome that had been assembled into chromosomes (You and Cloutier 2019) and the variants between the two genomes determined.

The following are the metrics from the Bowtie alignment of Stormont cirus Illumina whole genome sequence to the Bethune reference genome. The Pl data contained 237,824,651 reads each of 100 bases. Of these, 237,824,851 (100.00%) were paired, and of these, 83,822,572 (35.25%) aligned concordantly 0 times, 89,383,929 (37.58%) aligned concordantly exactly 1 time, and 64,618,350 (27.17%) aligned concordantly >1 times. Of the 83,822,572 pairs that aligned concordantly 0 times, 25,90,679 (3.09%) aligned discordantly 1 time. In addition, of the 81,231,893 pairs that aligned 0 times concordantly or discordantly 136,079,842 (83.76%) aligned 0 times, 14,406,415 (8.87%) aligned exactly 1 time and 11,977,529 (7.37%) aligned >1 times.

The overall data indicated that there was a 71.39% overall alignment rate of the Stormont cirus data to the Bethune reference sequence, that is, nearly 30% of the Stormont cirus reads that were not present in the reference sequence. Since the Stormont cirus genome has not yet been assembled, it is not possible to identify

Table 6.1 The sizes of the genomes of some *Linum usitatissimum* varieties, determined by flow cytometry, arranged from largest to smallest (Todd Michael, Personal communication)

Location	GRIN PI number	Genome (Mb)
India	523580	575
Canada	566850	518
Egypt	250092	512
India	524239	497
Germany	523497	489
Syria	181774	428
Afghanistan	426204	381

the nature of the sequences present in Stormont cirus that are absent from Bethune. Similar data has been obtained for alignments of derivatives from the original Stormont cirus variety, the genotrophs that are described in Chap. 14.

6.3   Comparison of Whole Genome Sequences of Bethune and Stormont Cirus

The differences between the Bethune and Stormont cirus genomes, as identified from the reference sequence, include deletions and single nucleotide polymorphisms as well as rearrangements. As noted in the alignment statistics, about 3% of the reads aligned discordantly, that is, the Stormont cirus genome for these sequences was not organized as the Bethune reference genome, which is most likely due to insertions in the Stormont cirus genome.

6.4   Single Nucleotide Polymorphisms Between Stormont Cirus and Bethune

The alignment of the Stormont cirus sequence reads with the Bethune genome was subject to mpileup to determine the SNPs between Bethune and Stormont cirus. The total number of SNPs identified between the two genomes was 149,997. The distribution of the polymorphisms across the genome is shown in Figs. 6.1, 6.2, and 6.3. What is notable is that the variation is not equally distributed across the genome.

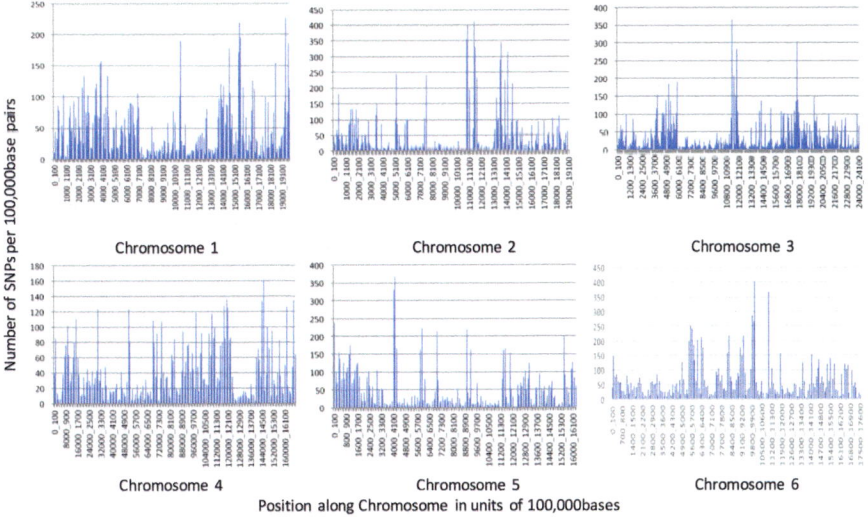

Fig. 6.1 Distribution of SNPs between Bethune and Stormont cirus across chromosomes 1–6

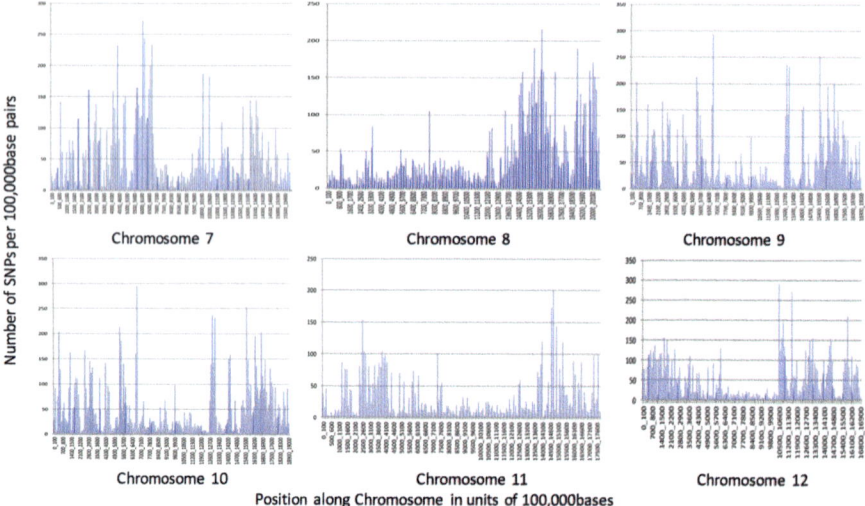

Fig. 6.2 Distribution of SNPs between Bethune and Stormont cirus across chromosomes 7–12. The SNPs have been placed in bins of 100,000 bp

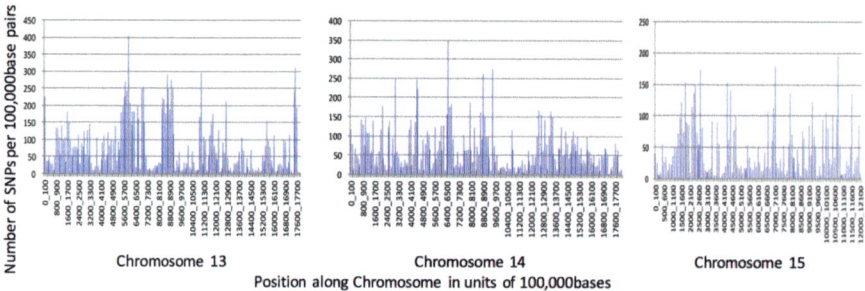

Fig. 6.3 Distribution of SNPs between Bethune and Stormont cirus across chromosomes 13–15. The SNPs have been placed in bins of 100,000 bp

There are regions that have a high density of SNPs while others are virtually devoid of variants. Chromosome 2 is particularly instructive where there are regions of 100 kbp (such as 7.9–8.0 Mb, 8.9–9.0 Mb and 9.8–9.9 Mb) in which only 3 or 4 SNPs occur.

These differences are well illustrated in Figs. 6.4 and 6.5, and 6.6. The region in Fig. 6.4 shows a region where there is a single SNP in an 18 kb region. As seen in Figs. 6.1 and 6.2 the density of SNPs can be very low, in some cases with <1 SNP per 100 kbp.

Figure 6.5 illustrates a region of high SNP density, where there are 134 SNPs in just over 6 kbp, a frequency of 1 SNP every 50 bp. Regions with high frequency of SNPs can reach 1 SNP every 20 bp. In determining the regions of high SNP variation, those regions that are present in more than one copy, and where there are

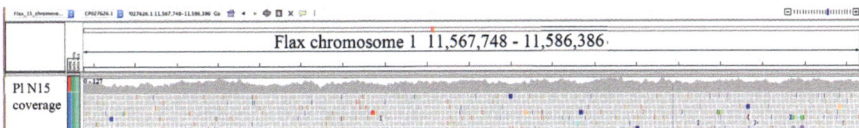

Fig. 6.4 Region of conservation between Stormont cirus and Bethune with only a single SNP in more than 18 kbp

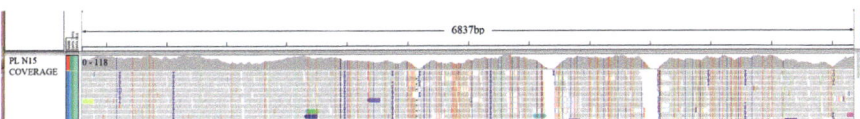

Fig. 6.5 Region of high SNP frequency between Bethune and Stormont cirus

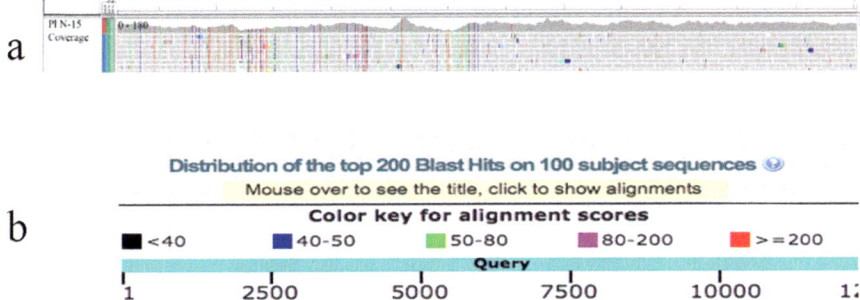

Fig. 6.6 (**a**) Transition between a region of high SNP frequency to one of high conservation on chromosomes 5. (**b**) The identification of possible genes from the Blastx analysis of the region

multiple nucleotides at specific positions, have been excluded, on the assumption that these are diverged relics of the ancient polyploidization.

The transition between regions of high diversity and regions of high conservation can be abrupt. In the case of Fig. 6.6, 69 SNPs are present in the first ~6 kbp of the region with none in the subsequent 6 kbp.

The region shown in Fig. 6.6 was subject to Blastx to determine if there were any genes within this region. Two genes were identified, one in the variable region and the other in the conserved region. The best match in the variable region was to a retroviral polyprotein, while that for the conserved region was for a putative gene, a region that has been identified as a putative gene in many species. This single example epitomizes the generality that many of the regions with high SNP frequency appear to be sequences related to retroviruses, while those that are

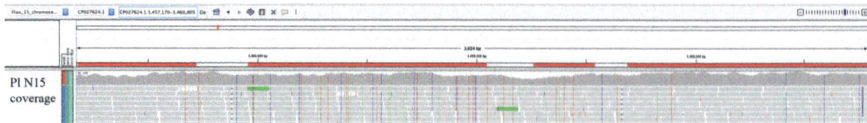

Fig. 6.7 The comparison between the N1-A rust resistance genes in Bethune and Stormont cirus. The shaded regions denote the coding regions, while the clear regions are the introns

conserved are frequently identified as structural genes. Considering the time of separation, and the selection for very different growth habits, of the oil-seed and fiber varieties, this high conservation at the nucleotide level, rather than just a conservation of protein structure, for structural genes as opposed to retrotransposons is somewhat unusual.

However, not all genic sequences are conserved, especially those for which variation should be the norm. One such family would be the genes responsible for disease resistance, where Bethune and Stormont cirus have different resistance profiles for flax rust. The region shown in Fig. 6.7 is that for the N1-A rust resistance gene with the exons and introns highlighted. It is clear that this gene is very different between the two varieties. However, in general, where there are regions of high genic content, the variability between the two genomes is low at the nucleotide level, as well as at the protein level.

6.5 Regions Present in Bethune But Absent from Stormont Cirus

As noted earlier, not all the sequencing reads from the Stormont cirus DNA aligned to the Bethune reference sequence. Therefore, the data do not allow the identification of the information present in Stormont cirus that is absent from Bethune. The inverse of that question, the information that is present in Bethune but absent in Stormont cirus can be addressed. The alignments can be interrogated to determine those regions for which there are no aligned reads in Stormont cirus. Such an example is shown in Fig. 6.8, which is a region of the Bethune genome for which there are no reads from Stormont cirus. Therefore, the information present here in Bethune is absent from Stormont cirus. A small number of these regions have been further characterized by PCR. In these cases, primers across the apparent gap in Stormont cirus were designed and used to amplify the region from both varieties. In most cases, the expected sizes of the fragments were obtained, as shown in Fig. 6.8, with Stormont cirus having the expected, shorter amplified fragment than Bethune.

In some cases, the amplified region from Stormont cirus was longer than the expected size based on the alignment. A subset of these have been sequenced, and information present generating the unexpected length was not present in the Bethune genome.

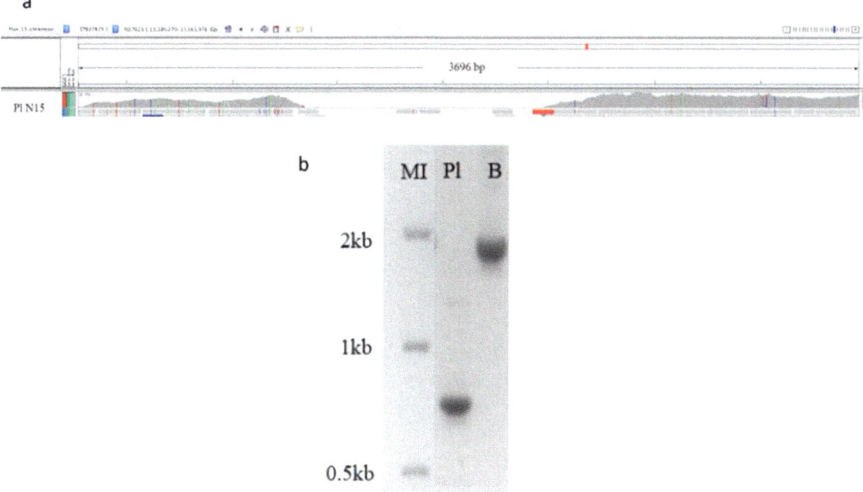

Fig. 6.8 (**a**) The region amplified from Pl visualized by IGV. Pl has a gap of 814 bases compared with the reference Bethune sequence. (**b**) The gel separation of the amplification of the regions shown in (a) from Bethune (B) and Pl. The size of the amplified fragment from the Bethune template is as expected from the reference sequence. The amplified fragment from the Pl template is expected to be 815 bases shorter than the Bethune-amplified fragment. This is shown by the size of the amplified band from Pl and further confirmed by sequencing of the two products

6.6   Repetitive Sequences

Since about half of the flax genome is comprised of repeated sequences, significant differences in nuclear genome size could arise from differences copy number variants. This might be especially important for the transposable element fraction of the genome, which could have undergone rounds of amplifications and deletions in the progeny lineages of Bethune and Stormont cirus. The observation that many of the regions of high nucleotide diversity appear to be transposon-related sequences is consistent with this fraction of the genome being the most rapidly evolving.

The number of 25S +18S ribosomal RNA genes (rDNA) repeat units varies between Bethune and Stormont cirus. Alignment of the complete repeat unit with the Illumina next-generation reads for the two varieties showed that 5.3% of the Bethune genome was made up of this family, while 3.62% of the Stormont cirus genome was attributable to the large rDNA (Wang 2018). Accounting for the difference in genome size, these proportions translate to 2900 genes in Stormont cirus and 2300 genes in Bethune.

The overall pattern of similarity between the oil-seed and fiber varieties ties in with the earlier identification of the distribution of repetitive sequences with the flax genome (Cullis 1981). Here the data indicated long stretches of low copy sequences not frequently interrupted by repeats. This pattern is consistent with the data shown in Figs. 6.1, 6.2, and 6.3, where there are regions of high diversity, frequently the sites of repeated sequences, and long stretches of low variation. One region epitomizing the variability within a chromosome is on chromosome 2 where the region from 7.6 Mb to 10.8 Mb has an average SNP frequency of 13 per 100,000 bp, while the region from 10.8 Mbp to 11.0 Mbp has 754 SNPs. This comparison can now be extended for specific genes that are involved in the differences in form and function between the two ideotypes of flax.

References

Cloutier S, You FM, Soto-Cerda BJ (2019) *Linum* genetic markers, maps and QTL discovery. Chapter 7. In: Cullis CA (ed) Linum genetics and genomics. Springer, Cham.

Cullis CA (1977) Molecular aspects of environmental induction of heritable changes in flax. Heredity 38:129–154

Cullis CA (1981) DNA sequence organization in the flax genome. Biochim Biophys Acta 652:1–15

Cullis CA (1983) Environmentally induced DNA changes in plants. Crit Rev Plant Sci 1(2):117–131

Deyholos MK (2019) The first flax genome assembly. Chapter 4. In: Cullis CA (ed). Linum genetics and genomics. Springer, Cham.

Durrant A (1958) Environmental conditioning of flax. Nature 181(4613):928–929

Durrant A (1962a) Environmental induction of heritable change in linum. Heredity 17:27

Durrant A (1962b) Induction, reversion and epitrophism of flax genotrophs. Nature 196(4861):1302

Durrant A (1965) Genotrophic change in linum. Heredity 20:647

Durrant A (1971) Induction and growth of flax genotrophs. Heredity 27:277

Durrant A (1974) Association of induced changes in flax. Heredity 32:133–143

Durrant A (1981) Unstable genotypes. Philos Trans R Soc Lond Ser B Biol Sci 292(1062):467–474

Durrant A, Tyson H (1960) Conditioned lines of flax. Nature 185(4705):60–60

Evans GM, Durrant A, Rees H (1966) Associated nuclear changes in induction of flax genotrophs. Nature 212(5063):697

Evans GM, Rees H, Snell CL, Sun S (1972) The relationship between nuclear DNA amount and the duration of the mitotic cycle. Chromosomes Today 3:24–31

Wang H (2018) The potential inducing pattern of the flax genome. MS thesis, Case Western Reserve University

Wang ZW, Hobson N, Galindo L, Zhu SL, Shi DH, McDill J, Yang LF, Hawkins S, Neutelings G, Datla R, Lambert G, Galbraith DW, Grassa CJ, Geraldes A, Cronk QC, Cullis C, Dash PK, Kumar PA, Cloutier S, Sharpe AG, Wong GKS, Wang J, Deyholos MK (2012) The genome of flax (Linum usitatissimum) assembled de novo from short shotgun sequence reads. Plant J 72(3):461–473

You FM, Cloutier S (2019) The assembly of the flax genomes into chromosomes. Chapter 5. In: Cullis CA (ed). Linum genetics and genomics. Springer, Cham.

You FM, Xiao J, Li P, Yao Z, Jia G, He L, Zhu T, Luo MC, Wang X, Deyholos MK, Cloutier S (2018) Chromosome-scale pseudomolecules refined by optical, physical and genetic maps in flax. Plant J 95:371–384. https://doi.org/10.1111/tpj.13944. Epub 2018 May 21

Chapter 7
Linum Genetic Markers, Maps, and QTL Discovery

Sylvie Cloutier, Frank M. You, and Braulio J. Soto-Cerda

7.1 Introduction

Linum usitatissimum (L.), commonly referred to as flax, flaxseed, linseed, or fiber flax, has been cultivated for millennia for its oil-rich seeds and its high-quality stem fibers (Zohary and Hopf 2000). Like most crops, commercial plant breeding activities started at the end of the nineteenth century, but molecular markers were not developed until the end of the twentieth century because, as a marginal crop, the funding necessary for this research was insufficient in the 1980s and early 1990s (Cloutier 2016). Since then, however, molecular resources for flax have quickly expanded as the small size of the flax genome made it readily amenable to fast-evolving technologies, and there were significant investments in large-scale projects. These markers have been applied to understand and mine the genetic diversity of the species and its relatives and to assist in breeding efforts.

7.2 Marker Development

Over the last three decades, molecular markers such as restriction fragment length polymorphism (RFLP), random amplified polymorphic DNA (RAPD), amplified fragment length polymorphism (AFLP), sequence-specific amplified polymorphism

S. Cloutier (✉) · F. M. You
Ottawa Research and Development Centre, Agriculture and Agri-Food Canada,
Ottawa, ON, Canada
e-mail: SylvieJ.Cloutier@Canada.ca

B. J. Soto-Cerda
Agriaquaculture Nutritional Genomic Center, Temuco, Chile

© Springer Nature Switzerland AG 2019   97
C. A. Cullis (ed.), *Genetics and Genomics of Linum*, Plant Genetics and
Genomics: Crops and Models 23, https://doi.org/10.1007/978-3-030-23964-0_7

(SSAP), inter-simple sequence repeat (ISSR), simple sequence repeat (SSR or microsatellite), and single nucleotide polymorphism (SNP) have been developed in succession and often supplanted one another.

7.2.1 Early Generation Markers

Molecular research in *Linum* was practically non-existent during the RFLP era, and few markers of this type were published (Cullis et al. 1995; Spielmeyer et al. 1998; Oh et al. 2000). The first publications of *Linum* molecular markers were based on RAPD and AFLP technologies (Aldrich and Cullis 1993; Chen et al. 1998; Cullis et al. 1999; Spielmeyer et al. 1998). RAPD markers were used extensively in several early and even recent genetic diversity studies (Bolsheva et al. 2015; Diederichsen and Fu 2006; El-Nasr and Mahfouze 2013; Fu et al. 2002a, b, 2003a, b, c; Fu 2005, 2006; Muravenko et al. 2003, 2009, 2010; Ottai et al. 2012; Stegnii et al. 2000; Yurkevich et al. 2013). ISSR markers were similarly applied (Pali et al. 2015; Rajwade et al. 2010; Uysal et al. 2010; Wiesner and Wiesnerová 2004). Despite the fact that the first molecular-based genetic maps comprised mostly AFLP markers (Spielmeyer et al. 1998) and that an attempt at optimizing the AFLP technique in flax (van Treuren 2001) was made, with few exceptions, this marker type was not broadly applied in this species (Chandrawati et al. 2014; Everaert et al. 2001). Similarly, genetic diversity studies using retrotransposon-based SSAP markers are rare (Melnikova et al. 2014). The more recent development of a substantial number of SSR markers and, shortly thereafter, of a considerably larger number of SNP markers is now enabling large-scale genome-wide studies. As a consequence, a more extensive discussion on SSR and SNP markers is warranted. Further, the parallel development of other molecular resources, such as the sequence of the flax genome (Wang et al. 2012; You et al. 2018b) and large-scale expressed sequence tag (EST) collections (Venglat et al. 2011), are all converging to enhance and fuel flax genomics and genetics research on an unprecedented scale.

7.2.2 Simple Sequence Repeat

An SSR marker is created by the amplification of a locus that harbors a variable number of short tandem repeats, hence producing amplicons of different lengths. The first report of a dinucleotide SSR was in humans (Litt and Luty 1989), but SSRs have since been shown to be ubiquitous in eukaryotes. The advantages of SSRs include their frequent co-dominance, genome-wide distribution, multi-allelism, heritability, polymorphism, and robustness (Hwang et al. 2009; Powell et al. 1996). At first, degenerate primers with SSR tails (Fisher et al. 1996) were used in flax (Wiesner et al. 2001), but this approach was quickly discarded because it generated few markers (Table 7.1). Identification of flax-specific SSRs was performed by

Table 7.1 Simple sequence repeat markers developed from *Linum usitatissimum* (L.)

SSR source	Method	Genotypes tested	Polymorphic primer pairs	Loci detected	Loci per primer pair	References
Genomic	Degenerate primers	8	10	–	–	Wiesner et al. (2001)
Genomic	SSR-enriched libraries	93	23	28	1.22	Roose-Amsaleg et al. (2006)
ESTs	Sequence Data mining	23	248	275	1.11	Cloutier et al. (2009)
Genomic	Sequence Data mining	60	60	66	1.10	Soto-Cerda et al. (2011a)
Genomic	SSR-enriched libraries	8	35	37	1.06	Deng et al. (2010)
ESTs	Sequence Data mining	61	23	23	1.00	Soto-Cerda et al. (2011b)
Genomic	SSR-enriched libraries	8	38	38	1.00	Deng et al. (2011)
Genomic	SSR-enriched libraries	27	9	–	–	Kale et al. (2012)
Genomic	SSR-enriched libraries	19	42	42	1.00	Bickel et al. (2011)
ESTs	Sequence Data mining	16	145	149	1.03	Cloutier et al. (2012a)
BESs	Sequence Data mining	16	673	720	1.07	Cloutier et al. (2012a)
Total			1306	1378	1.07	

constructing SSR-enriched libraries using a variety of methods (Bickel et al. 2011; Deng et al. 2008; Kale et al. 2012; Roose-Amsaleg et al. 2006). These methods were tedious and time-consuming and yielded small numbers of unique SSRs and even fewer polymorphic ones (Table 7.1; Bickel et al. 2011; Deng et al. 2010, 2011). In 2011, Next Generation Sequencing (NGS) was used to identify SSRs, but even with this high-throughput sequencing platform, only nine SSRs were discovered and their polymorphism was not assessed (Kale et al. 2012). However, starting in 2006, greater numbers of *Linum usitatissimum* DNA sequences became available in GenBank (Fig. 7.1), and mining this data revealed a powerful and efficient method to identify SSRs (Cloutier et al. 2009, 2012a; Soto-Cerda et al. 2011a, b). Two large datasets were particularly useful. The first was a collection of 274,278 ESTs generated from libraries from 14 different tissues (Venglat et al. 2011) and the second, a collection of 80,337 Bacterial Artificial Chromosome (BAC) end sequences (Ragupathy et al. 2011). A total of 416 SSR assays detecting 447 loci were derived from ESTs (Cloutier et al. 2009, 2012a; Soto-Cerda 2011b), and 733 SSR assays detecting 786 loci were mined from genomic sequences (Cloutier et al. 2012a; Soto-Cerda 2011a). Overall, the mining of ESTs and genomic survey sequences (GSSs)

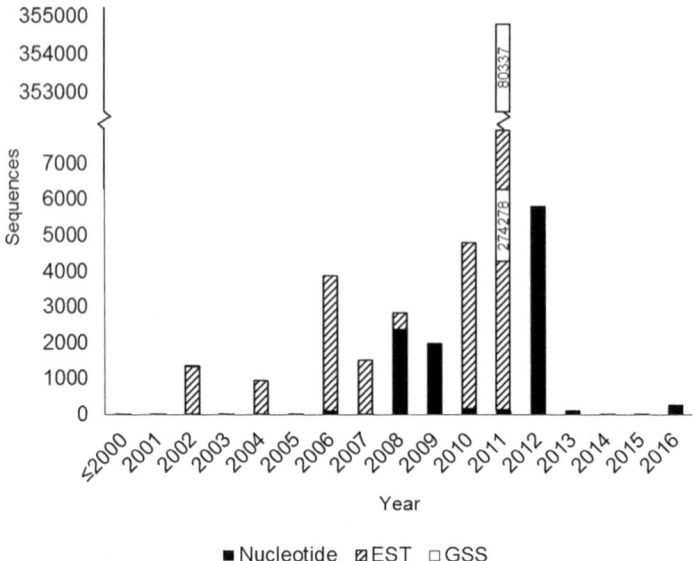

Fig. 7.1 *Linum usitatissimum* L. sequences deposited in GenBank up to June 2016. The entries are either nucleotide, expressed sequence tag (EST), or genomic survey sequence (GSS)

produced 88% of the SSR markers publically available to date for flax (Table 7.1). These markers were used to characterize segregating populations (Cloutier et al. 2011; Kumar et al. 2015) and germplasm collections (Soto-Cerda et al. 2013) including other *Linum* species (Fu and Peterson 2010; Soto-Cerda et al. 2011b, 2014a).

7.2.3 Single Nucleotide Polymorphism

Whole Genome Sequencing (WGS) and the resequencing of other genotypes revealed an abundance of SNPs in plant genomes. The advent of Next Generation Sequencing (NGS) platforms and the decreased cost of genome sequencing associated with them enabled SNP discovery for many plant species. SNPs are distributed throughout the genome and generally bi-allelic, but it is their abundance that makes them a desirable marker type for many applications, hence redefining the meaning of high-density maps. A total of 1067 SNPs were discovered in flax using 454 pyrosequencing combined to genomic reduction (Fu and Peterson 2012). At the same time, Illumina sequencing of reduced representation libraries was performed on eight genotypes of flax toward the discovery of 55,465 SNPs, of which, an estimated 96.8% could be validated in a biparental population (Kumar et al. 2012). However, considering the relatively small size of the flax genome, i.e., ~370 Mb (Ragupathy et al. 2011; Wang et al. 2012), usage of reduced representation libraries was deemed unnecessary, and direct resequencing using NGS was attempted. The entire Canadian core collection of 407 genotypes was directly resequenced, and a total of 1,775,461 high-quality SNPs were discovered (He et al. 2019; You et al. 2016).

7.3   Genetic, Cytogenetic, Physical, and Optical Maps

A linkage or genetic map represents the relative position of gene loci or markers on chromosomes or linkage groups which is estimated by their recombination frequencies. A physical map similarly displays gene loci and marker order, but the distance between them is generally expressed in base pairs. As such, the genetic to physical distance can vary greatly along the 15 chromosome pairs of the diploid flax genome.

The construction of a genetic map requires a mapping population and a set of markers. Together, the number of markers and the size of the population determine the resolution power of the map, i.e., its usefulness in positional cloning. Any of the marker types described above can be used, but some, such as SSRs, are preferred for their robustness, and others, such as SNPs, are preferred for their abundance.

7.3.1   First Generation Genetic Maps

The first reported genetic map of flax comprised 19 RFLP and 69 RAPD markers (Cullis et al. 1995). The next one comprised 2 morphological traits, 213 AFLP and 8 RFLP markers spanning 1400 centiMorgans (cM), and 18 linkage groups positioned on a biparental population of 59 doubled haploid lines (Spielmeyer et al. 1998). Because RFLPs were labor intensive and AFLPs suffered from suboptimal reproducibility (Fry et al. 2009), no other genetic map of flax was produced using these marker types. Shortly thereafter, Oh et al. (2000) published an improved genetic map comprising one sequence tag site (STS), 13 RFLP and 80 RAPD markers covering 1000 cM, and 15 linkage groups from their earlier CI1303/Stormont Cirrus and Leona/Koto F_2 populations of approximately 50 individuals (Cullis et al. 1995). The inherent reproducibility problem was the main argument against RAPD markers, but other disadvantages such as their dominance and co-migration issues also contributed to their near disappearance (Kumari and Thakur 2014).

In 2009, Cloutier et al. reported the first significant number of SSR markers in flax (Table 7.1). These and other publically available SSRs (Table 7.1) enabled the construction of a few single population SSR-based linkage maps. The first, derived from the mapping of a doubled haploid population of 78 individuals, comprised one morphological trait, five fatty acid desaturase gene-specific loci, five SNPs and 114 SSRs spanning 834 cM, and 24 linkage groups (Cloutier et al. 2011). The second was obtained from mapping 143 SSRs onto a population of 300-F_2 individuals to produce a 15-linkage group map of 1241 cM (Asgarinia et al. 2013).

7.3.2 Second Generation Genetic Maps

Second generation genetic maps are denser than first generation maps with approximately 700–800 markers. Cloutier et al. (2012b) constructed an integrated consensus map covering 15 linkage groups and totaling 1551 cM. Mapping information was combined for 389 markers from a recombinant inbred line (RIL) population of 243 individuals from a CDC Bethune/Macbeth (BM) cross, for 443 markers from a RIL population of 90 individuals from a E1747/Viking (EV) cross, and for 477 markers of a DH population from an SP2047/UGG5-5 (SU) cross. The consensus map included 770 markers, of which, 114 were shared by all three populations and 257 were common to any two populations. Of the mapped 770 markers, all were SSRs except for five SNPs, one morphological trait (yellow seed coat color), and six gene loci (*dgatA*, *dgatB*, *fad2A*, *fad2B*, *fad3A*, and *fad3B*). The marker density was estimated at one marker per 2 cM. More recently, Kumar et al. (2015) reported a single population genetic map of similar density but that combined 329 SNPs and 362 SSRs. The 1266 cM map, developed using the previously described CDC Bethune/Macbeth RIL population, also spanned all 15 linkage groups with a similar average marker density of one marker per 1.9 cM. The consensus map and the SNP/SSR CDC Bethune/Macbeth genetic maps are a good resource for the community because they can serve to anchor other markers to create fine maps for positional cloning.

7.3.3 High-Density Genetic Maps

High-density genetic maps are usually based on abundant SNP markers. You et al. (2018a) applied the genotyping by sequencing strategy to the three biparental populations used in the second generation genetic maps (Cloutier et al. 2012b) and identified 78,312 SNPs from the BM population, 90,265 SNPs from the EV population, and 103,100 SNPs from the SU population. This SNP data was analyzed to identify recombination bins, i.e., blocks of markers without recombination, for each population. A total of 1581, 2608, and 2283 recombination bins were defined for the BM, EV, and SU populations, respectively. SNP markers in each recombination bin were selected to construct a high-density genetic map. Segregating data of the single bin SNPs was merged to the SSR marker data from the second generation genetic maps to generate a high-density consensus map of 4594 SNP and 661 SSR markers that spanned all 15 chromosomes (You et al. 2015). The second generation and high-density genetic maps have been used in genomic studies and genome scaffolding (You et al. 2018a, b).

7.3.4 Cytogenetic, Physical, and Optical Maps

A karyotype represents the complete set of chromosomes of an individual which is visually described as a karyogram. Stains and fluorescent dyes are used to create characteristic banding patterns for each chromosome; hence, a karyogram displays the physical position of the telomeres, centromere, and the specific banding patterns of each chromosome. Such karyotypes were produced for several *Linum* species (Bolsheva et al. 2015; Yurkevich et al. 2013) including *L. usitatissimum* (Muravenko et al. 2003, 2009, 2010; Rachinskaya et al. 2011). Muravenko et al. (2003) first described the flax karyotype using C-banding and silver staining of the nucleolus organizer region. They later improved the karyotyping with C/DAPI-banding and fluorescent *in situ* hybridization with 5S and 26S ribosomal DNA (Muravenko et al. 2009, 2010; Rachinskaya et al. 2011). Such cytogenetic maps have been used to infer taxonomic relationships within the genus *Linum*.

The sensu stricto definition of a cytogenetic map is a low-resolution physical map because it depicts the physical organization of banding or stained markers on chromosomes. At the other end of the spectrum, the complete genomic sequence represents a high-resolution physical map because it describes the physical organization of genetic features along the chromosome at the base pair level. An intermediate level of resolution could be a physical map constructed from the analysis of a BAC library where a collection of BAC clones are restricted, fingerprinted, and assembled into overlapping fragments called contigs. Such a physical map of flax was constructed from the BAC library of CDC Bethune, producing 416 contigs spanning an estimated 368 Mb (Ragupathy et al. 2011). The BAC-end sequences were also obtained, and this information was used to anchor 204 contigs of the physical map to 670 markers of the genetic map (Cloutier et al. 2012b), corresponding to approximately 74% of the total genome. Ultimately, this information was used to order, orient, and refine the reference genome sequence of flax (Wang et al. 2012; You et al. 2018b).

An optical map is a high-resolution, whole genome restriction map assembled from single molecules sized by optical means. Long stained single genomic DNA molecules are nicked (or restricted) using one or more nicking (or restriction) endonucleases and fluorescently labelled (Dong et al. 2013; Lam et al. 2012). Such labelled long molecules are individually resolved by an automated high-throughput whole genome-mapping system that captures them as images which are subsequently converted to human-readable, single-molecule restriction maps using an image processing software. These restriction maps are then assembled into a consensus genome map comprising contigs of several megabases. Optical maps facilitate validation, ordering, and orientation of draft genome sequences. A high-resolution optical map of cv. CDC Bethune was constructed using the BioNano genome (BNG) mapping technology (You et al. 2018b). A total of 82 Gb of raw DNA molecules representing more than 200x genome equivalent with an N50 of 249 Kb was assembled into 251 BNG contigs with an improved N50 of 2.15 Mb and a total size of 317 Mb. This map has been used to validate scaffolds, detect and correct misassemblies, and generate chromosome-based pseudomolecules (You et al. 2018b, see chap. 5 for more details).

7.4 QTL Mapping Using Biparental Populations

Quantitative Trait Locus (QTL) mapping is performed to identify one or several regions of DNA that contain or are linked to at least one genetic feature controlling a trait of interest and which could be the first step toward the map-based identification of the feature(s). Biparental populations where the two parents differ for the phenotype(s) under study are commonly used. Mapping of F_2 populations followed by phenotyping of F_3 families can be used, but the high degree of heterozygosity in these populations can hinder the detection of QTL. F_1-derived doubled haploid populations or advanced single seed descent derived populations at the F_6 or later generations are advantageous because they are homozygous or nearly so. The downside of these populations is the time and labor required to develop them. On the other hand, because most loci are homozygous, genotyping is simplified to the two parental allelic forms at most loci. Also, phenotyping can be easily replicated because the lines are fairly stable and a sufficient quantity of seeds can usually be obtained. Diverse germplasm collections can also be used for QTL mapping using genome-wide association studies (GWASs). Such association mapping panels and studies are discussed in the last two sections of this chapter.

7.4.1 Disease Resistance

To date, only three reports of mapping disease resistance genes in flax using biparental populations exist (Spielmeyer et al. 1998; Bo et al. 2008; Asgarinia et al. 2013). The first one reports the mapping of QTL for fusarium wilt, a disease caused by the soil-borne fungal pathogen *Fusarium oxysporum* f. sp. *lini* Schlecht. This disease is best controlled through the use of resistant cultivars and crop rotations. Spielmeyer et al. (1998) mapped 59 doubled haploid lines from a cross between fusarium wilt resistant genotype CRZY8/RA91 and the Australian susceptible cv. Glenelg. Their analysis identified two major QTL for fusarium wilt resistance. The linkage group 6 QTL had a LOD score of 7.3 and accounted for 38% of the genetic variability for the trait in this cross, while the linkage group 10 QTL had a LOD score of 3.4 and represented 26% of the genetic variability. Further, an additive effect of both loci for resistance to wilt was reported.

The gene-for-gene concept described by Flor (1955) was advanced while studying the plant-pathogen interactions between *L. usitatissimum* and *Melampsora lini*, the fungal pathogen responsible for rust infections on flax. Not surprisingly, some of the first plant disease resistance genes cloned were flax rust resistance genes *L6*

(Lawrence et al. 1995) and *M* (Anderson et al. 1997). Bo et al. (2008), attempting to find markers close to the *M4* gene, did not construct a complete genetic map of flax but used a bulk segregant analysis to identify marker OPA18. They then confirmed its location to be 2.1 cM away from *M4* using a biparental population of 96 F_2 plants derived from a narrow cross between cv. Bison and the *M4* near-isogenic line NM4. Based on the segregation ratio, they postulated that *M4* was a single dominant gene.

More recently, Asgarinia et al. (2013) mapped a biparental population of 300 F_2 individuals derived from a cross between powdery mildew susceptible cv. NorMan and resistant cv. Linda. Powdery mildew in flax, caused by *Oidium lini* Skoric, is a foliar disease that affects most production areas of the world. In this study, phenotyping of F_3 and F_4 families was performed, and three consistent QTL for powdery mildew resistance were identified on linkage groups 1, 7, and 9. The LOD scores ranged between 21.2 and 24.2 in the F_3 and 4.5 and 7.8 in the F_4 families. The QTL on linkage group 1 explained 27 and 10% of the phenotypic variation for powdery mildew resistance in the F_3 and F_4 families, respectively, while the one on linkage group 7 explained 22 and 39% and the one on linkage group 9 explained 43 and 48%.

7.4.2   Seed Traits

Linseed refers specifically to oilseed flax, a crop that is recognized for its high oil content (45–50% in current cultivars) and its high linolenic acid content which is an essential fatty acid (Cloutier 2016). Oil from conventional linseed varieties will generally contain 55–57% alpha-linolenic acid, an omega-3 fatty acid. In comparison, canola (or rapeseed) has approximately 12% alpha-linolenic acid, while all other major oilseed crops have 0–10%. Cloutier et al. (2011) mapped an F_1-derived doubled haploid population of 78 individuals derived from a cross between the yellow-seeded low-linolenic acid line SP2047 and the brown-seeded high-linolenic acid line UGG5-5. They performed QTL analysis for eight seed traits including oil, palmitic acid, stearic acid, oleic acid, linoleic acid, and linolenic acid contents as well as iodine value (a measurement of the degree of unsaturation) and seed coat color. An inverse correlation existed between linoleic acid content and both linolenic acid content and iodine value as a consequence of linoleic acid accumulation in low-linolenic germplasm because they have non-functional fatty acid desaturase 3 (FAD3A and FAD3B). Consequently, the two QTL on linkage groups 7 and 16 were concomitant for the three traits. Further, the QTL on linkage group 7 included the *fad3A* locus encoding one of the two main delta-15 fatty acid desaturases. In this study, phenotyping was performed at two locations, and these QTL accounted for 13–29% of the phenotypic variations depending on the trait. A single QTL accounting for 42% of the phenotypic variation for palmitic acid was identified on linkage group 9. A QTL for color brightness (L∗) accounting for 72% of the phenotypic variation overlapped with the yellow or brown phenotypic marker on linkage group 22.

The recent study by Kumar et al. (2015) is the most comprehensive biparental QTL mapping study to date in flax. The genetic map is well covered with 329 SNP and 362 SSR markers, and the phenotyping was performed for 18 traits on 243 individuals grown in two locations for 4 years. The RIL population was derived from a cross between two Canadian linseed cultivars, namely, CDC Bethune and Macbeth, where the former was the genotype that was sequenced and which serves as the reference genome for flax (Wang et al. 2012). Here, palmitic acid, stearic acid, oleic acid, linoleic acid, and linolenic acid contents, as well as oil and protein contents and iodine value, were evaluated. One QTL each for palmitic acid, linoleic acid, oil, and protein contents, two each for linoleic acid content and iodine value, and three each for stearic and oleic acid were identified for a total of 14 seed-related QTL. The location of the QTL for fatty acid composition differed between the two studies because the SP2047/UGG5-5 cross identified major genes such as *fad3A* which was a null mutant in SP2047 while the second cross did not differ greatly for major genes such as stearoyl-ACP desaturases (*sad1* and *sad2*) and fatty acid desaturases (*fad2A*, *fad2B*, *fad3A*, *fad3B*). This finding is also supported by R^2 values (% phenotypic variation) for individual QTL that were lower than the SP2047/UGG5-5, i.e., ranging from 6% to 13%. Both of the above populations (SP2047/UGG5-5 and CDC Bethune/Macbeth) also served in the construction of the consensus map (Cloutier et al. 2012b).

7.4.3   Agronomic and Fiber-Related Traits

While QTL for agronomic traits were studied using association panels (see below), a single report has been published to date on their discovery using a biparental population. In addition to the 14 QTL for seed traits, six QTL for agronomic and fiber-related traits were identified using the CDC Bethune/Macbeth RIL population (Kumar et al. 2015). For the fiber-related traits, one QTL for cell wall and one for straw weight accounted for 14 and 30% of the phenotypic variance, respectively. Also, one QTL each for thousand seed weight and seeds per boll, which are yield components, and one for yield per se were identified. Finally, one QTL for days to maturity was identified. Interestingly, the QTL for cell wall, straw weight, seeds per boll, yield per se, and days to maturity co-located on linkage group 4. A much higher-density map and larger-scale phenotyping would be required to determine whether this is caused by pleiotropy or linkage. Further, correlations were observed between thousand seed weight, seeds per boll, and yield per se, indicating that yield improvement may be achieved through selection for the simpler yield components.

7.5   Association Mapping Panels

Association mapping (AM) relies upon the identification of associations between phenotype and allele frequencies (Soto-Cerda and Cloutier 2012). It is a powerful genetic mapping tool for crops that provides high-resolution, broad allele coverage and cost-effective gene tagging for the evaluation of ex situ conserved natural genetic diversity of crops. Multiple kinds of populations can be used as AM panels including gene bank collections, elite breeding materials, and specialized populations such as nested association mapping (NAM) and multiparent advanced generation intercross (MAGIC) populations (Sehgal et al. 2016). AM studies have often investigated the variation residing in core collections or elite breeding materials which bypass the expense and time associated with the construction of designer populations. However, potential population and family structures and unbalanced allele frequencies often caused by the diverse geographic origin of the plant material can create spurious associations and overlook the contribution of low frequency alleles (Soto-Cerda et al. 2014b). To overcome these limitations, NAM and MAGIC populations have been developed in many crops (McMullen et al. 2009; reviewed by Huang et al. 2015). NAM populations are developed by crossing a set of diverse founder lines to a single reference line followed by several cycles of selfing to create multiple RILs. MAGIC populations are created by several generations of intercrossing among multiple founder lines. These populations are advantageous from the point of view of increasing rare allele frequency and balancing the overall allele frequencies, thus permitting the capture of favorable QTL alleles of low frequency (Soto-Cerda et al. 2014b).

7.5.1 Core Collections of L. usitatissimum

Worldwide ex situ collections of cultivated flax maintain approximately 10,000 unique accessions (Diederichsen 2007). Some efforts have been made to assemble genetically diverse flax accessions. The N.I. Vavilov Institute established two collections where the first consists of 50 accessions representing donors for specific genes, mostly disease resistance genes (Kutuzova 2000), and where the second comprises 250 lines that captures the variation for morphological, agronomic, and disease resistance characters (Brutch 2002). The Centre for Genetic Resources at Wageningen in the Netherlands has established a fiber flax core collection of 84 individuals from 506 accessions (van Soest and Bas 2002). Core collections consist of a manageable number of accessions expected to represent most of the genetic variability. The process for selecting a minimum sample size with maximum variation has a normalizing effect that is expected to reduce population structure and decrease linkage disequilibrium (LD), thus creating a situation favorable for AM studies (Sehgal et al. 2016).

In Canada, a world collection of 3378 accessions of cultivated flax is maintained by Plant Gene Resources of Canada (PGRC). Since 1998, this collection has been assessed for morphological, agronomic, and quality characters (reviewed by Diederichsen and Fu 2008). This wealth of information permitted the assembly of the Canadian flax core collection of 407 accessions encompassing both the oil and fiber morphotypes (Diederichsen et al. 2013). Its further molecular characterization using 448 SSRs showed its abundant genetic diversity (5.32 alleles per locus), high allelic richness (5.68), weak population structure represented by two major sub-populations (F_{ST} = 0.094), weak family structure (0.023), and fast genome-wide LD decay (<1 cM), all positive attributes for AM studies (Soto-Cerda et al. 2013). Recently, Soto-Cerda et al. (2014a) showed that the Canadian flax core collection captures a similar allelic richness and genetic variability as that observed in 125 accessions of its wild progenitor, pale flax. The presence of primarily cultivars and breeding materials (~85%) in the Canadian flax core collection is advantageous considering that QTL mapping and commercial variety development can be conducted simultaneously (Yu and Buckler 2006).

7.5.2   Linum bienne *Collections*

Pale flax (*Linum bienne* Mill.) is the wild progenitor of cultivated flax (*L. usitatissimum* L.) as supported by morphological, cytological, and molecular characterizations (Diederichsen and Hammer 1995; Fu and Allaby 2010; Fu and Peterson 2010; Gill and Yermanos 1967; Tammes 1928; Soto-Cerda et al. 2011b, 2014a). Both species share the same chromosome number (*n* = 15) and are interfertile. Pale flax is a winter annual or perennial plant with narrow leaves, dehiscent capsules and, that usually displays large variation in the vegetative plant parts and growth habit (Diederichsen and Hammer 1995; Uysal et al. 2010, 2012). In total, 314 pale flax accessions have been stored in gene banks around the world (Diederichsen 2007; Habibollahi et al. 2015; Uysal et al. 2010). The largest collection of 120 accessions is located at the Leibniz Institute of Plant Genetics and Crop Plant Research (IPK). Previous phenotypic and molecular characterizations have been carried out using a small number of pale flax genotypes (Allaby et al. 2005; Diederichsen and Hammer 1995; Fu 2011; Uysal et al. 2010, 2012). The most comprehensive molecular characterization of pale flax was reported by Soto-Cerda et al. (2014a) where genetic diversity, population structure, LD decay, and AM analyses for nine traits were conducted with 112 genome-wide SSRs on 125 accessions of diverse European origins. This report corroborated the long held view that pale flax is a potential source of novel variation for cultivated flax. Indeed, pale flax contains not only high genetic diversity (8.03 alleles per locus) but it is also characterized by a fast LD decay (<1.8 cM) and unique QTL alleles for traits such as thousand seed weight (TSW), seeds per boll (SPB), plant height (PH), and flowering time (FT). While these QTL have potential to improve cultivated flax, they still need to be validated by marker-assisted backcrossing or interspecific QTL mapping.

7.6   Genome-Wide Association Studies

Association mapping generally falls into two broad categories. Candidate gene association studies relate polymorphisms in selected candidate genes that have purported roles in controlling phenotypic variation of a trait, while genome-wide

Fig. 7.2 Genome-wide linkage disequilibrium in the Canadian flax core collection based on one million single nucleotide polymorphism (SNP) markers. The blue dotted line indicates the cutoff value where the LD decays to half of the maximum value ($r^2 = 0.22$) used to determine the genome-wide LD block size

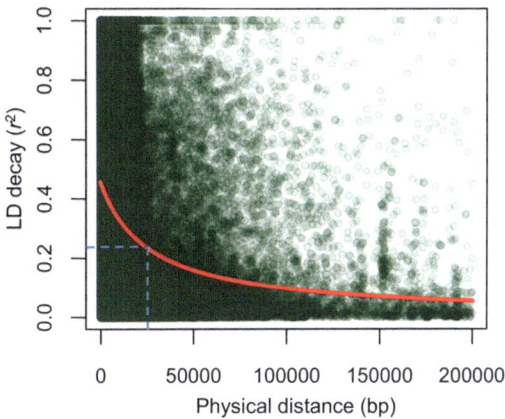

association studies (GWASs) survey genetic variation in the whole genome to find signals of association for complex traits (Zhu et al. 2008). Prior to the advent of NGS technologies, whole genome surveys were conducted with a few hundreds to a couple of thousands of mostly SSR markers (reviewed by Soto-Cerda and Cloutier 2012). The discovery of millions of SNP markers, through the application of NGS technologies, has allowed highly dense GWASs in many crops, including orphan and marginal crops (reviewed by Huang and Han 2014, Soto-Cerda and Cloutier 2012), with the potential not only to identify and map QTL, but also to identify causal polymorphisms at the single nucleotide level. In flax, the first GWASs were conducted using 464 mapped SSR loci (Soto-Cerda et al. 2014b, c), and more recently, the resequencing of the Canadian flax core collection permitted the identification of ~1.7 million SNPs which have been used to fine-tune the genome-wide LD decay, estimated at ~30 kb (Fig. 7.2), to better describe its genetic structure (Fig. 7.3) and to better understand the genetic architecture of yield-related traits (Soto-Cerda et al. 2015).

7.6.1 Agronomic Traits

Nowadays, the suite of genomic resources available for flax molecular breeding provide an opportunity to speed up the development of superior flax cultivars through the identification of QTL for agronomic traits such as TSW, PH, FT, plant branching (PB), lodging resistance (LDG), number of bolls per plant (BPP), bolls per unit area (BPA), and SPB. Soto-Cerda et al. (2014c) using the Canadian flax core collection identified five stable additive QTL for TSW across up to eight environments which explained ~30% of the phenotypic variation. In the same study, a seemingly pleiotropic QTL controlling FT and PH, identified on linkage group 1, accounted for 7.4 and 4.6% of the phenotypic variation, respectively. Other QTL

Fig. 7.3 Neighbor-joining tree constructed using 1.7 million single nucleotide polymorphism (SNP) markers showing the genetic relationships observed in the Canadian flax core collection. The two main morphotypes are color-coded: fibre types are blue and oilseed types are green

pertaining to PB and LDG were also identified (Soto-Cerda et al. 2014c). More recently, a GWAS of yield-related traits in the same flax core collection genotyped with 1.7 M SNPs permitted the identification of 6, 13, and 2 quantitative trait nucleotides (QTNs) affecting the variation of SPB, TSW, and yield per se, respectively (Soto-Cerda et al. 2015). Most of these QTNs were environment-specific and were distributed across ten of the 15 linkage groups. The six QTNs associated to SPB explained on average 7.4% of the phenotypic variation. QTN Lu5-2695662 had the highest positive effect (8.9%, 3.5 seeds/boll) on this trait. The 13 QTNs associated with TSW accounted on average for 7.82% of the phenotypic variation with an average positive effect of 1.3 grams. Two of these QTNs were stable at least in two environments: marker Lu3-30403311 on linkage group 3 accounted for 8%, and marker Lu12-6380748 on linkage group 12 accounted for 5.8% of the phenotypic variation for TSW. The two QTNs associated with yield explained 8.5% of the phenotypic variation and could increase yield by an average of 180 kg/ha. Candidate genes were identified in the vicinity (~30 kb) of these QTNs. For example, in the vicinity of marker Lu5-2695662, associated with SPB, a 1-aminocyclopropane-1-carboxylate synthase 4 involved in ethylene biosynthesis and fruit ripening is an interesting candidate. A stearoyl-ACP desaturase 1 (*sad1*) involved in the fatty acid metabolic pathway was located in the neighboring area of marker Lu10-2429571, a QTN for TSW (Soto-Cerda et al. 2015).

7.6.2   Seed Traits

Seed trait QTL in flax have been previously identified using biparental QTL mapping (Cloutier et al. 2011; Kumar et al. 2015). The first GWAS was conducted for seven seed traits including oil content, palmitic acid, stearic acid, oleic acid, linoleic acid, linolenic acid, and iodine value in the Canadian flax core collection assessed across six environments and genotyped with 460 SSRs (Soto-Cerda et al. 2014b). Nine stable QTL that were significant in at least four environments were identified, varying from one for oil to three for linoleic and linolenic acids. QTL on linkage group 12 for linoleic and linolenic acids co-localized with QTL previously identified by conventional biparental QTL mapping (Cloutier et al. 2011) and some mapped near genes known to be involved in the fatty acid biosynthesis pathway. The QTL for oil content explained 7.6% of the phenotypic variation, and the favorable QTL allele could increase oil content by 1.3%. The three QTL for linoleic and linolenic acids co-located on linkage groups 3, 5, and 12 and explained an overall 7.8% and 6.9% of the traits' variation, respectively. The QTL for iodine value mapped to linkage group 8 and explained 9.4% of the phenotypic variation. Interestingly, 58% of the QTL alleles identified were absent in the Canadian cultivars, suggesting that the Canadian flax core collection possesses alleles potentially useful to improve seed traits (Soto-Cerda et al. 2014b). The broad allelic diversity present in the Canadian flax core collection and the somewhat narrow genetic base of Canadian cultivars partly explain the reason for their lack of detection in the biparental RIL population derived from a cross between the Canadian linseed cultivars CDC Bethune and Macbeth reported by Kumar et al. (2015). However, further validation of QTL identified by both approaches must be conducted in order to accurately apply them in marker-assisted breeding.

7.7   Conclusion

Here, we described the history of the development of molecular markers for flax, starting with RAPDs in the mid to late 1990s up until the discovery of ~1.7 M SNPs approximately 20 years later. Such tremendous progress was possible as a consequence of major investments in flax genomics such as the Genome Canada funded project entitled Total Utilization of Flax GENomics (TUFGEN) and as a consequence of the advent of NGS and its constantly decreasing cost. Years 2011 and 2012 were a turning point for flax genomics resources with the release of major sequence datasets such as the BAC end sequences and physical map (Ragupathy et al. 2011), large-scale ESTs (Venglat et al. 2011), and the whole genome sequence of CDC Bethune (Wang et al. 2012). The resequencing of the 407 accessions of the Canadian flax core collection yielded ~1.7 M SNPs which complemented the earlier efforts in the development of more than 1300 SSRs. Taken together, these resources have fueled flax research, allowing the construction of medium- to high-density genetic maps and precise QTL mapping using either biparental populations or association panels. Numerous QTL for important agronomic and seed quality traits were thus identified in both cultivated flax and its wild progenitor. Some of these QTL are currently being validated in breeding programs using marker-assisted selection to screen early generation germplasm and to introgress novel alleles. The recent refinement of the flax genome reference assembly (You et al. 2018b) promises to be the next major keystone in genomics-assisted breeding for flax.

References

Aldrich J, Cullis CA (1993) RAPD analysis in flax: optimization of yield and reproducibility using *KlenTaq* 1 DNA polymerase, chelex 100, and gel purification of genomic DNA. Plant Mol Biol Report 11:128–141

Allaby RG, Peterson GW, Merriwether DA, Fu YB (2005) Evidence of the domestication history of flax (*Linum usitatissimum*) from genetic diversity of the *sad2* locus. Theor Appl Genet 112:58–65

Anderson PA, Lawrence GJ, Morrish BC, Ayliffe MA, Finnegan EJ, Ellis JG (1997) Inactivation of the flax rust resistance gene *M* associated with loss of a repeated unit within the leucine-rich repeat coding region. Plant Cell 9:641–651

Asgarinia P, Cloutier S, Duguid S, Rashid K, Mirlohi AF, Banik M, Saeidi G (2013) Mapping quantitative trait loci for powdery mildew resistance in flax (*Linum usitatissimum* L.). Crop Sci 53:2462–2472

Bickel CL, Gadani S, Lukacs M, Cullis CA (2011) SSR markers developed for genetic mapping in flax (*Linum usitatissimum* L.). Res Rep Biol 2:23–29

Bo TY, Ma JJ, Chen JX, Miao TY, Zhai WX (2008) Identification of specific molecular markers linked to the rust resistance gene *M4* in flax. Australas Plant Pathol 37:417–420

Bolsheva N, Zelenin AV, Nosova IV, Amosova AV, Samatadze TE, Yurkevich OY, Melnikova NV, Zelenina DA, Volkov AA, Muravenko OV (2015) The diversity of karyotypes and genomes within section *Syllinum* of the genus *Linum* (Linaceae) revealed by molecular cytogenetic markers and RAPD analysis. PLoS One 10:e0122015

Brutch NB (2002) The flax genetic resources collection held at the N.I. Vavilov Institute, Russian Federation. In: Maggioni LM, Pavelek M, van Soest LJM, Lipman E (eds) Flax genetic resources in Europe. IPGRI, Maccarese Rome, pp 61–65

Chen Y, Hausner G, Kenaschuk E, Procunier D, Dribnenki P, Penner G (1998) Identification of microspore-derived plants in anther culture of flax (*Linum usitatissimum* L.) using molecular markers. Plant Cell Rep 18:44–48

Cloutier S (2016) Chapter 31: Linseed. In: Corke H, Faubion J, Seetharaman K, Wrigley C (eds) Encyclopedia of food grains, 2nd edn. Elsevier, Oxford, pp 259–264

Cloutier S, Niu Z, Datla R, Duguid S (2009) EST-derived microsatellites from flax (*Linum usitatissimum* L.). Theor Appl Genet 119:53–63

Cloutier S, Ragupathy R, Niu Z, Duguid S (2011) SSR-based linkage map of flax (*Linum usitatissimum* L.) and mapping of QTLs underlying fatty acid composition traits. Mol Breed 28:437–451

Cloutier S, Miranda E, Ward K, Radovanovic N, Reimer E, Walichnowski A, Datla R, Rowland G, Duguid S, Ragupathy R (2012a) Simple sequence repeat marker development from bacterial artificial chromosome end sequences and expressed sequence tags of flax (*Linum usitatissimum* L.). Theor Appl Genet 125:685–694

Cloutier S, Ragupathy R, Miranda E, Radovanovic N, Reimer R, Walichnowski A, Ward K, Rowland G, Duguid S, Banik M (2012b) Integrated consensus genetic and physical maps of flax (*Linum usitatissimum* L.). Theor Appl Genet 125:1783–1795

Cullis CA, Oh TJ, Gorman MB (1995) Genetic mapping in flax (*Linum usitatissimum*). Proc 3rd Meeting Int Flax Breed Res Group, St Valéry en caux, France, 7–8 November 1995, pp 161–169

Cullis CA, Swami S, Song Y (1999) RAPD polymorphisms detected among the flax genotrophs. Plant Mol Biol 41:795–800

Deng X, Chen XB, Long SH, Wang XC, Gao Y, He DF, Wang J, Wang YF (2008) Microsatellite marker enrichment with magnetic beads in flax. Acta Agron Sin 34:2099–2105

Deng X, Long SH, He DF, Li X, Wang YF, Liu J, Chen XB (2010) Development and characterization of polymorphic microsatellite markers in *Linum usitatissimum*. J Plant Res 123:119–123

Deng X, Long SH, He DF, Li X, Wang YF, Hao DM, Qiu CS, Chen XB (2011) Isolation and characterization of polymorphic microsatellite markers from flax (*Linum usitatissimum* L.). Afr J Biotechnol 10:734–739

Diederichsen A (2007) Ex situ collections of cultivated flax (*Linum usitatissimum* L.) and other species of the genus *Linum* L. Genet Resour Crop Evol 54:661–678

Diederichsen A, Fu YB (2006) Phenotypic and molecular (RAPD) differentiation of four infraspecific groups of cultivated flax (*Linum usitatissimum* L. subsp. *usitatissimum*). Genet Resour Crop Evol 53:77–90

Diederichsen A, Fu YB (2008) Flax genetic diversity as the raw material for future success. In: FAO/ESCORENA (eds) Proceedings of the 2008 international conference on flax and other bast plants. Saskatoon, Canada, July 21–23, pp 270–280

Diederichsen A, Hammer K (1995) Variation of cultivated flax (*Linum usitatissimum* L. subp. *usitatissimum*) and its wild progenitor pale flax (subsp. *angustifolium* (Huds.) Thell.). Genet Resour Crop Evol 42:263–272

Diederichsen A, Kusters PM, Kessler D, Bainas Z, Gugel RK (2013) Assembling a core collection from the flax world collection maintained by Plant Gene Resources of Canada. Genet Resour Crop Evol 60:1479–1485

Dong Y, Xie M, Jiang Y, Xiao N, Du X, Zhang W, Tosser-Klopp G, Wang J, Yang S, Liang J, Chen W, Chen J, Zeng P, Hou Y, Bian C, Pan S, Li Y, Liu X, Wang W, Servin B, Sayre B, Zhu B, Sweeney D, Moore R, Nie W, Shen Y, Zhao R, Zhang G, Li J, Faraut T, Womack J, Zhang Y, Kijas J, Cockett N, Xu X, Zhao S, Wang J, Wang W (2013) Sequencing and automated whole-genome optical mapping of the genome of a domestic goat (*Capra hircus*). Nat Biotechnol 31:135–141

El-Nasr THSA, Mahfouze JA (2013) Genetic variability of golden flax (*Linum usitatissimum* L.) using RAPD markers. World Appl Sci J 26:851–856

Everaert I, De Riek J, De Loose M, Van Waes J, Van Bockstaele E (2001) Most similar variety grouping for distinctness evaluation of flax and linseed (*Linum usitatissimum* L.) varieties by means of AFLP and morphological data. Plant Var Seeds 14:69–87

Fisher PJ, Gardner RC, Richardson TE (1996) Single locus microsatellites isolated using 5' anchored PCR. Nucleic Acids Res 24:4369–4371

Flor HH (1955) Host-parasite interaction in flax rust – its genetics and other implications. Phytopathology 45:680–685

Fry NK, Savelkoul PH, Visca P (2009) Amplified fragment-length polymorphism analysis. Methods Mol Biol 551:89–104

Fu YB (2005) Geographic patterns of RAPD variation in cultivated flax. Crop Sci 45:1084–1091

Fu YB (2006) Redundancy and distinctness in flax germplasm as revealed by RAPD dissimilarity. Plant Genet Resour 4:117–124

Fu YB (2011) Genetic evidence for early flax domestication with capsular dehiscence. Genet Resour Crop Evol 58:1119–1128

Fu YB, Allaby RG (2010) Phylogenetic network of *Linum* species as revealed by non-coding chloroplast DNA sequences. Genet Resour Crop Evol 57:667–677

Fu YB, Peterson GW (2010) Characterization of expressed sequence tag-derived simple sequence repeat markers for 17 *Linum* species. Botany 88:537–543

Fu YB, Peterson GW (2012) Developing genomic resources in two *Linum* species via 454 pyrosequencing and genomic reduction. Mol Ecol Resour 12:492–500

Fu YB, Diederichsen A, Richards KW, Peterson G (2002a) Genetic diversity within a range of cultivars and landraces of flax (*Linum usitatissimum* L.) as revealed by RAPDs. Genet Resour Crop Evol 49:167–174

Fu YB, Peterson G, Diederichsen A, Richards KW (2002b) RAPD analysis of genetic relationships of seven flax species in the genus *Linum* L. Genet Resour Crop Evol 49:253–259

Fu YB, Rowland GG, Duguid SD, Richards KW (2003a) RAPD analysis of 54 North American flax cultivars. Crop Sci 43:1510–1515

Fu YB, Guérin S, Peterson G, Carlson JE, Richards KW (2003b) Assessment of bulking strategies for RAPD analyses of flax germplasm. Genet Resour Crop Evol 50:743–746

Fu YB, Guérin S, Peterson GW, Diederichsen A, Rowland GG, Richards KW (2003c) RAPD analysis of genetic variability of regenerated seeds in the Canadian flax cultivar CDC Normandy. Seed Sci Technol 31:207–211

Gill KS, Yermanos DM (1967) Cytogenetic studies on the genus *Linum* I. Hybrids among taxa with 15 as the haploid chromosome number. Crop Sci 7:623–627

Habibollahi H, Noormohammadi Z, Sheidai M, Farahani F (2015) SSR and EST-SSR-based population genetic structure of *Linum* L. (Linaceae) species in Iran. Genet Resour Crop Evol 63:1127. https://doi.org/10.1007/s10722-015-0306-7

He L, Xiao J, Rashid KY, Yao Z, Li P, Jia G, Wang X, Cloutier S, You FM (2019) Genome-wide association studies for pasmo resistance in flax (Linum usitatissimum L.). Front Plant Sci 9:1982

Huang X, Han B (2014) Natural variations and genome-wide association studies in crop plants. Annu Rev Plant Biol 65:531–551

Huang BE, Verbyla KL, Verbyla AP, Raghavan C, Singh VK, Gaur P, Leung H, Varshney RK, Cavanagh CR (2015) MAGIC populations in crops: current status and future prospects. Theor Appl Genet 128:999–1017

Hwang T-Y, Sayama T, Takahashi M, Takada Y, Nakamoto Y, Funatsuki H, Hisano H, Sasamoto S, Sato S, Tabata S, Kono I, Hoshi M, Hanawa M, Yano C, Xia Z, Harada K, Kitamura K, Ishimoto M (2009) High-density integrated linkage map based on SSR markers in soybean. DNA Res 16:213–225

Kale SM, Pardeshi VC, Kadoo NY, Ghorpade PB, Jana MM, Gupta VS (2012) Development of genomic simple sequence repeat markers for linseed using next-generation sequencing technology. Mol Breed 30:597–606

Kumar S, You FM, Cloutier S (2012) Genome wide SNP discovery in flax through next generation sequencing of reduced representation libraries. BMC Genomics 13:684

Kumar S, You FM, Duguid S, Booker H, Rowland G, Cloutier S (2015) QTL for fatty acid composition and yield in linseed (*Linum usitatissimum* L.). Theor Appl Genet 128:965–984

Kumari N, Thakur SK (2014) Randomly amplified polymorphic DNA – a brief review. Am J Anim Vet Sci 9:6–13

Kutuzova SN (2000) Katalog mirovoj kolekcii VIR, Vypusk 714, Donory chozjajstvenno cennych priznakov dja selekcii l'na-dolgunca. [Catalogue of the world collection at the VIR, volume 714, donors of economically important characters for breeding of fibre flax]. VIR, St. Petersburg, p 50

Lam ET, Hastie A, Lin C, Ehrlich D, Das SK, Austin MD, Deshpande P, Cao H, Nagarajan N, Xiao M, Kwok PY (2012) Genome mapping on nanochannel arrays for structural variation analysis and sequence assembly. Nat Biotechnol 30:771–776

Lawrence GJ, Finnegan EJ, Ayliffe MA, Ellis JG (1995) The *L6* gene for flax rust resistance is related to the Arabidopsis bacterial resistance gene *RPS2* and the tobacco viral resistance gene *N*. Plant Cell 7:1195–1206

Litt M, Luty JA (1989) A hypervariable microsatellite revealed by *in vitro* amplification of a dinucleotide repeat within the cardiac muscle actin gene. Am J Hum Genet 44:397–401

Maurya R, Singh PK, Ranade SA, Yadav HK (2014) Diversity analysis in Indian genotypes of linseed (*Linum usitatissimum* L.) using AFLP markers. Gene 549:171–178

McMullen MD, Kresovich S, Villeda HS, Bradbury P, Li H, Sun Q, Flint-Garcia S, Thornsberry J, Acharya C, Bottoms C, Brown P, Browne C, Eller M, Guill K, Harjes C, Kroon D, Lepak N, Mitchell SE, Peterson B, Pressoir G, Romero S, Oropeza Rosas M, Salvo S, Yates H, Hanson M, Jones E, Smith S, Glaubitz JC, Goodman M, Ware D, Holland JB, Buckler ES (2009) Genetic properties of the maize nested association mapping population. Science 325:737–740

Melnikova NV, Kudryavtseva AV, Zelenin AV, Lakunina VA, Yurkevich OY, Speranskaya AS, Dmitriev AA, Krinitsina AA, Belenikin MS, Uroshlev LA, Snezhkina AV, Sadritdinova AF, Koroban NV, Amosova AV, Samatadze TE, Guzenko EV, Lemesh VA, Savilova AM, Rachinskaia OA, Kishlyan NV, Rozhmina TA, Bolsheva NL, Muravenko OV (2014) Retrotransposon-based molecular markers for analysis of genetic diversity within the genus *Linum*. Biomed Res Int 2014:231589

Muravenko OV, Lemesh VA, Samatadze TE, Amosova AV, Grushetskaya ZE, Popov KV, Semenova OY, Khotyuleva LV, Zelenin AV (2003) Genome comparisons with chromosomal and molecular markers for three closely related flax species and their hybrids. Russ J Genet 39:414–421

Muravenko OV, Yurkevich OY, Bolsheva NL, Samatadze TE, Nosova IV, Zelenina DA, Volkov AA, Popov KV, Zelenin AV (2009) Comparison of genomes of eight species of sections *Linum* and *Adenolinum* from the genus *Linum* based on chromosome banding, molecular markers and RAPD analysis. Genetica 135:245–255

Muravenko OV, Bolsheva NL, Yurkevich OY, Nosova IV, Rachinskaya OA, Samatadze TE, Zelenin AV (2010) Karyogenomics of species of the genus *Linum* L. Russ J Genet 46:1182–1185

Oh TJ, Gorman M, Cullis CA (2000) RFLP and RAPD mapping in flax (*Linum usitatissimum*). Theor Appl Genet 101:590–593

Ottai MES, Al-Kordy MAA, Hussein RM, Hassanein MS (2012) Genetic diversity among Romanian fiber flax varieties under Egyptian conditions. Aust J Basic Appl Sci 6:162–168

Pali V, Mehta N, Verulkar SB, Xalxo MS, Saxena RR (2015) Molecular diversity in flax (*Linum usitatissimum* L.) as revealed by DNA based markers. Int J Plant Res 28:157–165

Powell W, Machray GC, Provan J (1996) Polymorphism revealed by simple sequence repeats. Trends Plant Sci 1:215–222

Rachinskaya OA, Lemesh VA, Muravenko OV, Yurkevich OY, Guzenko EV, Bol'sheva NL, Bogdanova MV, Samatadze TE, Popov KV, Malyshev SV, Shostak NG, Heller K, Hotyleva LV, Zelenin AV (2011) Genetic polymorphism of flax *Linum usitatissimum* based on the use of molecular cytogenetic markers. Russ J Genet 47:56–65

Ragupathy R, Rathinavelu R, Cloutier S (2011) Physical mapping and BAC-end sequence analysis provide initial insights into the flax (*Linum usitatissimum* L.) genome. BMC Genomics 12:217

Rajwade AV, Arora RS, Kadoo NY, Harsulkar AM, Ghorpade PB, Gupta VS (2010) Relatedness of Indian flax genotypes (*Linum usitatissimum* L.): an inter-simple sequence repeat (ISSR) primer assay. Mol Biotechnol 45:161–170

Roose-Amsaleg C, Cariou-Pham E, Vautrin D, Taversier R, Solignac M (2006) Polymorphic microsatellite loci in *Linum usitatissimum*. Mol Ecol Notes 6:796–799

Sehgal D, Singh R, Rani Rajpal V (2016) Chapter 2: Quantitave trait loci mapping in plants: concepts and approaches. In: Rani Rajpal V, Rama Rao S, Raina SN (eds) Molecular breeding for sustainable crop improvement, 2nd edn. Springer, Cham, pp 31–59

Soto-Cerda BJ, Cloutier S (2012) Chapter 3: Association mapping in plant genomes. In: Caliskan M (ed) Genetic diversity in plants. InTech, Rijeka, pp 29–54

Soto-Cerda BJ, Carrasco RA, Aravena GA, Urbina HA, Navarro CS (2011a) Identifying novel polymorphic microsatellites from cultivated flax (*Linum usitatissimum* L.) following data mining. Plant Mol Biol Report 29:753–759

Soto-Cerda BJ, Urbina Saavedra H, Navarro Navarro C, Ortega PM (2011b) Characterization of novel genic SSR markers in *Linum usitatissimum* (L.) and their transferability across eleven *Linum* species. Electron J Biotechnol 14:4. https://doi.org/10.2225/vol14-issue2-fulltext-6

Soto-Cerda BJ, Diederichsen A, Ragupathy R, Cloutier S (2013) Genetic characterization of a core collection of flax (*Linum usitatissimum* L.) suitable for association mapping studies and evidence of divergent selection between fiber and linseed types. BMC Plant Biol 13:78

Soto-Cerda BJ, Diederichsen A, Duguid S, Booker H, Rowland G, Cloutier S (2014a) The potential of pale flax as a source of useful genetic variation for cultivated flax revealed through molecular diversity and association analyses. Mol Breed 34:2091–2107

Soto-Cerda BJ, Duguid S, Booker H, Rowland G, Diederichsen A, Cloutier S (2014b) Association mapping of seed quality traits using the Canadian flax (*Linum usitatissimum* L.) core collection. Theor Appl Genet 127:881–896

Soto-Cerda BJ, Duguid S, Booker H, Rowland G, Cloutier S (2014c) Genomic regions underlying agronomic traits in linseed (*Linum usitatissimum* L.) as revealed by association mapping. J Integr Plant Biol 56:75–87

Soto-Cerda BJ, Duguid S, Booker H, You F, Rowland G, Cloutier S (2015) Genome-wide association study (GWAS) of yield-related traits in flax (*Linum usitatissimum* L.). International Plant & Animal Genome XXIII, San Diego, January 10–14

Spielmeyer W, Green AG, Bittisnich D, Mendham N, Lagudah ES (1998) Identification of quantitative trait loci contributing to Fusarium wilt resistance on an AFLP linkage map of flax (*Linum usitatissimum*). Theor Appl Genet 97:633–641

Stegnii VN, Chudinova YV, Salina EA (2000) RAPD analysis of flax (*Linum usitatissimum* L.) varieties and hybrids of various productivity. Russ J Genet 36:1149–1152

Tammes T (1928) The genetics of the genus *Linum*. Bibliogr Genet 4:1–36

Uysal H, Fu YB, Kurt O, Peterson GW, Diederichsen A, Kusters P (2010) Genetic diversity of cultivated flax (*Linum usitatissimum* L.) and its wild progenitor pale flax (*Linum bienne* Mill.) as revealed by ISSR markers. Genet Resour Crop Evol 57:1109–1119

Uysal H, Kurt O, Fu YB, Diederichsen A, Kusters P (2012) Variation in phenotypic characters of pale flax (*Linum bienne* Mill.) from Turkey. Genet Resour Crop Evol 59:19–30

van Soest LJM, Bas N (2002) Current status of the CGN *Linum* collection. In: Maggioni LM, Pavelek M, van Soest LJM, Lipman E (eds) Flax genetic resources in Europe. IPGRI, Maccarese Rome, pp 44–48

Van Treuren R (2001) Efficiency of reduced primer selectivity and bulked DNA analysis for the rapid detection of AFLP polymorphisms in a range of crop species. Euphytica 117:27–37

Venglat P, Xiang D, Qiu S, Stone S, Tibiche C, Cram D, Alting-Mees M, Nowak J, Cloutier S, Deyholos M, Bekkaoui F, Sharpe A, Wang E, Rowland G, Selvaraj G, Datla R (2011) Gene expression analysis of seed development. BMC Plant Biol 11:74

Wang Z, Hobson N, Galindo L, Shilin Z, Daihu S, McDill J, Yang L, Hawkins S, Neutelings G, Datla R, Lambert G, Galbraith DW, Grassa CJ, Geraldes A, Cronk QC, Cullis C, Dash PK, Kumar PA, Cloutier S, Sharpe A, Wong GKS, Wang J, Deyholos MK (2012) The genome of flax (*Linum usitatissimum*) assembled *de novo* from short shotgun sequence reads. Plant J 72:461–473

Wiesner I, Wiesnerovà D (2004) Statistical correlations of primer thermodynamic stability $\Delta G°$ for enhanced flax ISSR-PCR cultivar authentication. J Agric Food Chem 52:2568–2571

Wiesner I, Wiesnerovà D, Tejklovà E (2001) Effect of anchor and core sequence in microsatellite primers on flax fingerprinting patterns. J Agric Sci 137:37–44

You FM, Li P, Ragupathy R, Kumar S, Zhu T, Luo M-C, Duguid SD, Rashid KY, Booker HM, Deyholos MK, Fu YB, Sharpe AG, Cloutier S (2016) The draft flax genome pseudomolecules. In: The 66th flax Institute of the United States. Fargo, pp 17–24

You F, Li P, Kumar S, Ragupathy R, Banik M, Duguid S, Booker H, Deyholos M, Fu Y-B, Sharpe AG, Cloutier S (2015) The refined flax genome, its evolution and application. Proc 23rd Plant and Animal Genome meeting, San Diego, CA, January 10–15, P1039

You FM, Xiao J, Li P, Yao Z, Jia G, He L, Kumar S, Soto-Cerda BJ, Duguid SD, Booker HM, Rashid KY, Cloutier S (2018a) Genome-wide association study and selection signatures detect genomic regions associated with seed yield and fatty acid composition traits in bi-parental populations of flax (Linum usitatissimum L.). Int J Mol Sci 19:2303

You FM, Xiao J, Li P, Yao Z, Jia G, He L, Zhu T, Luo MC, Deyholos M, Cloutier S (2018b) Chromosome-scale pseudomolecules refined by optical, physical and genetic maps in flax. Plant J 95:371–384

Yu J, Buckler E (2006) Genetic association mapping and genome organization of maize. Curr Opin Biotechnol 17:155–160

Yurkevich OY, Naumenko-Svetlova AA, Bolsheva NL, Samatadze TE, Rachinskaya OA, Kudryavtseva AV, Zelenina DA, Volkov AA, Zelenin AV, Muravenko OV (2013) Investigation of genome polymorphism and seed coat anatomy of species of section *Adenolinum* from the genus *Linum*. Genet Resour Crop Evol 60:661–676

Zhu C, Gore M, Buckler E, Yu J (2008) Status and prospects of association mapping in plants. Plant Genome 1:5–20

Zohary D, Hopf M (2000) Domestication of plants in the Old World : the origin and spread of cultivated plants in West Asia, Europe and the Nile Valley, 3rd edn. Oxford University Press, Oxford

Chapter 8
Genetic Potential and Gene Expression Landscape in Flax

Jonathan S. Griffiths and Raju S. S. Datla

8.1 Introduction

Flax (*Linum usitatissimum* L.) as a domesticated plant has a rich history and is uniquely poised for further agricultural development. First domesticated in the Fertile Crescent as a Neolithic founder crop, flax was cultivated for its seeds and the nutritional oils they contain (Allaby et al. 2005). In southwest Germany, during the late Neolithic (4000–2400 B.C.), flax seeds underwent a dramatic reduction in size, most likely resulting from intensive changes in the use of flax toward fibers (Herbig and Maier 2011). Linen produced from flax fibers was one of the most important textiles in the world until the industrial revolution and the rise of cotton. More recently, flax production has shifted back to seeds and oils due to reduced use of linen relative to other and less expensive fibers and the nutritional benefits of the unique and rare oils produced by flax. The seed and the fiber are two important characteristics of flax that make it a valuable multi-purpose agricultural crop (Soto-Cerda et al. 2014). Despite these valuable traits, genomic resources for flax have developed slowly and have become available only recently compared to other crops. Furthermore, flax as a crop is facing many challenges including low productivity and returns to the growers. This combined with relatively few modern breeding programs primarily supported by academic and public agencies makes flax an excellent target for improvement and development to meet the current and projected future demands. Genomic and transcriptomic analyses provide a foundational resource and establish baseline for further detailed and targeted functional studies and can help identify key genes involved in seed and fiber quality and quantity. Here we review the major advances in flax gene expression and transcriptomics, specifically genes involved in fiber development and seed associated traits.

J. S. Griffiths · R. S. S. Datla (✉)
National Research Council, Saskatoon, SK, Canada
e-mail: raju.datla@nrc-cnrc.gc.ca

© Springer Nature Switzerland AG 2019
C. A. Cullis (ed.), *Genetics and Genomics of Linum*, Plant Genetics and
Genomics: Crops and Models 23, https://doi.org/10.1007/978-3-030-23964-0_8

8.2   Flax Production and Nutritional Qualities

Flax is produced worldwide, and Canada is among the highest producers and exporters (FAOSTAT database 2014). In Canada, flax is primarily produced for its seed. Flaxseed, or linseed, has a number of unique properties that make it an excellent agricultural product. Flax seed oil is used for a number of industrial products including linoleum, paint, varnish, and more. Aside from industrial applications, many studies have suggested that flaxseed can promote multiple different aspects of human health. For instance, flax has been suggested to support weight loss (Wu et al. 2010), lower cholesterol (Kristensen et al. 2012), reduce cancer growth (Thompson et al. 2005), and combat cardiovascular disease (Rodriguez-Leyva et al. 2010). Flax seed is enriched with a number of diverse phytochemicals that can modulate human health including cyclinopepetides, flavonoids, lignans, and the highest content of the essential omega-3 fatty acid α-linolenic acid (ALA) (Dabrowski and Sosulski 1984; Gui et al. 2012). These industrial and health applications make flax a promising crop plant that can be further developed for improved yields and phytochemical production.

The phloem (bast) fibers of flax are an excellent source of cellulose, which is what made them useful in the production of highly valued linens. Bast fibers are composed of up to 90% cellulose and contain relatively little lignin (Day et al. 2005b; del Rio et al. 2011). Bast fibers help to provide mechanical support to the plant and are located between the vascular cambium and the epidermis; they are easily removed from the plant, providing the raw material for linen production. Aside from cellulose, bast fibers also have an unusual lignin composition, further highlighting their use in investigating cell wall biosynthesis (Day et al. 2005b).

Flax seeds are also myxospermous, meaning that they produce seed mucilage. Mucilage is a specialized secondary (deposited after cell expansion has ceased) cell wall that could play roles in germination or seed dispersal (Haughn and Western 2012; Saez-Aguayo et al. 2014). Flax mucilage is rich in pectins and arabinogalactan proteins, which have been linked with immunomodulating properties. Mucilage can compose up to 41% of flax meal and represents a significant target for modifying seed composition (Ray et al. 2013). Mucilage is a major carbon sink for myxospermous species, and reducing mucilage biosynthesis can dramatically increase oil yields (Shi et al. 2012). Identifying mucilage biosynthetic genes through transcriptional analysis could identify major targets useful for modifying or increasing oil composition in flax seeds.

Cyclinopeptides are an interesting group of naturally occurring small cyclic peptides found in seeds and roots of flax (Bruhl et al. 2007; Gui et al. 2012). Composed of eight or nine amino acids, these small peptides could potentially protect against liver damage and suppress the human immune system (Kessler et al. 1986; Wieczorek et al. 1991; Gaymes et al. 1997; Górski et al. 2001). Interestingly, six unique cyclinopeptides are produced from only two genes. Cyclinopeptides amounts in seeds vary considerably between flax cultivars (Gui et al. 2012). Cyclinopeptides can affect oil viscosity and bitterness of flaxseed oil. Cyclinopeptides could therefore be targeted in breeding approaches to alter oil properties or for specific isolation of cyclinopeptides for health applications (Gui et al. 2012; Lao et al. 2014). Further analysis of specific expression patterns for each cyclinopeptide linked to physical properties of flaxseeds would be beneficial.

Many of these salient aspects of flax biology are primed for modification to either improve their quality and quantity or to tweak certain aspects of their production. In order to do so, genes involved in these processes need to be identified and functionally characterized. This can be accomplished through genomic and transcriptomic analysis to identify important genes implicated in unique flax biological processes.

8.3 Flax Genomics

Flax is well suited to breeding and genetic studies. It is self-pollinating and diploid and has a relatively small genome that was recently sequenced. Flax contains a total of $2n = 30$ chromosomes, with an average genome size of ~750 Mb (Wang et al. 2012). Following publication of a detailed genetic and physical map of the genome (Cloutier et al. 2012a, b), a full genome sequence was determined, and this key resource was now available for exploring the genetic and agronomic potential of this important ancient crop (Wang et al. 2012). The genetic potential of flax genome is estimated to encode ~43,000 genes (Wang et al. 2012). Moreover, the full genome sequence is an excellent foundational resource for further breeding efforts. The flax genome has also attracted attention due to a number of unusual structural features within its genome (Cullis 1973, 1976, 1981, 2005; Cullis and Cleary 1986; Schneeberger and Cullis 1991; Wang et al. 2012). In the offspring of a selfed-individual, flax can show up to 15% variation in genome size due to changes in the copy number of certain genetic elements such as a retrotransposon (Chen et al. 2005). Retrotransposons have been shown to induce heritable genomic changes resulting in phenotypic variations and plant growth under different environmental factors (Schneeberger and Cullis 1991; Chen et al. 2005; Johnson et al. 2011). Flax also has a high number of duplicated genes that could be contributing to rapid genomic changes (Wang et al. 2012). Flax has undergone multiple whole genome duplication events, one very ancient at approximately 30 million years ago and another 5–9 million years ago (Wang et al. 2012; Sveinsson et al. 2014). These genome plasticity properties could prove to be useful for further domestication and selection of important genes involved in oil and fiber development.

Association mapping has already been applied toward identifying key regions in the flax genome that are related to yield (Soto-Cerda et al. 2014). Focusing on Canadian flax varieties, simple yield-related traits were analyzed for their segregation with various molecular markers throughout the flax genome (Soto-Cerda et al. 2014). Given the complexity of yield, simplifying association mapping to yield-related traits like the number of seeds or seed size helps to improve the identification of associations. This study was able to isolate 12 unique markers that are associated with six agronomic traits (Soto-Cerda et al. 2014). These markers will prove to be valuable resources for further improving yield.

Recently, the powerful genome editing tool, CRISPR (Clustered Regularly Interspaced Short Palindromic Repeats)/Cas9 (CRISPR-associated system 9), was successfully used to edit a flax gene (Sauer et al. 2016). Through mutation of a single targeted gene, resistance to the broad spectrum herbicide glyphosate was introduced into flax (Sauer et al. 2016). This novel trait could be useful in agricultural settings given the widespread use of glyphosate in agriculture. Furthermore, development of herbicide resistant flax cultivars is a major and significant advance to address the crop losses caused by competing weeds. Recent advances in CRISPR/Cas9-based technologies combined with the newly published genome sequence open up many possibilities to modifying flax to improve yields. Further identification of gene targets through insights gained from gene expression analysis in flax would complement these studies.

8.4   Gene Expression in Flax

Analyzing gene expression during agriculturally important developmental processes can identify the key genes involved more easily than a functional genomic approach. Gene expression studies in flax have focused on important agricultural traits like seed and bast fiber development, and drought and pathogen stress responses. Multiple studies have also examined gene regulation through looking at the distribution of microRNAs (miRNA) and their gene targets in flax. While not extremely numerous yet, transcriptome profiling of flax has largely been performed using expressed sequence tag approaches and in-house microarrays (Day et al. 2005a; Cloutier et al. 2009). Transcriptomic approaches in flax have recently switched to newer and more powerful RNA sequencing (RNA-seq) technologies (Dmitriev et al. 2016; Galindo-Gonzalez and Deyholos 2016; Gorshkov et al. 2016; Zhang and Deyholos 2016). Now with the availability of full flax genome sequence and rapid advances in RNA-seq-based transcriptome tools and technologies, profiling gene expression studies offer a quick and powerful way to identify key genes involved in agronomical important traits.

Bast fibers have emerged as an excellent system to explore secondary cell wall deposition. The plant cell wall is composed of five different components including cellulose, hemicellulose and pectins which link cellulose microfibrils together, lignins, and proteins. Flax bast fibers are highly enriched in cellulose and contain little lignin, which is why they are so useful in linen production (Chantreau et al. 2015). Cellulose-rich bast fibers can be easily separated from lignin-rich xylem cells in the interior of the flax stem, making tissue isolation for expression profiling simple (Mokshina et al. 2014; Gorshkov et al. 2016). Aside from having enriched cellulose, other interesting and unique cell wall polysaccharides are also present and required for normal bast fiber development, like extremely long-chain polysaccharides of galactans, which can be of interest to cell wall biologists (Roach et al. 2011). Overall, profiling gene expression during bast fiber development is an excellent way to learn more about cell wall deposition and identify gene targets to modify bast fiber properties.

Gene expression during bast fiber development has been explored using expressed sequence tag approaches (Roach and Deyholos 2007), microarrays (Roach and Deyholos 2008), and RNA-seq (Gorshkov et al. 2016). Developing bast fibers are enriched in expression of *CELLULOSE SYNTHASE* (*CESA*) genes, arabinogalactan proteins, β-galactosidases, and chitinase-like proteins, all related to cell wall structure and cellulose crystallinity, and are important targets for further improving bast fiber quality (Roach and Deyholos 2007, 2008). As many as 16 putative *CESA* genes have been identified in flax (Chantreau et al. 2015). Surprisingly, few *CESA*s have been identified as being significantly upregulated in association with bast fiber development (Roach and Deyholos 2008; Mokshina et al. 2014; Gorshkov et al. 2016). Despite this, viral-induced gene silencing of flax *CESA* genes resulted in reduced bast fiber number and structure (Chantreau et al. 2015). Many other genes related to cell wall metabolism are upregulated including xyloglucan endotransglucosylases, rhamnogalacturonate lyases, beta-galactosidases, and chitinases (Roach and Deyholos 2008; Gorshkov et al. 2016). These genes encode proteins that degrade components of the plant cell wall including pectins and hemicelluloses, in a process that is required for correct secondary cell wall deposition. Degradation of galactans is required for normal bast fiber development, further illustrating the complex factors at work (Roach et al. 2011). Other notable genes involved in hemicellulose and lignin biosynthesis are downregulated, which is in accord with the high proportion of cellulose to other cell wall components (Gorshkov et al. 2016). Multiple transcription factors have also been shown to be upregulated during bast fiber development, potentially identifying a number of key regulators of secondary wall deposition (Gorshkov et al. 2016). Together, these fiber-focused studies have made significant advances and produced new insights into this agronomically important target in this crop.

Expression analysis of flax seeds has focused on the important economical traits of the seed, like ALA production (Rajwade et al. 2014). Seed ALA content is a major target for improving flax seed major economical attributes. Genes involved in ALA content have unsurprisingly been a major target for gene expression studies. A number of desaturase genes show significant variation correlated with ALA production (Rajwade et al. 2014). FATTY ACID DESATURASE 2 (FAD2), FAD3A, and FAD3B were observed to be highly correlated with ALA accumulation in flaxseeds (Rajwade et al. 2014). This study further suggests that multiple genes and even gene families are required for ALA production (Rajwade et al. 2014). During flax seed development, a number of mucilage-related genes are highly expressed specifically in the seed coat (Venglat et al. 2011). These genes, including *RHAMNOSE SYNTHASE*, a member of a gene family required for mucilage biosynthesis, are important targets for potentially increasing oil yields (Western et al. 2004; Shi et al. 2012; Venglat et al. 2011). Expression analysis studies have already identified key genes involved in oil production, and further expression modification approaches could help to provide tailored oil contents which would be beneficial to production of health-related oils or oils used in industrial settings.

In response to drought, to which flax is susceptible, many genes were identified as differentially regulated (Dash et al. 2014). Unsurprisingly, many of the genes were related to signaling, including hormone biosynthesis and response factors, kinases, and transcription factors (Dash et al. 2014). The most highly affected pathway appeared to be photosynthesis, with many genes downregulated in response to drought (Dash et al. 2014). This study also showed that for the most part, the same genes were downregulated in the shoot and in the root, suggesting a global coordinated response to reduced water availability (Dash et al. 2014). This study helped identify useful targets to help develop resistance to drought stress in flax, a potentially useful trait in environmentally unstable times.

Gene expression analyses have also been used to examine the response of flax to the pathogen *Fusarium oxysporum*, a fungus that causes wilt (Galindo-Gonzalez and Deyholos 2016). This detailed study helped to identify multiple different classes of genes that are differentially regulated in response to *Fusarium*. A variety of responses were identified including upregulation of many different types of defense-related genes including transcription factors, chitinases, peroxidases, and other cell-wall modification genes (Galindo-Gonzalez and Deyholos 2016). Taken together, this study demonstrates the usefulness of transcriptomic approaches to understanding plant-pathogen interactions in flax and also helps identify a number of key genes that could be used to help develop flax lines with improved resistance to pathogens and other environmental stresses.

Beyond direct expression studies, many reports have identified a number of miRNAs that are involved in regulating gene expression in flax (Neutelings et al. 2012; Barvkar et al. 2013). Similar to other plant species, flax contains a number of miRNAs that can regulate the expression of genes that in turn function in diverse developmental aspects of flax (Barvkar et al. 2013). As many as 116 miRNAs have been identified in flax, and most of the identified flax miRNAs target transcription factors and are involved in flower development (Barvkar et al. 2013). Genes involved in cell wall biosynthesis or modifications were also shown to be targeted by miRNAs (Neutelings et al. 2012). Overall, many different transcriptomic approaches have identified multiple different cell wall modification genes that are involved in stress responses and are also regulated by miRNAs, suggesting that further expression profiling of miRNAs could provide further insights into development of agriculturally important traits.

8.5   Challenges and Opportunities

Great progress has been made on identifying gene targets for agricultural improvement in flax, but much remains to be done. Initial studies have demonstrated that flax is very amenable to modern molecular genetics techniques, and that future advances in flax genomic studies will benefit greatly from reduced sequencing costs and the emerging novel approaches. There remain many major areas where flax can be further developed to improve our understanding of this ancient crop and to ideally increase yields.

Many important gene targets involved in key processes like bast fiber and seed development have been identified through transcriptome analysis; however, major gaps are still present that conclusively link specific genes to these processes. Functional data for many genes is lacking, and very few loss-of-function gene mutations are available for flax. Some major approaches to exploring gene function involve using virus-induced gene silencing approaches to downregulate CESA genes, which was the first successful gene knock-down in flax (Chantreau et al. 2015). Other resources include a TILLING (Targeting Induced Local Lesions in Genomes) platform, and mutant collection is now developed and available for flax (Chantreau et al. 2013). One study identified a number of mutant alleles in six desaturase genes in flax, including two alleles that had a significant effect on linoleic acid and ALA content (Thambugala et al. 2013). Beyond these initial studies, the flax TILLING platform and mutant collection have not been widely exploited as yet. An Ion Torrent-based sequencing tool has recently been developed for identifying mutations in flax, which would be useful to identify loss-of-function mutations in this species (Galindo-Gonzalez et al. 2015).

Aside from bast fiber development and quality, seed development, oil production, and health-modulating phytochemicals of flax are arguably the most economically important traits in flax. Transcriptomic analysis of flax seed development is still underdeveloped and lags behind progress in bast fiber development. There are many more publications on the health-modulating properties of several flax seed compounds than there are on expression of the genes involved in the biosynthesis of these compounds. Addressing these could potentially lead to comprehensive understanding of flax seed development and discovery of novel gene targets for further improvements in flax seed's nutritional and health-promoting aspects.

References

Allaby RG, Peterson GW, Merriwether DA, Fu YB (2005) Evidence of the domestication history of flax (Linum usitatissimum L.) from genetic diversity of the sad2 locus. Theor Appl Genet 112:58–65

Barvkar VT, Pardeshi VC, Kale SM, Qui S, Rollins M, Datla R, Gupta VS, Kadoo NY (2013) Genome-wide identification and characterization of microRNA genes and their targets in flax (*Linum usitatissimum*): characterization of flax miRNA genes. Planta 237:1149–1161

Bruhl L, Matthaus B, Fehling E, Wiege B, Lehmann B, Luftmann H, Bergander K, Quiroga K, Scheipers A, Frank O, Hofmann T (2007) Identification of bitter off-taste compondes in the stored cold pressed linseed oil. J Agric Food Chem 55:7864–7868

Chantreau M, Grec S, Gutierrez L, Dalmais M, Pineau C, Demailly H, Paysant-Leroux C, Tavernier R, Trouve JP, Chatterjee M, Guillot X, Brunaud V, Chabbert B, van Wuytwinkel O, Bendahmane A, Thomasset B, Hawkins S (2013) PT-Flax (phenotyping and TILLinG of flax): development of a flax (Linum usitatissimum L.) mutant population and TILLinG platform for forward and reverse genetics. BMC Plant Biol 13:159

Chantreau M, Chabbert B, Billiard S, Hawkins S, Neutelings G (2015) Functional analyses of cellulose synthase genes in flax (*Linum usitatissimum*) by virus-induced gene silencing. Plant Biotechnol J 13:1312–1324

Chen Y, Schneeberger RG, Cullis CA (2005) A site-specific insertion sequence in flax genotrophs induced by environment. New Phytol 167:171–180

Cloutier S, Niu Z, Datla R, Duguid S (2009) Development and analysis of EST-SSRs for flax (Linum usitatissimum L.). Theor Appl Genet 119:53–63

Cloutier S, Miranda E, Ward K, Radovanovic N, Reimer E, Walichnowski A, Datla R, Rowland G, Duquid S, Ragupathy R (2012a) Simple sequence repeat marker development from bacterial artificial chromosome end sequences and expressed sequence tags of flax (Linum usitatissimum L.). Theor Appl Genet 125:685–694

Cloutier S, Ragupathy R, Miranda E, Radovanvic N, Reimer E, Walichnowski A, Ward K, Rowland G, Duguid S, Banik M (2012b) Integrated consensus genetic and physical maps of flax (*Linum usitatissimum* L.). Theor Appl Genet 125:1783. https://doi.org/10.1007/s00122-012-1953-0

Cullis CA (1973) DNA differences between flax genotypes. Nature 243:515–516

Cullis CA (1976) Environmentally induced changes in ribosomal RNA cistron number in flax. Heredity 36:73–80

Cullis CA (1981) DNA sequence organization in the flax genome. Biochim Biophys Acta 652:1–15

Cullis CA (2005) Mechanisms and control of rapid genomic changes in flax. Ann Bot 95:201–206

Cullis CA, Cleary W (1986) Rapidly varying DNA sequences in flax. Can J Genet Cytol 28:252–259

del Rio JC, Rencoret J, Gutierrez A, Nieto L, Jimenez-Barbero J, Martinez AT (2011) Structural characterization of guaiacyl-rich lignins in flax (Linum usitatissimum) fibers and shives. J Agric Food Chem 59:11088–11099

Dabrowski KJ, Sosulski FW (1984) Composition of free and hydrolysable phenolic acids in defatted flours of ten oilseeds. J Am Oil Chem Soc 32:128–130

Dash PK, Cao Y, Jailani AK, Gupta P, Venglat P, Xiang D, Rai R, Sharma R, Thirunavukkarasu N, Abdin MZ, Yadava DK, Singh NK, Singh J, Salvaraj G, Deyholos M, Kumar PA, Datla R (2014) Genome-wide analysis of drought induced gene expression changes in flax (*Linum usitatissimum*). GM Crops Food 5(2):106–119

Day A, Addi M, Kim W, David H, Bert F, Mesnage P, Rolando C, Chabbert B, Neutelings G, Hawkins S (2005a) ESTs from the fibre-bearing stem tissues of flax (Linum usitatissimum L.): expression analysis of sequences related to cell wall development. Plant Biol (Stuttg) 7:23–32

Day A, Ruel K, Neutelings G, Cronier D, David H, Hawkins S, Chabbert B (2005b) Lignification in the flax stem: evidence for an unusual lignin in bast fibers. Planta 222:234–245

Dmitriev AA, Kudryavtseva AV, Krasnow GS, Koroban NV, Speranskaya AS, Krinitsina AA, Belenikin MS, Snezhkina AV, Sadritdinova AF, Kishlyan NV, Rozhmina TA, Yurkevich OY, Muravenko OV, Bolsheva NL, Melnikova NV (2016) Gene expression profiling of flax (Linum usitatissimum L.) under edaphic stress. BMC Plant Biol 16(Suppl 3):237. https://doi.org/10.1186/s12870-016-0927-9

Food and Agriculture Organization of the United Nations (2014) FAOSTAT database. Rome: FAO. Retrieved March 3, 2017 from: http://www.fao.org/faostat/en/#data/QC

Galindo-Gonzalez L, Deyholos MK (2016) RNA-seq transcriptome response of flax (Linum usitatissimum L.) to the pathogenic fungus *Fusarium oxysporum* f. sp. lini. Front Plant Sci 7:1766. eCollection 2016

Galindo-Gonzalez L, Pinzon-Latorre D, Bergen EA, Jensen DC, Deyholos MK (2015) Ion Torrent sequencing as a tool for mutation discovery in the flax (Linum usitatissimum L.) genome. Plant Methods 11:19

Gaymes TJ, Cebrat M, Siemion IZ, Kay JE (1997) Cyclolinopeptide A (CLA) mediates its immunosuppressive activity through cyclophilin-dependent calcineurin inactivation. FEBS Lett 418:224–227

Gorshkov O, Mokshina N, Gorshkov V, Chemikosova S, Gogolev Y, Gorshkova T (2016) Transcriptome portrait of cellulose-enriched flax fibers at advanced stage of specialization. Plant Mol Biol. https://doi.org/10.1007/s11103-016-0571-7

Górski A, Kasprzycka M, Nowaczyk M, Wieczoreck Z, Siemion IZ, Szelejewski W, Kutner A (2001) Cyclolinopeptide: a novel immunosuppressive agent with potential anti-lipemic activity. Transplant Proc 33:553–553

Gui B, Shim YY, Datla RS, Covello PS, Stone SL, Reaney MJ (2012) Identification and quantification of cyclolinopeptides in five flaxseed cultivars. J Agric Food Chem 60:8571–8579

Haughn GW, Western TL (2012) Arabidopsis seed coat mucilage as a specialized cell wall that can be used as a model for genetic analysis of plant cell wall structure and function. Front Plant Sci 3:64

Herbig C, Maier U (2011) Flax for oil or fibre? Morphometric analysis of flax seeds and new aspects of flax cultivation in late Neolithic wetland settlements in southwest Germany. Veg Hist Archaeobotany 20:527–533

Johnson C, Moss T, Cullis C (2011) Environmentally induced heritable changes in flax. J Vis Exp. pii: 2332. https://doi.org/10.3791/2332

Kessler H, Klein M, Müller A, Bats JW, Ziegler K, Frimmer M (1986) Conformation prerequisites for the in vitro inhibition of cholate uptake in hepatocytes by cyclic analogs of antamanide and somatostatin. Angew Chem Int Ed Engl 25:997–999

Kristensen M, Jensen MG, Aarestrup J, Petersen KEN, Sondergaard L, Mikkelsen MS, Astrup A (2012) Flaxseed dietary fibers lowers cholesterol and increase fecal excretion, but magnitude of effect depend on food type. Nutr Metab (Lond) 9:8–16

Lao YW, Mackenzie K, Vincent W, Krokhin OV (2014) Characterization and complete separation of major cyclolinopeptides in flaxseed oil by reversed-phase chromatography. J Sep Sci 34:1788–1796

Mokshina N, Gorshkova T, Deyholos MK (2014) Chitinase-like (CTL) and cellulose synthase (CESA) gene expression in gelatinous-type cellulosic walls of flax (Linum usitatissimum L.) bast fibers. PLoS One 9:e97949

Neutelings G, Fenart S, Lucau-Danila A, Hawkins S (2012) Identification and characterization of miRNAs and their potential targets in flax. J Plant Physiol 169:1754–1766

Rajwade AV, Kadoo NY, Borikar SP, Harsulkar AM, Ghorpade PB, Gupta VS (2014) Differential transcriptional activity of SAD, FAD2 and FAD3 desaturase genes in developing seeds of linseed contributes to varietal variation in the a-linolenic acid content. Phytochemistry 98:41–53

Ray S, Paynel F, Morvan C, Lerouge P, Driouich A, Ray B (2013) Characterization of mucilage polysaccharides, arabinogalactanproteins and cell-wall hemicellulosic polysaccharides isolated from flax seed meal: a wealth of structural moieties. Carbohydr Polym 93:651–660

Roach MJ, Deyholos MK (2007) Microarray analysis of flax (Linum usitatissimunm L.) stems identified transcripts enriched in fibre-bearing phloem tissues. Mol Gen Genomics 278:149–165

Roach MJ, Deyholos MK (2008) Microarray analysis of developing flax hypocotyls identifies novel transcripts correlated with specific stages of phloem fibre differentiation. Ann Bot 102:317–330

Roach MJ, Mokshina NY, Badhan A, Snegireva AV, Hobson N, Deyholos MK, Gorshkova TA (2011) Development of cellulosic secondary walls in flax fibers requires beta-galactosidase. Plant Physiol 156:1351–1363

Rodriguez Leyva D, Dupasquier CM, McCullough R, Pierce GN (2010) The cardiovascular effects of flaxseed and its omega-3 fatty acid, alpha-linolenic acid. Can J Cardiol 26:489–496

Saez-Aguayo S, Rondeau-Mouro C, Macquet A, Kronholm I, Ralet MC, Berger A, Salle C, Poulain D, Granier F, Botran L, Loudet O, de Meaux J, Marion-Poll A, North HM (2014) Local evolution of seed flotation in Arabidopsis. PLoS Genet 10:e1004221

Sauer NJ, Narvaez-Vasquez J, Mozoruk J, Miller RB, Warburg ZJ, Woodward MJ, Mihiret YA, Lincoln TA, Segami RE, Sanders SL, Walker KA, Beetham PR, Schopke CR, Gocal GFW (2016) Oligonucleotide-mediated genome editing provides precision and function to engineered nucleases and antibiotics in plants. Plant Physiol 170:1917–1928

Schneeberger RG, Cullis CA (1991) Specific DNA alterations associated with the environmental induction of heritable changes in flax. Genetics 128:619–630

Shi L, Katavic V, Yu Y, Kunst L, Haughn G (2012) Arabidopsis glabra2 mutant seeds deficient in mucilage biosynthesis produce more oil. Plant J 69:37–46

Soto-Cerda BJ, Duquid S, Booker H, Rowland G, Diederichsen A, Cloutier S (2014) Genomic regions underlying agronomic traits in linseed (Linum usitatissimum L.) as revealed by association mapping. J Integr Plant Biol 56:75–87

Sveinsson S, McDill J, Wong GK, Li J, Li X, Deyholos MK, Cronk QC (2014) Phylogenetic pinpointing of a paleopolyploidy events within the flax genus (Linum) using transcriptomics. Ann Bot 113:753–761

Thambugala D, Duguid S, Loewen E, Rowland G, Booker H, You FM, Cloutier S (2013) Genetic variation of six desaturase genes in flax and their impact on fatty acid composition. Theor Appl Genet 126:2627–2641

Thompson LU, Chen JM, Li T, Strasser-Weippl K, Goss PE (2005) Clin Cancer Res 11:3828–3836

Venglat P, Xiang D, Qui S, Stone SL, Tibiche C, Cram D, Alting-Mees M, Nowak J, Cloutier S, Deyholos M, Bekkaoui F, Sharpe A, Wang E, Rowland G, Selvaraj G, Datla R (2011) Gene expression analysis of flax seed development. BMC Plant Biol 11:74. https://doi.org/10.1186/1471-2229-11-74

Wang Z, Hobson N, Galindo L, Zhu S, Shi D, McDill J, Yang L, Hawkins S, Neutelings G, Datla R, Lambert G, Galbraith DW, Grassa C, Geraldes A, Cronk QC, Cullis C, Dash PK, Kumar PA, Cloutier S, Sharpe AG, Wong GKS, Wang J, Deyholos MK (2012) The genome of flax (Linum usitatissimum) assembled de novo from short shotgun sequence reads. Plant J 72:461–473

Western TL, Young DS, Dean GH, Tan WL, Samuels AL, Haughn GW (2004) Plant Physiol 134:296–306

Wieczorek Z, Bengtsson B, Trojnar J, Siemion IZ (1991) Immunosuppressive activity of cyclo-linopeptide A. Pept Res 4:275–283

Wu H, Pan A, Yu Z, Lu L, Zhang G, Yu D, Zong G, Zhou Y, Chen X, Tang L, Feng Y, Zhou H, Chen X, Li H, Demark-Wahnefried W, Hu FB, Lin X (2010) Lifestyle counseling and supplementation with flaxseed or walnuts influence the management of metabolic syndrome. J Nutr 140:1937–1942

Zhang N, Deyholos MK (2016) RNASeq analysis of the shoot apex of flax (Linum usitatissimum) to identify phloem fiber specification genes. Front Plant Sci 7:950

Chapter 9
Flax Small RNAs

Alexey A. Dmitriev, Anna V. Kudryavtseva, and Nataliya V. Melnikova

9.1 The Role of Small RNAs in Plants

Small RNAs are the key regulators of numerous biological processes in plants: development, reproduction, defense against viruses and transposable elements, response to different stresses, etc. (Li and Zhang 2016; Couzigou and Combier 2016; Shriram et al. 2016; Ito 2013; Ruiz-Ferrer and Voinnet 2009). Plant small RNAs participate in post-transcriptional gene regulation via transcript cleavage or translational repression, transcriptional gene silencing, and RNA-directed DNA methylation (Borges and Martienssen 2015; Bartel 2004; Wu et al. 2010; Schwab et al. 2005; Xie et al. 2015). Most of small RNAs have 20–24 nucleotides in length and initially produced as double-stranded duplexes from RNA precursors. Small RNAs in plants are divided into hairpin RNAs (hpRNAs), which are processed from single-stranded hairpins, and small interfering RNAs (siRNAs), which are processed from double-stranded RNA precursors (Fig. 9.1). hpRNAs include microRNAs (miRNAs) and other hpRNAs, while siRNAs can be subdivided into heterochromatic, secondary, and natural antisense transcript siRNAs (Axtell 2013). miRNA is the most extensively studied class of plant small RNAs.

9.2 miRNA Biogenesis in Plants

miRNA-coding genes (*MIRs*) are transcribed by RNA polymerase II into primary miRNAs (pri-miRNAs). Pri-miRNAs are capped and polyadenylated RNA molecules, which fold into hairpin-like structures. Pri-miRNAs are cleaved by DICER-LIKE 1 (DCL1) with the involvement of accessory proteins into stem-loop precursor

A. A. Dmitriev · A. V. Kudryavtseva · N. V. Melnikova (✉)
Engelhardt Institute of Molecular Biology, Russian Academy of Sciences, Moscow, Russia

© Springer Nature Switzerland AG 2019   129
C. A. Cullis (ed.), *Genetics and Genomics of Linum*, Plant Genetics and
Genomics: Crops and Models 23, https://doi.org/10.1007/978-3-030-23964-0_9

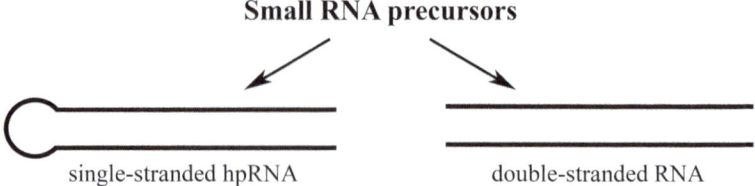

Small RNA precursors

single-stranded hpRNA   double-stranded RNA

Fig. 9.1 Classification of small RNAs

Fig. 9.2 Biogenesis of miRNA in plants. miRNA genes are transcribed by RNA polymerase II (Pol II) into hairpin-like polyadenylated pri-miRNAs. Pri-miRNAs are cleaved by DCL1 initially into a stem-loop pre-miRNAs and then into mature miRNA duplexes, consisting of the active miRNA and miRNA*. HEN1 adds methyl group onto 2′-OH at the 3′ terminal of plant miRNAs. Mature single-stranded miRNAs associated with the AGO are incorporated into RISC, which recognizes target mRNAs via complementarity

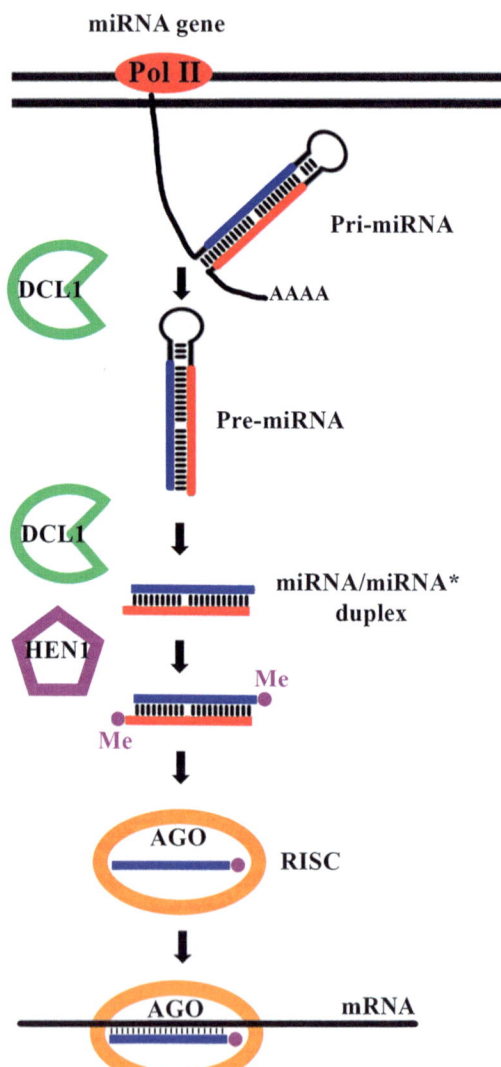

miRNAs (pre-miRNAs). Then pre-miRNAs are processed again by DCL1 and assisting proteins and form mature miRNA duplexes, which consist of the active miRNA strand and its opposite fragment (miRNA*). HUA ENHANCER 1 (HEN1) adds methyl group onto 2′-OH at the 3′ terminal of plant miRNAs that protect them from degradation. The duplexes are transferred into the cytoplasm with the help of HASTY, where miRNAs* are generally degraded, while active mature single-stranded miRNAs associated with the Argonaute (AGO) are incorporated into the RNA-induced silencing complex (RISC). RISC recognizes target mRNA via complementarity and then cleavages target or represses their translation (Fig. 9.2, Bartel 2004; Lee et al. 2004; He and Hannon 2004; Kurihara et al. 2006; Yu et al. 2005; Li et al. 2005; Bologna and Voinnet 2014).

9.3 Identification of miRNAs and Their Targets

For identification and validation of miRNAs and their targets, several approaches were developed. The most commonly used methods are computational prediction based on secondary structure, cloning and sequencing of small RNA libraries, and high-throughput sequencing (Sun et al. 2014). Bioinformatics prediction of plant miRNAs is based on the analysis of current databases to identify sequences with the secondary stem-loop structure, which is typical for miRNA precursors (Adai et al. 2005). The computational approach is widely used due to its low cost and high efficiency, but it failed to determine precise miRNA sequences as several candidate miRNA orthologs or paralogs might be predicted for the miRNA. miRNA-rapid amplification of cDNA ends (miR-RACE) allows one to overcome this problem. This approach includes preparation of miR-enriched library, 5′ and 3′ miRNA-rapid amplification of cDNA ends, and cloning and sequencing of the PCR products (Song et al. 2010). High-throughput sequencing is the most used tool for the study of plant miRNAs that simultaneously provides millions of reads. It allows identification and quantification of even low abundant miRNAs, which are expressed only in the specific tissue or under definite environments. Data on nucleotide sequences of the studied sample is not required for high-throughput sequencing of miRNAs, so this method can be used for less characterized plant species (Ma et al. 2015).

Computational prediction of miRNA targets is based on sequence complementarity of miRNA and its target mRNA (Schwab et al. 2005; Rhoades et al. 2002; Shao et al. 2013; Dai et al. 2011). Identification of truncated transcripts, which result from cleavages by miRNAs or other small RNAs, is used for experimental verification of miRNA targets in 5′ RACE (rapid amplification of cDNA ends) (Llave et al. 2002) and degradome sequencing methods (Steele 1991; German et al. 2008; Henderson and Jacobsen 2008). Another strategy for miRNA target identification is the search for correlation between miRNA expression and expression of its potential target gene using high-throughput sequencing, microarrays, and qPCR (Sun et al. 2014; Oulas et al. 2015; Thomson et al. 2011; Chou et al. 2016).

9.4   miRNAs in Flax

Development of high-throughput sequencing technologies allowed getting extensive data on flax genome and transcriptome. In 2010 and 2011, ESTs (expressed sequence tags) from different flax tissues were sequenced (Fenart et al. 2010; Venglat et al. 2011), and in 2012, whole-genome shotgun sequencing and assembly of flax genome were performed (Wang et al. 2012). These studies enabled a computational prediction of small RNAs in flax. As a result, 297 miRNA loci and 462 small nuclear RNA (snRNA) loci were identified on the basis of flax whole-genome sequencing (Wang et al. 2012). In the study of Neutelings et al., flax ESTs (Fenart et al. 2010) and miRBase database (http://www.mirbase.org/, Kozomara and Griffiths-Jones 2014) were used for miRNA prediction – 20 conserved miRNAs were identified (Neutelings et al. 2012). Conserved miRNAs are those which are common for different plant species. Moss and Cullis used flax genome (Wang et al. 2012) and unigene (Venglat et al. 2011) sequences and miRBase (Griffiths-Jones et al. 2006) and PMRD (Zhang et al. 2010) databases to screen for flax miRNAs. As a result, 12 conserved and 649 novel miRNAs were predicted in flax (Moss and Cullis 2012). In the study of Barvkar et al., 116 conserved miRNAs from 23 families were revealed in the flax genome using a computational approach (Barvkar et al. 2013). Twenty-six miRNAs from 19 families were identified in flax plants by Barozai (Barozai 2012). At present, 124 computationally predicted miRNAs from 24 families are published in miRBase for flax (Fenart et al. 2010; Barvkar et al. 2013).

Apart from bioinformatics prediction, identification of flax miRNAs that are expressed in specific tissues or under defined conditions is necessary. Using high-throughput sequencing of small RNAs, 96 conserved miRNAs from 21 families and 475 novel potential miRNAs were identified in flax leaves under normal, deficient, and excessive nutrition (Melnikova et al. 2016, 2015, 2014). Under alkaline-salt, neutral salt, and alkaline stresses and non-stressed control, 118 to 122 known and 212 to 233 novel miRNAs were revealed by high-throughput sequencing in flax seedlings (Yu et al. 2016).

Summarized data on conserved miRNA families identified in flax are presented in Table 9.1. Results of computational prediction and high-throughput sequencing of flax conserved miRNA families were close: 22 of 24 families were identified using both methods, while lus-miR165 and lus-miR828 were predicted only by bioinformatics analysis.

Most of identified flax miRNA genes are located in intergenic regions (Barvkar et al. 2013) that is in agreement with results obtained for other plants (Tang 2010). Clusters of miRNA genes were identified in flax (Moss and Cullis 2012). For instance, different miRNAs from miR169 family are encoded by the same gene, and they have a common precursor (Barvkar et al. 2013). The same was observed for miRNAs from miR398 family (Neutelings et al. 2012).

Table 9.1 Conserved miRNA families identified in flax

miRNA family	Method of identification	References
lus-miR156	CP/Seq	Wang et al. (2012), Neutelings et al. (2012), Barvkar et al. (2013), Barozai (2012), Melnikova et al. (2016), Yu et al. (2016)
lus-miR159	CP/Seq	Wang et al. (2012), Neutelings et al. (2012), Moss and Cullis (2012), Barvkar et al. (2013), Barozai (2012), Melnikova et al. (2016), Yu et al. (2016)
lus-miR160	CP/Seq	Wang et al. (2012), Barvkar et al. (2013), Melnikova et al. (2016), Yu et al. (2016)
lus-miR162	CP/Seq	Neutelings et al. (2012), Barvkar et al. (2013), Barozai (2012), Melnikova et al. (2016), Yu et al. (2016)
lus-miR164	CP/Seq	Wang et al. (2012), Moss and Cullis (2012), Barvkar et al. (2013), Melnikova et al. (2016), Yu et al. (2016)
lus-miR165	CP	Moss and Cullis (2012)
lus-miR166	CP/Seq	Wang et al. (2012), Neutelings et al. (2012), Barvkar et al. (2013), Barozai (2012), Melnikova et al. (2016), Yu et al. (2016)
lus-miR167	CP/Seq	Wang et al. (2012), Moss and Cullis (2012), Barvkar et al. (2013), Barozai (2012), Melnikova et al. (2016), Yu et al. (2016)
lus-miR168	CP/Seq	Wang et al. (2012), Neutelings et al. (2012), Barvkar et al. (2013), Melnikova et al. (2016), Yu et al. (2016)
lus-miR169	CP/Seq	Wang et al. (2012), Moss and Cullis (2012), Barvkar et al. (2013), Melnikova et al. (2016), Yu et al. (2016)
lus-miR171	CP/Seq	Wang et al. (2012), Neutelings et al. (2012), Barvkar et al. (2013) Barozai (2012), Melnikova et al. (2016), Yu et al. (2016)
lus-miR172	CP/Seq	Wang et al. (2012), Neutelings et al. (2012), Barvkar et al. (2013), Barozai (2012), Melnikova et al. (2016), Yu et al. (2016)
lus-miR319	CP/Seq	Neutelings et al. (2012), Moss and Cullis (2012), Barvkar et al. (2013), Barozai (2012), Melnikova et al. (2016), Yu et al. (2016)
lus-miR390	CP/Seq	Wang et al. (2012), Barvkar et al. (2013), Melnikova et al. (2016), Yu et al. (2016)
lus-miR393	CP/Seq	Neutelings et al. (2012), Barvkar et al. (2013), Melnikova et al. (2016), Yu et al. (2016)
lus-miR394	CP/Seq	Wang et al. (2012), Barvkar et al. (2013), Barozai (2012), Melnikova et al. (2016), Yu et al. (2016)
lus-miR395	CP/Seq	Wang et al. (2012), Barvkar et al. (2013), Melnikova et al. (2016), Yu et al. (2016)
lus-miR396	CP/Seq	Wang et al. (2012), Neutelings et al. (2012), Barvkar et al. (2013), Melnikova et al. (2016), Yu et al. (2016)
lus-miR397	CP/Seq	Barvkar et al. (2013), Barozai (2012), Melnikova et al. (2016), Yu et al. (2016)
lus-miR398	CP/Seq	Wang et al. (2012), Neutelings et al. (2012), Barvkar et al. (2013), Melnikova et al. (2016), Yu et al. (2016)
lus-miR399	CP/Seq	Wang et al. (2012), Barvkar et al. (2013), Melnikova et al. (2016), Yu et al. (2016)

(continued)

Table 9.1 (continued)

miRNA family	Method of identification	References
lus-miR408	CP/Seq	Wang et al. (2012), Neutelings et al. (2012), Moss and Cullis (2012), Barvkar et al. (2013), Barozai (2012), Melnikova et al. (2016), Yu et al. (2016)
lus-miR530	CP/Seq	Wang et al. (2012), Barvkar et al. (2013), Yu et al. (2016)
lus-miR828	CP	Wang et al. (2012), Barvkar et al. (2013)

Note: *CP* computational prediction, *Seq* high-throughput sequencing

9.5   Expression Analysis of miRNAs in Flax

Analysis of miRNA expression level in specific tissues under defined conditions is essential for the understanding of miRNA functions in plants (Sunkar and Zhu 2004; Zhang 2015; Datta and Paul 2015). In flax, miRNA expression evaluation was performed in different tissues under normal and stress conditions using high-throughput sequencing and qPCR. High-throughput sequencing was preferable for identification and preliminary expression evaluation, while qPCR was suitable for a more accurate evaluation of miRNA expression in extended sample sets.

qPCR is the most cost- and time-effective method for gene expression analysis (Die and Roman 2012). However, as miRNAs have a length of about 21–24 bp, special approaches for evaluation of miRNA levels using qPCR are necessary. Stem-loop is a widely used method for qPCR-based miRNA detection, which includes miRNA reverse transcription with specific stem-loop primer and further qPCR (Chen et al. 2005; Salone and Rederstorff 2015; Yang et al. 2014; Tong et al. 2015, Fig. 9.3). The other approach for qPCR quantification of miRNA levels was also developed (Shen et al. 2015).

In flax, miRNAs from the following families were highly expressed: in anthers – miR162, miR166, miR167, miR168, and miR408; in etiolated seedlings, stem tissues, and apex – miR172; in etiolated seedling and leaf – miR156; in apex – miR166 (Neutelings et al. 2012; Barvkar et al. 2013).

Under phosphate deficiency, upregulation of miR399 and downregulation of novel potential lus-miR-N1 were revealed, while under excessive nutrition, upregulation of miR395 was observed (Melnikova et al. 2016, 2015). Under alkaline, neutral salt, and alkaline-salt stresses, miRNA levels were evaluated using qPCR and high-throughput sequencing that resulted in the identification of numerous miRNAs with stress-associated expression alterations (Yu et al. 2016).

Thus, patterns of miRNA expression were determined in different flax tissues. Moreover, differentially expressed miRNAs were identified in flax plants under abiotic stresses. However, more genotypes with diverse characteristics under different stress conditions should be studied to understand the regulatory role of miRNAs in flax.

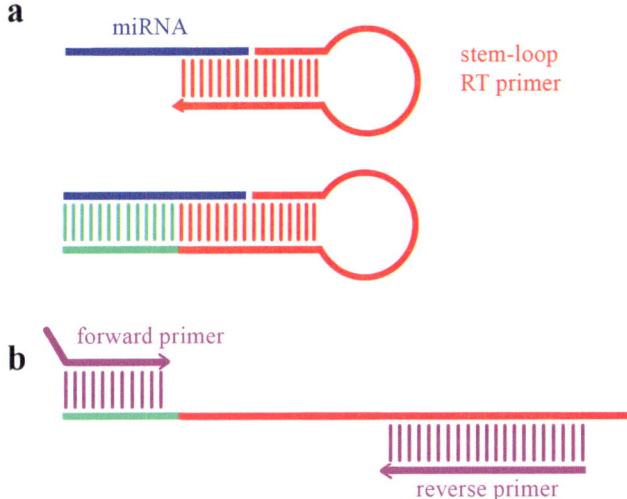

Fig. 9.3 qPCR analysis of miRNA expression using stem-loop RT primer. (**a**) Reverse transcription of miRNA with stem-loop primer; (**b**) qPCR with TaqMan probe (specific or universal) or DNA-binding dye (SYBR Green, EvaGreen)

9.6   Target Prediction of Flax miRNAs

miRNAs regulate gene expression through cleavage or translational repression of target transcripts to which they have high complementarity (Li et al. 2014). Therefore, identification of miRNA targets is necessary for understanding the role of miRNA in plant growth, development, and adaptation to different environments.

For prediction of miRNA targets, psRNATarget software (http://plantgrn.noble. org/psRNATarget/) (Dai and Zhao 2011) was used in the majority of flax studies (Neutelings et al. 2012; Moss and Cullis 2012; Barvkar et al. 2013; Melnikova et al. 2016; Yu et al. 2016). Computational target prediction for flax miRNAs enables the identification of genes, whose expression regulation could be implemented via miRNAs (Table 9.2). These data are useful for the determination of the processes that are controlled by miRNAs in flax plants.

Although potential miRNA targets were predicted in flax, experimental data verification was necessary. miRNAs repress the expression of their targets, so the negative correlation between miRNA expression and expression of predicted targets could indicate their interaction (Oulas et al. 2015; Thomson et al. 2011; Hausser and Zavolan 2014). Expression of miRNAs and their potential target genes were evaluated in flax using qPCR and degradome sequencing. The negative correlation was revealed for lus-miR156a and squamosa promoter-binding protein (for convenience, here and further the proteins encoded by the target genes are given); lus-miR166b and leucine zipper protein HOX32; lus-miR166d and homeodomain transcription factor; lus-miR168b and cytochrome P450; lus-miR172e and AP2 domain-containing transcription factor; lus-miR172 and AP2 domain-containing

Table 9.2 miRNA targets in flax

miRNA family	Proteins encoded by the predicted target genes	References
lus-miR156	Squamosa promoter-binding protein Sec61 transport protein Nonclathrin coat protein zeta2-cop F-box family protein LIGULELESS1 protein SPL domain class transcription factor E3 ubiquitin protein ligase upl2	Neutelings et al. (2012), Barvkar et al. (2013)
lus-miR159	Allyl alcohol dehydrogenase Phosphoglyceride transfer family protein Enoyl-CoA hydratase/protein tyrosine phosphatase E3 ubiquitin protein ligase DCP2 (DECAPPING 2) hydrolase GTP-binding protein Arginine-tRNA-protein transferase Ring finger protein Pm27 3-hydroxyacyl-CoA dehydratase mRNA-decapping enzyme subunit 2 Trafficking protein particle complex subunit 6b Myb domain protein Mitochondrial transcription termination factor family protein	Neutelings et al. (2012), Barvkar et al. (2013), Yu et al. (2016)
lus-miR160	Auxin response factor Arsenical pump-driving ATPase	Barvkar et al. (2013), Yu et al. (2016)
lus-miR162	Eukaryotic translation initiation factor Dicer-like 1 protein Initiation factor 3G	Neutelings et al. (2012), Barvkar et al. (2013)
lus-miR164	NAC domain-containing protein CCAAT-binding transcription factor subunit A S-like RNase NUC173 domain-containing protein Methionine sulfoxide reductase	Barvkar et al. (2013)

lus-miR166	Zinc finger (CCCH-type)/Nucleic acid BP	Neutelings et al. (2012), Barvkar et al. (2013)
	Pentatricopeptide repeat (PPR)-containing protein	
	UDP-glucuronic acid 4-epimerase	
	Homeodomain transcription factor	
	EARLY flowering 4 protein	
	Brassinosteroid LRR receptor kinase precursor	
	Electron transfer flavoprotein beta-subunit	
lus-miR167	LIM1	Barvkar et al. (2013)
	Corticosteroid 11-beta-dehydrogenase	
	Cysteine protease	
lus-miR168	an1-type zinc finger protein	Neutelings et al. (2012), Barvkar et al. (2013)
	20S proteasome beta subunit	
	Nodulin-like protein	
	3-ketoacyl-CoA synthase	
	Structural maintenance of chromosome	
	Putative ubiquitin-associated (UBA) protein	
	Cytochrome P450	
lus-miR169	Tetratricopeptide-like helical domain-containing protein	Barvkar et al. (2013), Yu et al. (2016)
	Transducin family protein	
	Jasmonate-zim-domain protein, GRAS family transcription factor	
lus-miR171	Methylenetetrahydrofolate dehydrogenase	Neutelings et al. (2012); Barvkar et al. (2013)
	Beta-1,3-glucanase-like protein	
	Potassium transporter	
	ADP-ribosylation factor GTPase-activating protein	
	Beta-glucosidase	
	FAD/NAD(P)-binding oxidoreductase family protein	
	Glucan endo-1,3-beta-glucosidase precursor	

(continued)

Table 9.2 (continued)

miRNA family	Proteins encoded by the predicted target genes	References
lus-miR172	DNA-binding protein Proton-exporting ATPase Xyloglucan endotransglycosylase/hydrolase Trehalose 6-phosphate synthase AP2 domain-containing transcription factor Polyadenylate-binding protein cdc2MsC Endoribonuclease/protein kinase IRE1 5-formyltetrahydrofolate cyclo-ligase ATP-dependent Clp protease proteolytic subunit Auxin-responsive protein IAA6 Phytochrome-associated protein 1 Aspartic proteinase precursor Sialin IAA-amino acid hydrolase ILR1 Calcium lipid-binding protein Ubiquitin-protein ligase	Neutelings et al. (2012), Barvkar et al. (2013)
lus-miR319	TCP 26S proteasome non-ATPase regulatory subunit Ribosomal protein L24 Protein hira hir1 Ubiquitin carboxyl-terminal hydrolase Myb domain protein	Neutelings et al. (2012), Moss and Cullis (2012), Barvkar et al. (2013), Yu et al. (2016)
lus-miR390	Serine-threonine protein kinase	Barvkar et al. (2013)
lus-miR393	Leucyl-tRNA synthetase Ribonucleotide reductase Sucrose phosphate synthase Auxin signaling F-box	Neutelings et al. (2012), Yu et al. (2016)

lus-miR394	ATP-binding protein Regulatory-associated protein of mTOR Plant sec1 FGGY family of carbohydrate kinase S-adenosyl-L-methionine-dependent methyltransferases superfamily protein Galactose oxidase/kelch repeat superfamily protein Signal transduction histidine kinase FGGY family of carbohydrate kinase Jasmonate-zim-domain protein	Barvkar et al. (2013), Yu et al. (2016)
lus-miR395	Sulfite reductase Sulfate adenylyltransferase R3h domain containing protein Nuclear matrix protein 1	Barvkar et al. (2013)
lus-miR396	Peroxin-3 family protein Prefoldin subunit Ceramidase family protein Coatomer protein complex Ribonuclease family protein Aspartic proteinase Phytoene dehydrogenase NADH-plastoquinone oxidoreductase 2-dehydro-3-deoxyphosphoheptonate aldolase Leucine-rich repeat transmembrane protein kinase Heat shock protein 70 MAP kinase ATP-binding protein Leunig-like protein GRL1 Putative growth-regulating factor 12 Aldo/keto reductase Acyl-CoA oxidase	Neutelings et al. (2012), Barvkar et al. (2013)

(continued)

Table 9.2 (continued)

miRNA family	Proteins encoded by the predicted target genes	References
lus-miR397	Laccase UDP-glucose 4-epimerase Adaptin ear-binding coat-associated protein Rac-GTP-binding protein	Barvkar et al. (2013)
lus-miR398	RNA helicase BHLH domain class transcription factor Cysteine desulfurylase MO25 protein Ras-related protein Putative transporter ABC transporter Uracil phosphoribosyltransferase Glucose-6-phosphate/phosphate translocator 1	Neutelings et al. (2012), Barvkar et al. (2013)
lus-miR399	CDK-activating kinase Ubiquitin-conjugating enzyme E2	Barvkar et al. (2013), Melnikova et al. (2016)
lus-miR408	Blue copper protein Geranylgeranyl diphosphate synthase Nuclear transcription factor Y Plantacyanin Chloroplast import apparatus	Neutelings et al. (2012), Yu et al. (2016)

transcription factor; and lus-miR396c and GREEN-RIPE LIKE 1 in different flax tissues (Barvkar et al. 2013). Inverse expression profiles were also observed for lus-miR408 and BCP; lus-miR156 and F-box protein; lus-miR159 and E3 ubiquitin ligase; lus-miR171 and K+ transporter; lus-miR319 and ribosomal protein L24; lus-miR393 and leucyl-tRNA synthetase; lus-miR396 and aspartic proteinase; and lus-miR398 and MO25 protein in different flax tissues and developmental stages (Neutelings et al. 2012). Further, the negative correlation was observed for lus-miR399 expression and expression of gene encoding ubiquitin-conjugating enzyme E2 and for lus-miR-N1 and gene encoding ubiquitin-activating enzyme E1 in flax under normal and imbalanced nutrition (Melnikova et al. 2016). Moreover, inverse expression changes were revealed in flax under saline, alkaline, or saline-alkaline stresses for lus-miR156 and tetratricopeptide repeat-like superfamily protein; lus-miR159 and cystatin/monellin superfamily protein; lus-miR160 and transducin/ WD40 repeat-like superfamily protein; lus-miR162 and heat shock protein 70; lus-miR164 and ARM repeat superfamily protein; lus-miR166 and pathogenesis-related gene 1; lus-miR167 and G-box binding factor 1; lus-miR168 and methionine gamma-lyase; lus-miR169 and transducin/WD40 repeat-like superfamily protein; lus-miR169 and profilin 5; lus-miR169 and alpha/beta-hydrolases superfamily protein; lus-miR171 and UDP-glucosyl transferase 85A3; lus-miR171 and jasmonate-zim-domain protein 3; lus-miR172 and deoxyxylulose-5-phosphate synthase; lus-miR319b and O-methyltransferase family protein; lus-miR390 and purine permease 3; lus-miR393 and F-box family protein; lus-miR394 and pyruvate dehydrogenase kinase; lus-miR395 and ATP sulfurylase 1; lus-miR396 and xylose isomerase family protein; lus-miR397 and REF4-related 1; lus-miR397 and 3-deoxy-d-arabino-heptulosonate 7-phosphate synthase; lus-miR398 and copper/zinc superoxide dismutase 2; lus-miR398 and myosin family protein with DiI domain; lus-miR399 and trigalactosyldiacylglycerol2; lus-miR399 and cofactor-independent phosphoglycerate mutase; lus-miR408 and cyclophilin 20-2; lus-miR530 and WRKY family transcription factor; and lus-miR828 and ribosomal protein L3 family protein using miRNA and transcriptome profiling and degradome sequencing (Yu et al. 2016).

Thus, bioinformatics and experimental identification of miRNA targets in flax enabled prediction of genes, which are regulated via miRNAs, and determination of their functions. However, compared to well-studied plant species, obtained data for flax is still limited, and future analysis of miRNA functions is necessary.

9.7 Conclusions

Small RNAs play an essential role in the regulation of numerous processes in plants, including growth and development and responses to different stresses. In flax, currently, only miRNA class of small RNAs is extensively studied. More than one hundred conserved and hundreds of novel potential miRNAs were identified in flax using bioinformatics methods and high-throughput sequencing. Expression profiles of miRNAs were determined in different flax tissues. Stress-responsive miRNAs with expression alterations under deficient and excessive nutrition and alkaline, neutral salt, and alkaline-salt conditions were revealed. Finally, miRNA targets were identified in flax using computational prediction, high-throughput sequencing, and qPCR analysis. Obtained results brought new insights into the regulation of flax cell processes via miRNA.

9.8   Future Perspectives

Although miRNAs and their targets were identified in flax, more detailed characterization of miRNA functions is necessary. Single miRNA can regulate the expression of several genes, and expression of a particular gene can be controlled by several miRNAs. Therefore, miRNA studies should be more complex and focused on miRNA-regulated networks. Moreover, in flax, only miRNA is in the focus of research, while other classes of small RNAs are almost unstudied. A comprehensive investigation of different small RNAs will provide new opportunities for understanding the mechanisms of growth, development, reproduction, and stress response in flax plants.

References

Adai A, Johnson C, Mlotshwa S, Archer-Evans S, Manocha V, Vance V, Sundaresan V (2005) Computational prediction of miRNAs in Arabidopsis thaliana. Genome Res 15(1):78–91

Axtell MJ (2013) Classification and comparison of small RNAs from plants. Annu Rev Plant Biol 64:137–159

Barozai MYK (2012) In silico identification of micrornas and their targets in fiber and oil producing plant flax (Linum usitatissimum L.). Pak J Bot 44:1357–1362

Bartel DP (2004) MicroRNAs: genomics, biogenesis, mechanism, and function. Cell 116(2):281–297

Barvkar VT, Pardeshi VC, Kale SM, Qiu S, Rollins M, Datla R, Gupta VS, Kadoo NY (2013) Genome-wide identification and characterization of microRNA genes and their targets in flax (Linum usitatissimum): characterization of flax miRNA genes. Planta 237(4):1149–1161

Bologna NG, Voinnet O (2014) The diversity, biogenesis, and activities of endogenous silencing small RNAs in Arabidopsis. Annu Rev Plant Biol 65:473–503

Borges F, Martienssen RA (2015) The expanding world of small RNAs in plants. Nat Rev Mol Cell Biol 16(12):727–741

Chen C, Ridzon DA, Broomer AJ, Zhou Z, Lee DH, Nguyen JT, Barbisin M, Xu NL, Mahuvakar VR, Andersen MR et al (2005) Real-time quantification of microRNAs by stem-loop RT-PCR. Nucleic Acids Res 33(20):e179

Chou CH, Chang NW, Shrestha S, Hsu SD, Lin YL, Lee WH, Yang CD, Hong HC, Wei TY, Tu SJ et al (2016) miRTarBase 2016: updates to the experimentally validated miRNA-target interactions database. Nucleic Acids Res 44(D1):D239–D247

Couzigou JM, Combier JP (2016) Plant microRNAs: key regulators of root architecture and biotic interactions. New Phytol 212(1):22–35

Dai X, Zhao PX (2011) psRNATarget: a plant small RNA target analysis server. Nucleic Acids Res 39(Web Server issue):W155–W159

Dai X, Zhuang Z, Zhao PX (2011) Computational analysis of miRNA targets in plants: current status and challenges. Brief Bioinform 12(2):115–121

Datta R, Paul S (2015) Plant microRNAs: master regulator of gene expression mechanism. Cell Biol Int 39(11):1185–1190

Die JV, Roman B (2012) RNA quality assessment: a view from plant qPCR studies. J Exp Bot 63(17):6069–6077

Fenart S, Ndong YP, Duarte J, Riviere N, Wilmer J, van Wuytswinkel O, Lucau A, Cariou E, Neutelings G, Gutierrez L et al (2010) Development and validation of a flax (Linum usitatissimum L.) gene expression oligo microarray. BMC Genomics 11:592

German MA, Pillay M, Jeong DH, Hetawal A, Luo S, Janardhanan P, Kannan V, Rymarquis LA, Nobuta K, German R et al (2008) Global identification of microRNA-target RNA pairs by parallel analysis of RNA ends. Nat Biotechnol 26(8):941–946

Griffiths-Jones S, Grocock RJ, van Dongen S, Bateman A, Enright AJ (2006) miRBase: microRNA sequences, targets and gene nomenclature. Nucleic Acids Res 34(Database issue):D140–D144

Hausser J, Zavolan M (2014) Identification and consequences of miRNA-target interactions – beyond repression of gene expression. Nat Rev Genet 15(9):599–612

He L, Hannon GJ (2004) MicroRNAs: small RNAs with a big role in gene regulation. Nat Rev Genet 5(7):522–531

Henderson IR, Jacobsen SE (2008) Sequencing sliced ends reveals microRNA targets. Nat Biotechnol 26(8):881–882

Ito H (2013) Small RNAs and regulation of transposons in plants. Genes Genet Syst 88(1):3–7

Kozomara A, Griffiths-Jones S (2014) miRBase: annotating high confidence microRNAs using deep sequencing data. Nucleic Acids Res 42(Database issue):D68–D73

Kurihara Y, Takashi Y, Watanabe Y (2006) The interaction between DCL1 and HYL1 is important for efficient and precise processing of pri-miRNA in plant microRNA biogenesis. RNA 12(2):206–212

Lee Y, Kim M, Han J, Yeom KH, Lee S, Baek SH, Kim VN (2004) MicroRNA genes are transcribed by RNA polymerase II. EMBO J 23(20):4051–4060

Li C, Zhang B (2016) MicroRNAs in control of plant development. J Cell Physiol 231(2):303–313

Li J, Yang Z, Yu B, Liu J, Chen X (2005) Methylation protects miRNAs and siRNAs from a 3′-end uridylation activity in Arabidopsis. Curr Biol 15(16):1501–1507

Li J, Reichel M, Li Y, Millar AA (2014) The functional scope of plant microRNA-mediated silencing. Trends Plant Sci 19(12):750–756

Llave C, Xie Z, Kasschau KD, Carrington JC (2002) Cleavage of Scarecrow-like mRNA targets directed by a class of Arabidopsis miRNA. Science 297(5589):2053–2056

Ma X, Tang Z, Qin J, Meng Y (2015) The use of high-throughput sequencing methods for plant microRNA research. RNA Biol 12(7):709–719

Melnikova NV, Belenikin MS, Bolsheva NL, Dmitriev AA, Speranskaya AS, Krinitsina AA, Samatadze TE, Amosova AV, Muravenko OV, Zelenin AV et al (2014) Flax inorganic phosphate deficiency responsive miRNAs. J Agric Sci 6(6):156–160

Melnikova NV, Dmitriev AA, Belenikin MS, Speranskaya AS, Krinitsina AA, Rachinskaia OA, Lakunina VA, Krasnov GS, Snezhkina AV, Sadritdinova AF et al (2015) Excess fertilizer responsive miRNAs revealed in Linum usitatissimum L. Biochimie 109:36–41

Melnikova NV, Dmitriev AA, Belenikin MS, Koroban NV, Speranskaya AS, Krinitsina AA, Krasnov GS, Lakunina VA, Snezhkina AV, Sadritdinova AF et al (2016) Identification, expression analysis, and target prediction of flax genotroph microRNAs under normal and nutrient stress conditions. Front Plant Sci 7:399

Moss TY, Cullis CA (2012) Computational prediction of candidate microRNAs and their targets from the completed Linum usitatissimum genome and EST database. J Nucleic Acids Investig 3:e2, 9–17

Neutelings G, Fenart S, Lucau-Danila A, Hawkins S (2012) Identification and characterization of miRNAs and their potential targets in flax. J Plant Physiol 169(17):1754–1766

Oulas A, Karathanasis N, Louloupi A, Pavlopoulos GA, Poirazi P, Kalantidis K, Iliopoulos I (2015) Prediction of miRNA targets. Methods Mol Biol 1269:207–229

Rhoades MW, Reinhart BJ, Lim LP, Burge CB, Bartel B, Bartel DP (2002) Prediction of plant microRNA targets. Cell 110(4):513–520

Ruiz-Ferrer V, Voinnet O (2009) Roles of plant small RNAs in biotic stress responses. Annu Rev Plant Biol 60:485–510

Salone V, Rederstorff M (2015) Stem-loop RT-PCR based quantification of small non-coding RNAs. Methods Mol Biol 1296:103–108

Schwab R, Palatnik JF, Riester M, Schommer C, Schmid M, Weigel D (2005) Specific effects of microRNAs on the plant transcriptome. Dev Cell 8(4):517–527

Shao C, Chen M, Meng Y (2013) A reversed framework for the identification of microRNA-target pairs in plants. Brief Bioinform 14(3):293–301

Shen Y, Tian F, Chen Z, Li R, Ge Q, Lu Z (2015) Amplification-based method for microRNA detection. Biosens Bioelectron 71:322–331

Shriram V, Kumar V, Devarumath RM, Khare TS, Wani SH (2016) MicroRNAs as potential targets for abiotic stress tolerance in plants. Front Plant Sci 7:817

Song C, Fang J, Wang C, Guo L, Nicholas KK, Ma Z (2010) MiR-RACE, a new efficient approach to determine the precise sequences of computationally identified trifoliate orange (Poncirus trifoliata) microRNAs. PLoS One 5(6):e10861

Steele AD (1991) Shift in genomic RNA patterns of human rotaviruses isolated from white children in South Africa. S Afr Med J 79(3):143–145

Sun X, Zhang Y, Zhu X, Korir NK, Tao R, Wang C, Fang J (2014) Advances in identification and validation of plant microRNAs and their target genes. Physiol Plant 152(2):203–218

Sunkar R, Zhu JK (2004) Novel and stress-regulated microRNAs and other small RNAs from Arabidopsis. Plant Cell 16(8):2001–2019

Tang G (2010) Plant microRNAs: an insight into their gene structures and evolution. Semin Cell Dev Biol 21(8):782–789

Thomson DW, Bracken CP, Goodall GJ (2011) Experimental strategies for microRNA target identification. Nucleic Acids Res 39(16):6845–6853

Tong L, Xue H, Xiong L, Xiao J, Zhou Y (2015) Improved RT-PCR assay to quantitate the pri-, pre-, and mature microRNAs with higher efficiency and accuracy. Mol Biotechnol 57(10):939–946

Venglat P, Xiang D, Qiu S, Stone SL, Tibiche C, Cram D, Alting-Mees M, Nowak J, Cloutier S, Deyholos M et al (2011) Gene expression analysis of flax seed development. BMC Plant Biol 11:74

Wang Z, Hobson N, Galindo L, Zhu S, Shi D, McDill J, Yang L, Hawkins S, Neutelings G, Datla R et al (2012) The genome of flax (Linum usitatissimum) assembled de novo from short shotgun sequence reads. Plant J 72(3):461–473

Wu L, Zhou H, Zhang Q, Zhang J, Ni F, Liu C, Qi Y (2010) DNA methylation mediated by a microRNA pathway. Mol Cell 38(3):465–475

Xie M, Zhang S, Yu B (2015) microRNA biogenesis, degradation and activity in plants. Cell Mol Life Sci 72(1):87–99

Yang LH, Wang SL, Tang LL, Liu B, Ye WL, Wang LL, Wang ZY, Zhou MT, Chen BC (2014) Universal stem-loop primer method for screening and quantification of microRNA. PLoS One 9(12):e115293

Yu B, Yang Z, Li J, Minakhina S, Yang M, Padgett RW, Steward R, Chen X (2005) Methylation as a crucial step in plant microRNA biogenesis. Science 307(5711):932–935

Yu Y, Wu G, Yuan H, Cheng L, Zhao D, Huang W, Zhang S, Zhang L, Chen H, Zhang J et al (2016) Identification and characterization of miRNAs and targets in flax (Linum usitatissimum) under saline, alkaline, and saline-alkaline stresses. BMC Plant Biol 16(1):124

Zhang B (2015) MicroRNA: a new target for improving plant tolerance to abiotic stress. J Exp Bot 66(7):1749–1761

Zhang Z, Yu J, Li D, Zhang Z, Liu F, Zhou X, Wang T, Ling Y, Su Z (2010) PMRD: plant microRNA database. Nucleic Acids Res 38(Database issue):D806–D813

Chapter 10
Development of Flax (*Linum usitatissimum* L.) Mutant Populations for Forward and Reverse Genetics

Sébastien Grec, Marion Dalmais, Manash Chatterjee, Abdelhafid Bendahmane, and Simon Hawkins

10.1 Introduction

Mutagenesis has historically been used to increase genetic diversity in a wide range of different crop species including cultivated flax (*Linum usitatissimum* L.) that has a limited genetic base, particularly for the fiber varieties (Allaby et al. 2005). Mutations can be introduced by using physical agents (X-rays, gamma radiation) or chemical treatments such as ethyl methane sulfonate (EMS) or sodium azide (NaN₃) (Dalmais et al. 2013). The generation of mutant populations has also been done by T-DNA insertion in model species for which genetic transformation is highly efficient such as *Arabidopsis* (Alonso et al. 2003) or *Brachypodium* (Hsia et al. 2017), as well as in crop species such as rice (An et al. 2005). While mutations sought are generally those resulting in a loss of gene function, gain of function mutants can also be obtained (Griffiths et al. 2005). The availability of mutant populations is also a valuable asset for the functional characterization of genes allowing the confirmation of their biological role(s). Flax can be genetically engineered, and different teams have up-/downregulated a limited number of genes in this species, thereby providing important functional information (Wróbel-Kwiatkowska et al. 2007; Day et al. 2009; Roach et al. 2011). Nevertheless, percentage transformation rates are generally low, and the availability of a chemically mutagenized (EMS) flax

S. Grec · S. Hawkins (✉)
Université de Lille, CNRS, UMR 8576 – UGSF – Unité de Glycobiologie Structurale et Fonctionnelle, Lille 5900, France
e-mail: simon.hawkins@univ-lille.fr

M. Dalmais · A. Bendahmane
Institute of Plant Sciences Paris Saclay IPS2, CNRS, INRA, Université Paris-Sud, Université Evry, Université Paris-Saclay, Orsay, France

M. Chatterjee
Bench Bio Pvt. Ltd, Vapi, India

© Springer Nature Switzerland AG 2019
C. A. Cullis (ed.), *Genetics and Genomics of Linum*, Plant Genetics and Genomics: Crops and Models 23, https://doi.org/10.1007/978-3-030-23964-0_10

population coupled with a TILLING (Targeting Induced Local Lesions In Genomes) approach represents an attractive alternative to accelerate gene function discovery.

10.2   Mutagenesis in Flax

As indicated above, different approaches can be used to generate mutant populations, and both radiation and chemical mutagenesis have been used in flax (Table 10.1). A large-scale T-DNA insertion mutant approach has not been used

Table 10.1 Examples of flax mutant populations created for increasing genetic variability and/or for TILLING

Ref[a]	Mutagens[b]	Dose/exposure[c]	Seed number per treatment	Phenotypes observed
(1)	γ-rays, EMS	30–90 KR (radiation), 0.3–0.4% (chemical)/22 h	7000	Increased plant height, albino, yellow seed color, blue flower color, deformed flowers, reduced linolenic acid/increased linoleic acid seed content
(2)	EMS	0.4%/22 h	20,000	Decreased linolenic acid seed content, changes in oleic, palmitic, and linoleic acid seed content
(3)	γ-rays, NEU, NMU, EMS, EI, DMS	1–150 KR (radiation), 0.006–0.4% (chemical)/12–18 h	500–1000	Chlorophyll loss, pathogen resistance, lodging resistance, fiber yield, seed size, sterility
(4)	EMS	0.4%/18 h	2000	Curley-stem, yellowing, albino, reduced height, increased sterility
(5)	EMS	0.3–0.75%/5–8 h	4000	Sterility, modifications in: stem height and diameter, leaf size and shape, branching, flower size and color, flowering time
(6)	EMS	0.5%/4 h	2000	Not assessed (TILLING population)
(7)	X-rays, γ-rays, HA, AA	10–15 KR (radiation), 0.1 and 0.2% (chemical)	100	Sterility, reduced shoot/root length, reduced dry weight
(8)	EMS	0.1–0.5%/18 h	200	Increased secondary branching, capsules and seed yield

[a]References: (1) Green and Marshall (1984); (2) Rowland and Bhatty (1990); (3) Bacelis (2001); (4) Tejklová (2002); (5) Chantreau et al. (2013); (6) Galindo-González et al. (2015); (7) Bhat et al. (2016): (8) Kulmi et al. (2017)

[b]Mutagens: radiation (x-/γ-rays), EMS, ethyl methane sulfonate; NEU, nitrosoethyl urea; NMU, nitrosomethyl urea; EI, ethylene imine; DMS, dimethyl sulfonate; HA, hydroxylamine; AA, 5-amino acridine

[c]Dose/exposure indicates radioactive intensity (in kiloroentgens) or chemical mutagen concentration (%) and soaking time (hours)

given the difficulty of genetically transforming flax with *Agrobacterium*. Following initial work in the early 1970s using gamma radiation (Srinivasachar and Malik 1971) and diethyl sulfate (Vereshchagin 1973) aimed at inducing increased variability in the proportion of unsaturated fatty acids, Green and Marshall (1984) used EMS to create a mutant population of the oil cultivar Glenelg. Visual phenotyping in the field of M1 and M2 generations identified a range of phenotypes characteristic of that observed with EMS treatments in other species. Detailed biochemical screening identified several mutant lines showing modified fatty acid profiles including reduced linolenic acid and increased linoleic acid content. Several years later Rowland and Bhatty (1990) used the same EMS protocol as Green & Marshall in a large-scale program to treat 20,000 seeds of the McGregor (oil) variety. Once again the objective of this research was to create mutants modified in seed fatty acid content, and screening identified 53 lines showing reduced levels of linolenic acid.

In addition to modifications in fatty-acid content, studies have shown that mutagenesis can lead to changes in other agronomically important features in flax. For example, Bacelis (2001) examined the effects of different doses of gamma rays and different concentrations of five chemical mutagens (Table 10.1). In this large-scale study, chlorophyll mutants were identified as well as families showing altered resistance to pathogens (*Fusarium*, *Melampsora*) and changes in lodging resistance, fiber yield, and seed size. As could be expected, the number of mutant phenotypes observed increased with increased radiation dose or mutagen concentration. In another study (Tejklová 2002), visual phenotyping of EMS mutants leads to the identification of a "curly-stem" phenotype, as well as yellowing, albino mutants, reduced height, and increased plant sterility.

More recently three different EMS concentrations (0.3%, 0.6%, and 0.75%) were used to generate a flax EMS population for both forward and reverse genetics in order to improve understanding of seed biology and cell wall metabolism in this species (Chantreau et al. 2013). The population consisted of 4894 M2 mutant seed families that, after sowing, produced 4033 families representing an overall percentage sterility of 17.6%. Sterility augmented with increasing EMS concentration as observed in other species. Analyses indicated that 38.5% (1522 families) showed visual phenotypes (Fig. 10.1) in stems (height and diameter), leaves (shape and color), plant architecture, and numerous flower-related phenotypes (flowering time, color, shape, fruit formation). Of interest is the fact that more than half of the mutants showed multiple phenotypes. The different phenotypes observed in this collection were integrated into the UTILLdb database (http://urgv.evry.inra.fr/UTILLdb). This is an open phenotypic and genomic mutant database that contains information on pea, *Brachypodium*, tomato, and flax mutant populations and allows breeders and scientists to search the database for different mutant families showing similar phenotypes.

While another study (Bhat et al. 2017) has reported sterility and reduced shoot/root size and dry mass, one recent work (Kulmi et al. 2017) identified two mutants showing higher capsule number that the authors associated with increased branching. Another 2000 strong EMS mutant population has also been generated by Canadian scientists but has mainly been used for reverse genetics (Galindo-González

Fig. 10.1 Examples of mutant flax phenotypes (Chantreau et al. 2013). (**a**) Mutant with altered leaf morphology (elongated leaves); (**b**) dwarf mutant with excessive branching; (**c**) flowering mutant with additional reproductive and non-reproductive organs; (**d**) mutant with altered leaf morphology (rolled leaves); (**e**) leafy mutant with reduced internodes; (**f**) pigment mutant showing stem and leaf chlorophyll loss

et al. 2015; Fofana et al. 2017b). Altogether, these studies clearly demonstrate the interest of using mutagenesis as a strategy to increase genetic diversity and improve agriculturally related traits in flax.

10.3 Production of Flax EMS Populations

A general overview of the production of a mutant population is given in Fig. 10.2. The first step involves soaking a batch of seeds in an EMS (ethyl methane sulfonate) solution to induce alkylation of guanine (G) bases in the DNA of cells in the apical meristem of the embryo. When these cells subsequently undergo DNA replication during the S phase of the mitotic cell cycle, alkylated G bases pair with thymine (T) bases instead of with cytosine (C) bases, thereby transforming 50% of a given GC pair containing an alkylated G into an AT pair. This induces point mutations giving rise to new single nucleotide polymorphisms/variations (SNPs, SNVs) distributed throughout the genome of the individual. SNPs/SNVs may be silent (no change in coded amino acid), missense (amino acid change), nonsense (truncation/premature stop signal), or splice-variant mutations. Other alkylating chemical mutagens that have been used to create mutant populations in flax include NEU (nitrosoethyl urea,

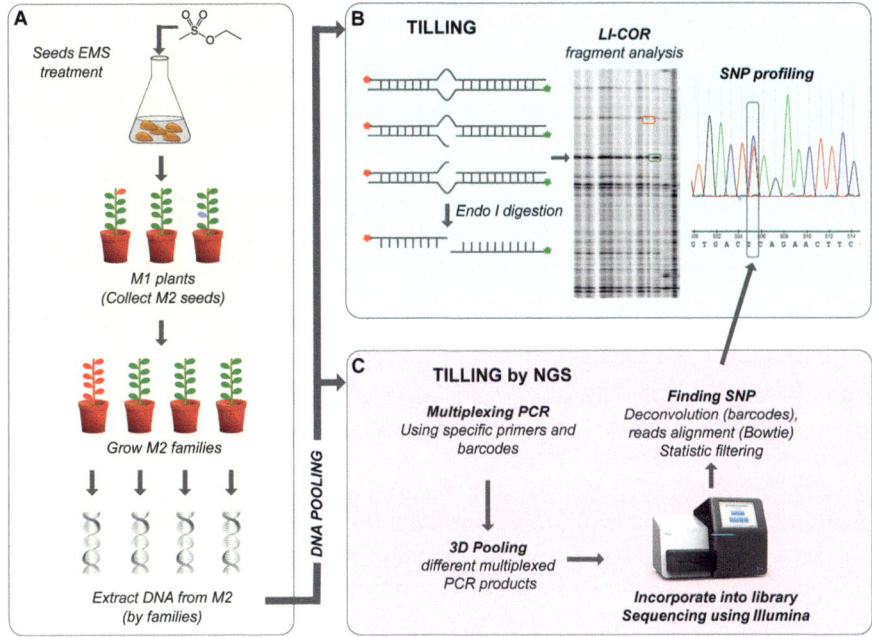

Fig. 10.2 Overview of flax mutant population creation and reverse genetics by TILLING. (**a**) Creation of EMS mutant population and extraction of DNA for TILLING by: (**b**) ENDO1 enzyme-based mismatch cleavage strategy or (**c**) NGS-based sequencing strategy

also known as N-ethyl-N-nitrosourea or ENU) and NMU (nitrosomethyl urea) that alkylate thymine leading to mis-pairing with guanine and T/A to G/C mutation, as well as EI (ethylene imine) and DMS (dimethyl sulfonate)(Puyo et al. 2014). Mismatch base pairing can also be induced by using hydroxylamine (HA) leading to the formation of hydroxyl cytosine and G/C to T/A transitions. Frameshift mutations can be induced by using amino acridines (AA) that intercalate into the DNA strand (Ferguson and Denny 1990).

The mutagen concentration and exposure time used to produce the best level of mutation is determined by performing preliminary "kill curve" experiments – if the EMS solution is too concentrated and/or the incubation time is too long, then seed viability will be low due to the accumulation of a high number of lethal mutations. If, on the other hand, the concentration/time is too low and short, then the mutation frequency will also be low, as will be the chances of either observing interesting phenotypes (forward genetics) or obtaining a sufficient number of mutations in the gene of interest that will have an effect on protein function (reverse genetics). The effective mutagen concentration and exposure time is also influenced by the ploidy level of the species targeted. Generally a higher EMS concentration and/or exposure time is necessary for species with duplicated genomes such as wheat or oat (see Table 6, Chantreau et al. 2013). Typical combinations successfully used to create flax EMS mutant populations include 0.3% EMS/5h, 0.6% EMS/6h, and 0.75%

EMS/8h (Chantreau et al. 2013) and 0.5% EMS/4h (Galindo-González et al. 2015; Fofana et al. 2017b).

The number of seeds used to generate the population depends upon the capacity of the team/structure to manage the collection. In general, the bigger the population, the better as this will increase the chances of obtaining interesting phenotypes and high numbers of target genes with deleterious mutations. However, a large collection of seeds implies a large collection of plants (that have to be grown and screened!). Generally population sizes are usually in the thousands (Chantreau et al. 2013; Galindo-González et al. 2015; Fofana et al. 2017b). Since not all seeds will germinate, nor all plants survive to produce seeds, it is better to start with a greater number of seeds.

In the next step, mutagenized seeds are sown, and the M1 plants are grown to maturity to produce the M2 seeds (cultivated flax is autogamous) that are then collected. At this stage, all the seeds from a single plant are pooled together and constitute a single mutant family. Since the EMS-generated mutations occur at random in different cells of the M1 embryo apical meristem, the derived cell lines and tissues from these initial cells will all present different mutations (see Box 10.1). As a result, the M1 plants are chimeric and heterozygote for any given mutation.

The seeds (M2 seeds) that were collected from the different M1 plants constitute the mutant population that is used for subsequent analyses. All the seeds obtained from a single M1 plant (usually several capsules) are pooled together and constitute a single M2 family. For the next step, M2 seeds are sown to produce M2 plants for phenotyping (forward genetics) and/or harvesting of plant material for DNA extraction (reverse genetics). Since segregation of the mutations occurs during M2 embryogenesis, the M2 seeds obtained from a single M1 family will be wild type, heterozygous, or homozygous for any given mutation (Box 10.2). It is for this reason that more than one (M2) seed per M1 family is planted in order to

---

Box 10.1

The extent of G base alkylation during EMS treatment depends on a combination of many different factors including EMS concentration, exposure time, seed size, architecture and seed coat thickness, DNA accessibility, etc. As a result, the genomes of the different cells in the M1 seed embryo apical meristem will show a wide range of different alkylation patterns. Usually several (hundreds of) genes (but not all) will be affected in any given cell; however, the genes affected in different cells will not necessarily be the same ones. This means that a given gene may be mutated in one cell, but not in another. In addition, since not all of the G bases in a given gene become alkylated, it is unlikely that a mutated gene in one cell will have the same mutations as the same gene in another cell. During plant development, mitotic divisions of the mutated cells (each of which shows different mutations) will give rise to a chimeric plant composed of cell lines and tissues derived from these cells.

Box 10.2

Flax is essentially autogamous and M2 embryos develop from the M2 zygote obtained after fusion of the M1-derived male and female gametes. Male/female gametes are haploid (n) and derived by mitotic divisions of the microspore/megaspore mother cells in the anther/ovule of the flax flower. In a mutant population, the microspore/megaspore mother cells themselves are derived from either (1) a cell containing a mutation (G base alkylated during EMS treatment at a specific locus in a given gene) or (2) a cell lacking a mutation at this locus. For case 1, the microspore/megaspore mother cell will be heterozygous for the mutation, and the derived haploid gametes will either carry a point mutation at this locus or be wild type (WT). For case 2, the microspore/megaspore mother cells are all WT as are the derived gametes. As a result, the M2 zygote (and embryo/plant) derived from the fusion of male and female gametes can be homozygous for the mutation (both gametes have the point mutation at the same locus), heterozygous (only one gamete carries the point mutation), or WT.

have a reasonable chance of obtaining homozygous mutants (most likely to give rise to a phenotype). Usually five seeds are sown per family, but this could be slightly reduced/increased depending upon the space and resources available. It is worthwhile remembering that five plants per family in a 3000-strong family will generate a total of 15,000 plants that will have to be phenotyped.

Leaves of plants can then be collected and dried/lyophilized for subsequent DNA extraction for TILLING (reverse genetics) and seeds (M3 seeds) collected from mature plants to maintain the collection. Although seeds can be stored for several years at 4 °C, their viability will decrease with time, and it is therefore necessary to periodically re-sow the seeds and collect new germplasm.

10.4   Reverse Genetics

As opposed to forward (or classical) genetics where one starts with a phenotype and then tries to identify the mutated gene(s) responsible for the observed modifications, reverse genetics identifies individuals with mutations in a specific gene for subsequent analysis. The use of reverse genetics has been greatly stimulated over the last 20 years by the introduction of high-throughput approaches such as TILLING (Targeting Induced Local Lesions In Genomes) that allows the identification of an allelic series of a targeted mutated gene in a mutagenized population (Henikoff et al. 2004; Wang et al. 2012a, b).

Genes can be targeted based upon scientific knowledge about the gene's function obtained from other studied species such as *Arabidopsis* or alternatively from transcriptomic studies in flax. Both developmental and organ and/or tissue-specific transcriptomics data exist for flax, as do expression data for plants grown under abiotic- and biotic-stress conditions, thereby providing a large number of potentially interesting genes that can be targeted (Roach and Deyholos 2007; Fenart et al. 2010; Huis et al. 2012; Dash et al. 2014; Mokshina et al. 2014, Galindo-González and Deyholos 2016; Zhang and Deyholos 2016). In addition to the important fiber and oil yield and quality, genes associated with stress tolerance and disease resistance would appear to be targets of choice in the context of climate change. Many genes of interest are members of multigene families, and it is therefore necessary to have a complete knowledge of the genome in order to be able to target the individual gene suspected to be associated with the biological process being studied. The genome of the flax variety Bethune has been sequenced and assembled into chromosomes (Wang et al. 2012a, b; You et al. 2018) and is accessible on the Phytozome database (https://phytozome.jgi.doe.gov/pz/portal.html), thereby allowing sequence data to be retrieved. Identification of the gene to be targeted is done by combining i) any available knowledge about the gene in question (e.g., expression data – is the gene expressed in the organ/tissue of interest, does the translated protein possess the adequate active site for interaction with the substrate/co-factor and/or other conserved motif, has the recombinant/purified protein been characterized, etc.?) with ii) phylogenetic analysis (http://www.ebi.ac.uk/Tools/msa/muscle/) of the flax proteins and homologous protein sequences from other species for which functional information is available. The target gene can then be chosen based upon available data and the sequence similarity to functionally characterized genes from other species (e.g., *Arabidopsis*).

Induced point mutations occur at random in the genome, and so not all families will have a mutation in the gene of interest. Furthermore, not all mutations will have a negative effect of the protein's function, and the challenge is therefore to identify the mutant families that contain harmful mutations. Mutations may be silent (base change but no amino-acid modification), missense (amino-acid change), or truncation (amino-acid changed to stop codon). The web-based CODDLE tool (Codons Optimized to Discover Deleterious Lesions) (http://blocks.fhcrc.org/proweb/coddle/) is generally used to identify those regions of the protein most likely to undergo mutations provoking a strong effect. Other tools available include SIFT (Sorting Tolerant From Intolerant; Kumar et al. 2009; http://sift.jcvi.org) that allows one to evaluate the impact of detected mutations on the protein and PROVEAN (**Pro**tein **V**ariation **E**ffect **An**alyzer, Choi et al. 2012) (http://provean.jcvi.org/index.php).

To detect induced point mutations in specific genomic targets, several mutation detection systems have been developed. All of them rely on polymerase chain reaction (PCR) amplification of multiple alleles in the selected genomic area (from M2 DNA families) that forms DNA heteroduplexes during the successive steps of heating and cooling. To reduce the numbers of PCRs from large DNA mutant collections (commonly composed of thousands of M2 families), DNA pooling is done (usually on robotic platforms) according to different strategies that allow subsequent deconvolution for identification of the family containing the mutated gene. The depth of DNA pooling depends on the sensitivity of the detection method chosen.

In all cases, DNA is extracted from the different M2 mutant families, quality-controlled and quantified. Either total genomic DNA can be used or a so-called "exome capture" approach in which only the coding DNA is sequenced, thereby greatly reducing the time and costs necessary for identifying the gene of interest, as well as simplifying the bioinformatic analyses. In this approach, the DNA extracted from the mutant population is pooled as standard and then hybridized to oligonucleotides corresponding to the transcriptome (i.e., expressed genes) of the organism. Such an approach was recently used for African rice, *Oryza glaberrima* (Henry et al. 2014).

One of the mutation detection systems most widely used in the last decades has been the enzymatic digestion system using specific mismatch endonucleases like ENDO1 (Triques et al. 2007), patented by the French National Institute for Agricultural Research (INRA) (Bendahmane et al. 2004) or *Cel 1* (Till et al. 2004) (Fig. 10.3). ENDO1 recognizes and cleaves all substitution and insertion types of mismatches. This technique was used in the flax TILLING screen (Chantreau et al. 2013) where target loci were PCR-amplified using nested PCR combined with labeled universal primers from 8-DNA pools (Dalmais et al. 2008). PCR products were digested with ENDO1 that cuts if a mutant is detected. Digestion products, labeled with infrared fluorescent both in 5′ and 3′, were detected on a DNA fragment analyzer using a polyacrylamide gel electrophoresis system. A deconvolution

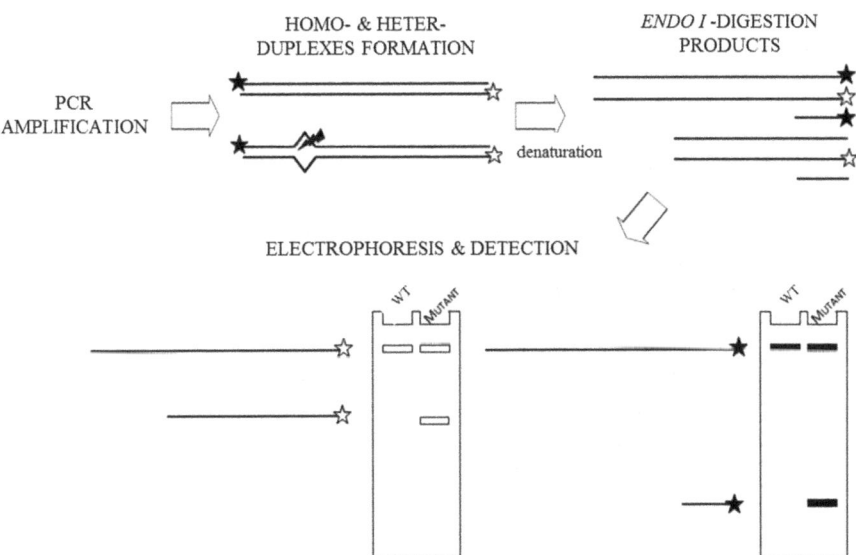

Fig. 10.3 Principle of the ENDO1 mutation detection system combined with a DNA fragment analyzer. Nested PCR is performed with fluorescent-labelled primers (both in 5′ and 3′), represented with black and white stars. DNA homo- or heteroduplexes occur during successive heating and cooling PCR cycles. If the PCR product carries a mutation (allele mutant), ENDO1 recognizes the mismatch and cuts the DNA (see lightning). Digestion products, labelled both in 5′ and 3′, are detected on a DNA fragment analyzer using a polyacrylamide gel electrophoresis system

step was performed following the same protocol to identify the family containing the mutated gene. PCR products from the mutated lines identified were characterized by Sanger sequencing so as to determine the position of the SNP detected. Although this method successfully enabled the detection of a relatively large number of mutants (see below), it is time-consuming and its capacity for high-throughput detection is quite limited.

The last decade has seen an explosion in the development of and the application of next-generation sequencing (NGS) technologies to many very different areas of biology. The detection of mutants is no exception, and the application of NGS has led to the establishment of new cost- and time-effective approaches. The use of NGS technology has required the development of new strategies of DNA pooling, library construction, and data analysis to identify mutations. Multiplexing of gene targets and genomes in the same sequencing run offers new perspectives in terms of productivity and sensitivity for SNP detection. All of the published protocols rely on PCR-amplification, from one- or multi-dimensional DNA pools, of the targeted region to generate DNA libraries with unique tracer sequences and barcodes followed by NGS sequencing and bioinformatic analyses (Tsai et al. 2011; Guo et al. 2015) (Fig. 10.4). Robust bioinformatics pipelines are necessary – often done by developing custom pipelines – to treat and analyze raw sequence data in order to identify induced mutations. The first step involves the alignment of reads with standard sequences to identify rare mutations. At this stage, it is important to clean the background and to avoid false positives. The importance of optimizing bioinformatics pipelines for this step was demonstrated in a recent study comparing different sequencing platforms (Roche 454, Ion Torrent PGM, and Illumina NextSeq)

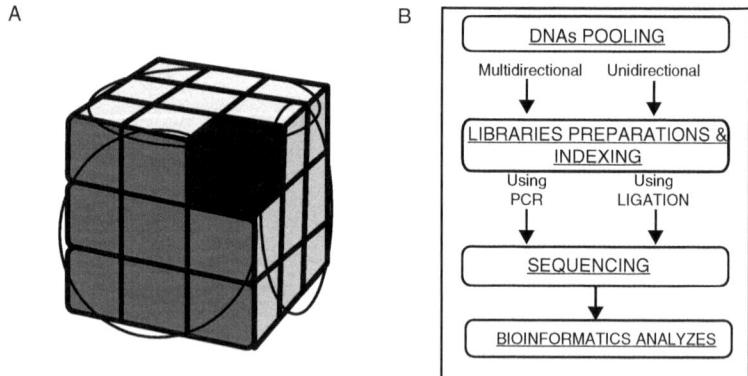

Fig. 10.4 NGS TILLING strategy and multi-dimensional pooling methods. (**a**) Principle of the three-dimensional pooling method, represented with a cube. The example displays 27 sub-cubes, representing 27 DNA samples. Each face of the cube (circle) is the equivalent of a one-dimensional pooling (3 pools), meaning that each sub-cube (sample) is represented three-fold in the 9 DNA-pools. For example, a mutation in the DNA-sample represented by the black sub-cube will be detected threefold in each 1D pools (corresponding to the 3 faces of the cube) after sequencing of only 9 DNA pools in place of 27 DNA samples. (**b**) General strategy of TILLING by NGS sequencing of PCR products from pooled genomic DNA

(Sandmann et al. 2017). In this study, the use of generalized linear models (GLMs), individually calibrated for the different platforms, eliminated up to 76% of all false-positive single nucleotide variations (SNVs) and 97% of all false-positive indels compared to the use of a standard analysis pipeline. Both throughput and sensitivity of the TILLING screens based on NGS have increased tremendously, compared with the previous enzymatic detection method and gel-based LICOR DNA fragment analyzer.

10.5 Reverse Genetics in Flax

Although flax mutant populations have existed since the 1970s, it is only recently that targeted reverse genetics approaches have been initiated in this species. As a proof of concept, TILLING was used to screen a flax mutant population for two genes (*CAD* (*Cinnamyl Alcohol Dehydrogenase*) and *C3H* (*Coumarate 3-Hydroxylase*)) involved in the biosynthesis of the cell wall polymer lignin (Chantreau et al. 2013)). In flax, as in other plant species possessing lignified tissues, *CAD* belongs to a multigene family, and before designing PCR primers, it was therefore important to identify the particular gene most likely involved in the biological process under study. Phylogenetic analyses identified the gene Lus10027864 (*LusCAD1*) as being the most closely related to the *Arabidopsis* genes *AtCAD4* and *AtCAD5* that are functionally responsible for converting hydroxycinnamaldehydes into the lignin hydroxycinnamyl alcohol monomers (Sibout et al. 2005), and this gene was therefore targeted. Subsequent characterization and expression studies of the flax lignin multigene family (Le Roy et al. 2017) confirmed that the targeted CAD gene belonged to the bona fide group of lignin genes.

The use of an ENDO1-based strategy led to the identification of 76 CAD mutants and 79 C3H mutants with an average mutation frequency of 1 mutation per 41 kb corresponding to approximately 9000 mutations per genome. As could be expected mutation frequency increased with higher EMS doses. Comparison with other species indicated that the observed mutation frequency was higher than that observed for many other diploid species and similar to that found in polyploid species such as wheat, rapeseed, and oats (Chantreau et al. 2013). Despite a higher mutation frequency, the sterility level was observed to be comparable to that of mutant populations in other species and is possibly related to the ancestral duplication of the flax genome resulting in a higher level of gene redundancy (Sveinsson et al. 2013). Although high mutation frequencies can cause difficulties for subsequent attempts to correlate the mutation(s) and phenotype(s) in classical outcrossing and mapping approaches (see below), they also reduce the number of families that have to be TILLED to identify mutants as well as generate a large number of independent mutants for functional studies.

Further analyses of the flax lignin mutations showed that 67 (CAD) and 74 (C3H) occurred in exons with 15 (CAD) and 24 (C3H) silent mutations, 51 (CAD) and 43 (C3H) missense mutations, and 1 (CAD) and 7 (C3H) truncation mutations.

Analyses showed that all mutations (except one) were G/C to A/T transitions as expected with the EMS mutagen (Sikora et al. 2011). While mutations were present throughout the CAD and C3H amplicons, they decreased in number in intronic regions reflecting the lower GC content. Preliminary work aimed at establishing a functional link between the tilled mutants and phenotype showed that 71% of the CAD missense and truncation mutants exhibited an orange-/brown-colored xylem characteristic of the "brown mid-rib" phenotype commonly observed in other CAD down-regulated species (Baucher et al. 1996; Sibout et al. 2005). Such an observation clearly establishes the interest of mutant populations for improving our knowledge of flax biology.

In 2014, an NGS-based methodology to identify mutants in Flax was developed in the INRA's Translational Research platform (Institute of Plant Sciences Paris-Saclay, France) to benefit from their TILLING-by-NGS pipeline and accelerate the identification of far higher numbers of mutants than that currently possible by a classical TILLING approach. This method relied on a two-dimensional DNA pools approach combined with processing of sequencing libraries using the MiSeq personal sequencer (Illumina®) and by unique tracer sequences and barcodes allowing flexibility in the number and pooling arrangement of targeted genes. A bioinformatic pipeline, called "Sentinel," was also developed to analyze the data sequences and to detect mutations and was patented by INRA (Bendahmane et al. 2016). Data analyses involved amplicon alignment to the reference sequence to identify possible single nucleotide changes, which were then evaluated for frequency, sequencing quality, intersection pattern in pools, and statistical relevance to produce a Bayesian score with an associated confidence threshold. This method is less time-consuming and more cost-effective in comparison with the ENDO1 enzymatic system. Moreover, experiments on other crop species have shown that it is also more sensitive with a gain of up to 2.55-fold. This methodology was used on a subset of the flax EMS collection generated by Chantreau and colleagues (Chantreau et al. 2013) to screen 15 genes involved in stress tolerance and cell wall metabolism. Over a total length of 5780 bp PCR-amplified, three hundred and eighty six mutations were detected from 2339 M2-DNA families. The success of this approach was validated by the observation that the calculated mutation detection frequency was assessed at 1 mutation detected each 35 kb screened, better than the previous assessment with ENDO1, at 1 mutation per 41 kb.

In another recent study, the Ion Torrent Personal Genome Machine (IT-PGM) was used for high-throughput sequencing (HTS) of another flax mutagenized population (Galindo-González et al. 2015). To evaluate the use of this innovative sequencing technology, the authors first confirmed in an elegant small-scale experiment that they would be able to detect mutant SNPs in pooled DNA. For this, they exploited known polymorphisms existing between the Macbeth and CDC cultivars and diluted (1:64 and 1:96) Macbeth DNA in CDC Bethune DNA to mimic the occurrence of a mutation in a WT background. Interestingly, this approach not only allowed detection of known SNPs but also led to the detection of previously unknown SNPs that were subsequently confirmed by Sanger sequencing.

Having confirmed that the IT-PGM could be successfully used to detect SNPs in DNA pools, the authors then went on to analyze DNA extracted from the leaves of 768 M2 individuals of the elite linseed cultivar CDC Bethune. As a proof of concept, the authors tilled four *PME* (*pectin methylesterase*) genes potentially involved in fiber cell wall metabolism (Pinzon-Latorre and Deyholos 2014). Although they identified 13 potential mutations using the 314 chip, only five of these were confirmed by Sanger sequencing, prompting the authors to change to the higher sensor count 316 chip for subsequent studies. The technological up-grade resulted in a five-fold increase in total read number after filtering and was associated with increased accuracy (11 Sanger-confirmed SNPs out of 16 putative SNPs identified by the IT-PGM). Subsequent analyses of read coverage for the four PME amplicons highlighted the fact that different factors (e.g., amplicon size, GC content, primer Gibbs energy) affected the number of mutations detected in the different genes. This study also underlined the necessity for NGS-based detection approaches to have sufficient read depth to avoid unacceptably high levels of false positives. As compared to Illumina sequencing technology, the IT-PGM has a lower throughput and generates a higher number of false positives. Nevertheless, the run time is considerably shorter and operating costs are generally lower.

A reverse genetics approach was also recently used to identify flax EMS mutants involved in seed lignan metabolism. Flax seeds are one of the richest natural sources of small bioactive molecules called lignans that have a number of beneficial effects on human health. The major flax lignan, secoisolariciresinol (SECO), is formed by the dimerization of two coniferyl alcohol molecules and accumulates in the seed coat under its diglucosylated form, SDG (secoisolariciresinol diglucoside). Since the uptake of flax lignans by human intestinal cells appears to be inversely related to the degree of glycosylation, a better control of lignan glycosylation could be an important factor in improving lignan bioavailability. Forfana and colleagues (Fofana et al. 2017b) had previously shown that the flax recombinant UDP-glycosyltransferase (UGT) enzyme UGT74S1 was able to glycosylate SECO to form the monoglucoside (SMG) and diglucoside (SDG). In a subsequent study (Fofana et al. 2017a), they screened a flax EMS mutant population consisting of almost 2000 M2 families and identified 93 families containing mutations in the exonic region of the UGT74S1 gene. In this case, mutations were detected using 454 pyrosequencing in contrast to the ENDO1 enzyme-based used by Chantreau and colleagues (Chantreau et al. 2013). The average frequency mutation was estimated at 1 mutation per 28 Kb. Of the 93 mutations, 18 were sense and 75 were missense and nonsense (truncation). Biochemical analyses of seed SDG lignan content in 69 of the 75 M2 families was determined by ultra-performance liquid chromatography-mass spectrometry (UPLC-MS) and compared to extracts from WT CDC-Bethune seeds. Results showed that 21 families showed reduced SDG levels, of which 8 contained almost no SDG, while 43 showed higher than WT content. Although attempts to correlate phenotype (SDG content) to mutation position were unsuccessful, the results clearly demonstrated that targeted mutation of the UGT74S1 gene could be correlated to changes in lignan content. Interestingly the results also revealed changes in the pool of phenolic glycosides as well as suggest that other UGTs may be able to glycosylate

SECO, thereby underlining the complexity of phenolic metabolism in flax and the importance of secondary metabolite glycosylation in plants (Huis et al. 2012; Le Roy et al. 2016). Overall this study is important as it demonstrates that TILLING can be used to successfully target traits of economic importance and provide new germplasm that breeders can introduce into flax improvement programs.

10.6 Forward Genetics in Flax

In contrast to a reverse genetics approach where plants containing mutations in a gene believed to be associated with a given biological process are targeted, the forward genetics approach involves screening the mutant population to identify plants showing a phenotype related to the biological process under study. Although plants showing a desired phenotype can be integrated directly into breeding programs, many fundamental studies also aim to identify the mutated gene(s) responsible for the new phenotype. Screening of a flax mutant population of the oil cultivar Glenelg led to the identification of two lines in which seed linolenic acid content was reduced to 29% total fatty acid content compared to 43% in WT plants (Green and Marshall 1984). Flax seeds (linseed) contain high amounts of the polyunsaturated fatty acid alpha-linolenic acid (ALA, C18:3) that are responsible for the industrial properties of linseed oil. Although beneficial for human health, the cooking properties of linolenic acid (C18:3) are less good than those of linoleic acid (C18:2), thereby driving research to breed low linoleic acid linseed varieties. Further analyses of seed fatty acid content in the two mutant families indicated that linoleic acid content increased to 30% compared to typical WT values of 18%. Fatty acid content was also the target of another large-scale biochemical screen of an EMS mutant population of the McGregor (oil) variety undertaken by Rowland and Bhatty (1990). Their work allowed the identification of 53 lines showing reduced levels of linolenic acid. Subsequent characterization of selected mutant lines (Ntiamoah and Rowland 1997; Vrinten et al. 2005) established that the low linolenic acid line solin 593–708 derived from the McGregor EMS mutants contained EMS-induced point mutations leading to premature stop codons in two fatty-acid desaturase genes (*LuFAD3A, LuFAD3B*) responsible for linoleic acid desaturation.

Reverse genetics was also recently used to successfully identify flax cell wall mutants (Chantreau et al. 2014). In this work, a two-stage screening approach (Fig. 10.5) based on (i) UV autofluorescence of lignified bast fibers and (ii) histochemical staining (Phloroglucinol-HCl) allowed the identification of 419 individual mutant families out of a total of 3391 families screened.

Subsequent characterization of one of these families – the *lignified bast fiber1* (*lbf1*) mutant – by a combination of wet chemistry, NMR, and immunolocalization proved that fiber cell walls contained more than three times the amount of lignin compared to WT plants (17% vs 5%, respectively). Since this level of lignification does not normally occur in cultivated flax plants, the availability of the *lbf1* mutant (and other *lbf* mutants) represents a unique opportunity to probe how increased

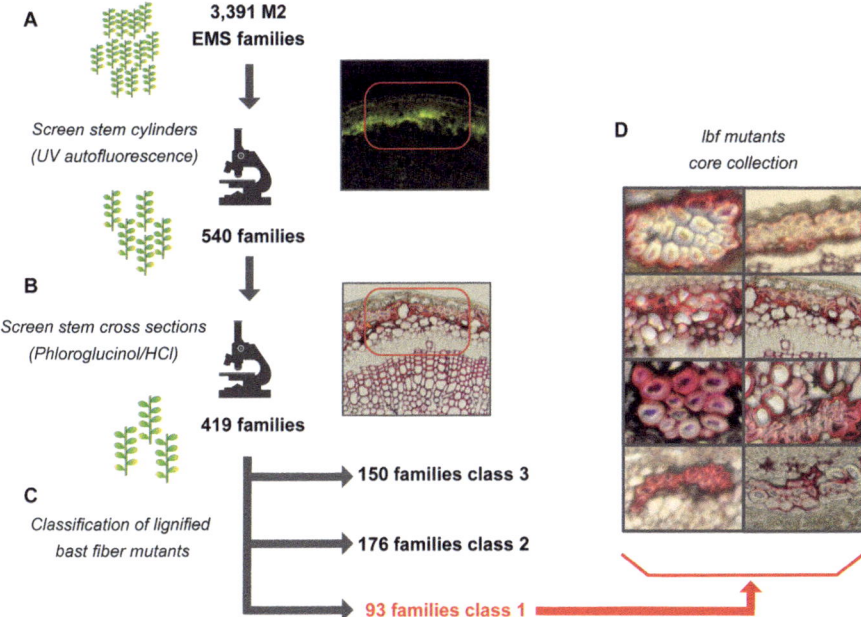

Fig. 10.5 Two-step screening strategy used to identify flax *lbf* mutants (Chantreau et al. 2014). (**a**) Stem cylinders from M2 plants were analyzed for bast fiber auto-fluorescence in multi-well plaques in a first round of screening. (**b**) Lignification of bast fiber cell walls was confirmed in a second round of screening using the Weisner reagent (Phloroglucinol/HCl). (**c**) *Lbf* mutants were classed into three families according to the intensity of the phenotype (class 1 = most heavily ligni-fied, class 3 = least heavily lignified). (**d**) Core collection of flax *lbf* mutants showing a range of lignified cell wall phenotypes

lignin affects cell wall mechanical and physical properties thereby contributing to our understanding of structure-function relationship important for improving fiber quality in textiles and composite materials. A similar strategy could also be used to identify flax mutants showing changes in other biological processes of interest (e.g., stress/disease resistance). As in all species, the challenge is to define and to put into place an efficient forward genetic screen that is as high throughput as possible.

Forward genetics has been used for many decades to determine gene function and to improve our understanding of plant biology. Although classical genetics involving outcrossing mutants to a polymorphic line followed by a fine mapping using a pool of individual F2 plants carrying the mutation is still successfully used, such an approach is more complicated with induced mutant populations (Jander et al. 2002). The high mutation rate generated by EMS treatment can give rise to thousands of mutations per genome (e.g., an average of 9000 mutations per genome were estimated in the EMS flax population generated by Chantreau et al. 2013), making this approach labor intensive and requiring the mapping of a pool of thou-sands of individual F2 plants. Additionally, some traits identified in mutant library screenings will probably be due to multiple mutations. In this case, the observed phenotype could be altered when mutants are crossed to other accessions, making phenotyping of the F2 population difficult. A variation of classical genetics is the use of bulked segregant analysis (BSA) strategy. In this approach, the allelic frequencies of genetic markers are determined in two groups of plants exhibiting contrasting phenotypes, leading to the identification of the genomic interval comprising the variation. Then, a sequencing of this restricted genomic region allows the identification of the causal mutations. This BSA approach can significantly accelerate the identification of a causal mutation and considerably reduce the costs of genotyping.

QTL (quantitative trait loci) mapping followed by molecular characterization can also be used to identify mutations associated to quantitative traits. For example, this approach had been successfully used to identify the Th498 to Ser498 substitution in the flax flavonoid 3′,5′ hydroxylase associated with yellow seed and white flower phenotypes (Sudarshan et al. 2017). This work is a good example of how the use of a flax fine genetic map (Cloutier et al. 2012) can be used to identify point mutations by a classical genetics approach.

Nevertheless, the biggest revolution in mutation identification comes from recent advances in sequencing technologies. The development of next-generation sequencing (NGS) technologies has considerably reduced the cost of performing whole genome sequencing, thereby making it technically and financially feasible to identify variants in the genome of a particular mutant (Schneeberger 2014). The emergence of high-throughput sequencing has been accompanied by the development of informatic tools designed to analyze a large data flow and in particular to detect sequence variations. Existing informatic tools dedicated to single nucleotide polymorphism (SNP) such as the GATK (https://software.broadinstitute.org/gatk/) suite or Varscan (http://varscan.sourceforge.net) can be adapted for mutation detection in EMS mutant genomes.

NGS-based strategies have been widely applied on *Arabidopsis thaliana* EMS populations (Schneeberger et al. 2014). As a proof of concept, Schneeberger identified causal mutations associated with slow growth and light green leaves resulting from lesions within a pool of 500 mutant F2 plants. Mutations were mapped using a bulked segregant analysis of an F2 population resulting from the cross between the mutant line (Col 0 Columbia reference) and a genetically divergent accession (Landsberg erecta). High-throughput sequencing (HTS) was performed on a bulked DNA sample from F2 individuals exhibiting the phenotype, and the relative frequency of each parental allele within the pool was determined using customized bioinformatic tools. Mutations were then validated using independent allelic mutants. This approach is known as the SHOREmap strategy. Subsequently, a computer pipeline (SHOREmap V3.0) was built to standardize and automate sequencing data processing procedures (Sun and Schneeberger 2015). The mutant can also be crossed directly to the original wild-type line and then selfed. Sequencing is then directly performed on a bulked DNA sample of individuals from an isogenic F2 showing the mutated phenotype. This strategy called MutMap minimizes the number of genetic crosses, thereby accelerating the identification of the causal SNP for the mutant phenotype (Abe et al. 2012). Similarly, Nordström et al. (2013) used

NGS to perform mapping by directly sequencing allelic rice mutants without crossing, and Fekih et al. (2013) sequenced segregating mutant rice progenies derived from a heterozygous mutant line. More recently the use of exome sequencing has been used to reduce the complexity of detecting mutations associated with a target phenotype. For example, this approach leads to the identification of a clear peak genomic region associated with increased plant height within a wheat EMS population (Mo et al. 2017). To date, direct sequencing mapping is mainly applied to model plant species such as *Arabidopsis* or major food crops such as rice and wheat that benefit from substantial genomic resources, but there is no doubt that the study of flax EMS mutant populations will benefit from these technological advances and strategies in the near future.

10.7 Conclusions and Future Perspectives

The creation and characterization of flax mutant populations is a powerful strategy for introducing genetic variability into this economically important species that suffers from a narrow genetic base, particularly for fiber varieties (Allaby et al. 2005). Currently, EMS-generated mutants are not considered as genetically modified organisms (GMOs), thereby providing another advantage over the use of classical genetic engineering approaches. The commercial importance of this point is well illustrated by the fact that, in 2015, Europe imported 68% of Canada's total linseed exports representing more than 400 thousand tonnes. The discovery in 2009 of traces of the GMO Triffid line (unauthorized in Europe) almost completely eliminated Canadian flaxseed exports to Europe over the following 2 years.

The rapid increase in the performance of NGS platforms has not only improved the efficiency of TILLING, thereby allowing reliable and rapid detection of mutants in targeted genes, but should also facilitate the identification of mutations responsible for an observed phenotype in a forward genetics approach. The latter point is particularly important given the recent explosion in genome editing (GE) technologies allowing scientists to directly modify nucleotides or sequences within a locus (Kumar et al. 2017). Techniques such as the CRISPR/Cas9 (clustered, regularly interspaced, short palindromic repeats) system allow direct intervention on the gene of interest, thereby potentially eliminating the need to follow a reverse genetics TILLING approach to obtain mutants in this gene. The recent demonstration that a combination of single-stranded oligonucleotides (ssODNs) and CRISPR/Cas9 could be used to generate herbicide-tolerant flax opens up new perspectives for both breeding purposes and fundamental biology in this species (Sauer et al. 2016). However, GE approaches are currently unable to create large mutant populations displaying a range of different, and often novel, phenotypes as illustrated by the generation of the flax series of *lbf* mutants characterized by a fiber cell wall composition not present in WT populations (Chantreau et al. 2014). As such, the creation of mutant populations will continue to be a powerful approach for improving our knowledge about flax biology.

Acknowledgments SH and SG gratefully acknowledge the support of the University of Lille, the Research Federation (Université de Lille, CNRS FR 3688, FRABio, Biochimie Structurale et Fonctionnelle des Assemblages Biomoléculaires), the Region "Hauts-de-France" (project ALIBIOTECH), and the French National Research Agency (ANR, project PT-Flax).

References

Abe A, Kosugi S, Yoshida K, Natsume S, Takagi H, Kanzaki H et al (2012) Genome sequencing reveals agronomically important loci in rice using MutMap. Nat Biotechnol 30(2):174

Allaby RG, Peterson GW, Merriwether DA, Fu YB (2005) Evidence of the domestication history of flax (Linum usitatissimum L.) from genetic diversity of the sad2 locus. Theor Appl Genet 112(1):58–65

Alonso JM, Stepanova AN, Leisse TJ, Kim CJ, Chen H, Shinn P et al (2003) Genome-wide insertional mutagenesis of Arabidopsis thaliana. Science 301(5633):653–657

An G, Lee S, Kim SH, Kim SR (2005) Molecular genetics using T-DNA in rice. Plant Cell Physiol 46(1):14–22

Bacelis K (2001) Experimental mutagenesis in fiber flax breeding. Biologia 1:40–43

Baucher M, Chabbert B, Pilate G, Doorsselaere JV, Tollier MT, Petit-Conil M, Cornu D, Monties B, Montagu MV, Inze D, Jouanin L, Boerjan W (1996) Red xylem and higher lignin extractability by down-regulating a cinnamyl alcohol dehydrogenase in poplar. Plant Physiol 112:1479–1490

Bendahmane A, Triques K, Sturbois B, Aubourg S, Caboche M (2004) Method for producing highly sensitive endonucleases, novel preparations of endonucleases and uses thereof. PCT/EP2005/009220

Bendahmane A, Marcel F, Dalmais M, Beaumont G, Mania B (2016) SENTINEL, SOFTWARE dedicated to TILLING by NGS Analysis. Certifier par l'Agence pour la Protection des programmes. Inter Deposit Digital Number.FR001.240004.000.R.P.2016.000.10000

Bhat IA, Pandit UJ, Sheikh IA, Hassan ZU (2017) Physical and chemical mutagenesis in Linum usitatissimum L. to induce variability in seed germination, survival and growth rate traits. Curr Bot 7:28–32

Chantreau M, Grec S, Gutierrez L, Dalmais M, Pineau C, Demailly H et al (2013) PT-flax (phenotyping and TILLinG of flax): development of a flax (Linum usitatissimum L.) mutant population and TILLinG platform for forward and reverse genetics. BMC Plant Biol 13(1):159

Chantreau M, Portelette A, Dauwe R, Kiyoto S, Crônier D, Morreel K et al (2014) Ectopic lignification in the flax lignified bast fiber1 mutant stem is associated with tissue-specific modifications in gene expression and cell wall composition. Plant Cell 26(11):4462–4482

Choi Y, Sims GE, Murphy S, Miller JR, Chan AP (2012) Predicting the functional effect of amino acid substitutions and indels. PLoS One 7(10):e46688

Cloutier S, Ragupathy R, Miranda E, Radovanovic N, Reimer E, Walichnowski A et al (2012) Integrated consensus genetic and physical maps of flax (Linum usitatissimum L.). Theor Appl Genet 125(8):1783–1795

Dalmais M, Schmidt J, Le Signor C et al (2008) UTILLdb, a *Pisum sativum in silico* forward and reverse genetics tool. Genome Biol 9(2):R43. https://doi.org/10.1186/gb-2008-9-2-r43

Dalmais M, Antelme S, Ho-Yue-Kuang S, Wang Y, Darracq O, d'Yvoire MB et al (2013) A TILLING platform for functional genomics in *Brachypodium distachyon*. PLoS One 8(6):e65503

Dash PK, Cao Y, Jailani AK, Gupta P, Venglat P, Xiang D et al (2014) Genome-wide analysis of drought induced gene expression changes in flax (Linum usitatissimum). GM Crops Food 5(2):106–119

Day A, Neutelings G, Nolin F, Grec S, Habrant A, Crônier D, Maher B, Rolando C, David H, Chabbert B, Hawkins S (2009) Caffeoyl coenzyme A O-methyltransferase down-regulation is associated with modifications in lignin and cell-wall architecture in flax secondary xylem. Plant Physiol Biochem PPB Société Française Physiol Végétale 47:9–19

Fekih R, Takagi H, Tamiru M, Abe A, Natsume S, Yaegashi H et al (2013) MutMap+: genetic mapping and mutant identification without crossing in rice. PLoS One 8(7):e68529

Fenart S, Ndong YPA, Duarte J, Rivière N, Wilmer J, van Wuytswinkel O et al (2010) Development and validation of a flax (Linum usitatissimum L.) gene expression oligo microarray. BMC Genomics 11(1):592

Ferguson LR, Denny WA (1990) Frameshift mutagenesis by acridines and other reversibly-binding DNA ligands. Mutagenesis 5(6):529–540

Fofana B, Ghose K, Somalraju A, McCallum J, Main D, Deyholos MK et al (2017a) Induced mutagenesis in UGT74S1 gene leads to stable new flax lines with altered secoisolariciresinol diglucoside (SDG) profiles. Front Plant Sci 8:1638

Fofana B, Ghose K, McCallum J, You FM, Cloutier S (2017b) UGT74S1 is the key player in controlling secoisolariciresinol diglucoside (SDG) formation in flax. BMC Plant Biol 17(1):35

Galindo-González L, Deyholos MK (2016) RNA-seq transcriptome response of flax (Linum usitatissimum L.) to the pathogenic fungus Fusarium oxysporum f. sp. lini. Front Plant Sci 7:1766

Galindo-González L, Pinzón-Latorre D, Bergen EA, Jensen DC, Deyholos MK (2015) Ion torrent sequencing as a tool for mutation discovery in the flax (Linum usitatissimum L.) genome. Plant Methods 11(1):19

Green AG, Marshall DR (1984) Isolation of induced mutants in linseed (Linum usitatissimum) having reduced linolenic acid content. Euphytica 33(2):321–328

Griffiths AJ, Wessler SR, Lewontin RC, Gelbart WM, Suzuki DT, Miller JH (2005) An introduction to genetic analysis. Macmillan, New York

Guo Y, Abernathy B, Zeng Y, Ozias-Akins P (2015) TILLING by sequencing to identify induced mutations in stress resistance genes of peanut (*Arachis hypogaea*). BMC Genomics 16(1):157. https://doi.org/10.1186/s12864-015-1348-0

Henikoff S, Till BJ, Comai L (2004) TILLING. Traditional mutagenesis meets functional genomics. Plant Physiol 135(2):630–636

Henry IM, Nagalakshmi U, Lieberman MC, Ngo KJ, Krasileva KV, Vasquez-Gross H et al (2014) Efficient genome-wide detection and cataloging of EMS-induced mutations using exome capture and next-generation sequencing. Plant Cell 26(4):1382–1397

Hsia MM, O'Malley R, Cartwright A, Nieu R, Gordon SP, Kelly S et al (2017) Sequencing and functional validation of the JGI Brachypodium distachyon T-DNA collection. Plant J 91:361

Huis R, Morreel K, Fliniaux O, Lucau-Danila A, Fénart S, Grec S et al (2012) Natural hypolignification is associated with extensive oligolignol accumulation in flax stems. Plant Physiol 158:1893

Jander G, Norris SR, Rounsley SD, Bush DF, Levin IM, Last RL (2002) Arabidopsis map-based cloning in the post-genome era. Plant Physiol 129(2):440–450

Kulmi MRM, Mogali SC, Patil KS, Leelavathi TM (2017) Isolation of high-yielding mutants through EMS-induced mutagenesis in linseed (Linum usitatissimum L.). Int J Curr Microbiol App Sci 6(8):278–285

Kumar P, Henikoff S, Ng PC (2009) Predicting the effects of coding non-synonymous variants on protein function using the SIFT algorithm. Nat Protoc 4(7):1073–1081

Kumar AP, McKeown PC, Boualem A, Ryder P, Brychkova G, Bendahmane A et al (2017) TILLING by sequencing (TbyS) for targeted genome mutagenesis in crops. Mol Breed 37(2):14

Le Roy J, Huss B, Creach A, Hawkins S, Neutelings G (2016) Glycosylation is a major regulator of phenylpropanoid availability and biological activity in plants. Front Plant Sci 7:735

Le Roy J, Blervacq AS, Créach A, Huss B, Hawkins S, Neutelings G (2017) Spatial regulation of monolignol biosynthesis and laccase genes control developmental and stress-related lignin in flax. BMC Plant Biol 17(1):124

Mo Y, Howell T, Vasquez-Gross H, de Haro LA, Dubcovsky J, Pearce S (2017) Mapping causal mutations by exome sequencing in a wheat TILLING population: a tall mutant case study. Mol Genet Genomics:293(2):463–477

Mokshina N, Gorshkova T, Deyholos MK (2014) Chitinase-like (CTL) and cellulose synthase (CESA) gene expression in gelatinous-type cellulosic walls of flax (Linum usitatissimum L.) bast fibers. PLoS One 9(6):e97949

Nordström KJ, Albani MC, James GV, Gutjahr C, Hartwig B, Turck F et al (2013) Mutation identification by direct comparison of whole-genome sequencing data from mutant and wild-type individuals using k-mers. Nat Biotechnol 31(4):325

Ntiamoah C, Rowland GG (1997) Inheritance and characterization of two low linolenic acid EMS-induced McGregor mutant flax (Linum usitatissimum). Can J Plant Sci 77(3):353–358

Pinzon-Latorre D, Deyholos MK (2014) Pectinmethylesterases (PME) and pectinmethylesterase inhibitors (PMEI) enriched during phloem fiber development in flax (Linum usitatissimum). PLoS One 9:e105386

Puyo S, Montaudon D, Pourquier P (2014) From old alkylating agents to new minor groove binders. Crit Rev Oncol Hematol 89(1):43–61

Roach MJ, Deyholos MK (2007) Microarray analysis of flax (Linum usitatissimum L.) stems identifies transcripts enriched in fibre-bearing phloem tissues. Mol Genet Genomics 278(2):149–165

Roach MJ, Mokshina NY, Badhan A, Snegireva AV, Hobson N, Deyholos MK, Gorshkova TA (2011) Development of cellulosic secondary walls in flax fibers requires B-galactosidase. Plant Physiol 156:1351–1363

Rowland GG, Bhatty RS (1990) Ethyl methanesulphonate induced fatty acid mutations in flax. J Am Oil Chem Soc 67(4):213–214

Sandmann S, de Graaf AO, van der Reijden BA, Jansen JH, Dugas M (2017) GLM-based optimization of NGS data analysis: a case study of Roche 454, ion torrent PGM and Illumina NextSeq sequencing data. PLoS One 12(2):e0171983

Sauer NJ, Narváez-Vásquez J, Mozoruk J, Miller RB, Warburg ZJ, Woodward MJ et al (2016) Oligonucleotide-mediated genome editing provides precision and function to engineered nucleases and antibiotics in plants. Plant Physiol 170(4):1917–1928

Schneeberger K (2014) Using next-generation sequencing to isolate mutant genes from forward genetic screens. Nat Rev Genet 15(10):662

Sibout R, Eudes A, Mouille G, Pollet B, Lapierre C, Jouanin L, Séguin A (2005) Cinnamyl alcohol dehydrogenase-C and -D are the primary genes involved in lignin biosynthesis in the floral stem of Arabidopsis. Plant Cell Online 17:2059–2076

Sikora P, Chawade A, Larsson M, Olsson J, Olsson O (2011) Mutagenesis as a tool in plant genetics, functional genomics, and breeding. Int J Plant Genom 2011:1

Srinivasachar D, Malik RS (1971) Gamma-ray induced variability in the iodine value of linseed oil. Curr Sci 11:298–299

Sudarshan GP, Kulkarni M, Akhov L, Ashe P, Shaterian H, Cloutier S et al (2017) QTL mapping and molecular characterization of the classical D locus controlling seed and flower color in Linum usitatissimum (flax). Sci Rep 7(1):15751

Sun H, Schneeberger K (2015) SHOREmap v3. 0: fast and accurate identification of causal mutations from forward genetic screens. In Plant Functional Genomics (pp. 381–395). Humana Press, New York, NY

Sveinsson S, McDill J, Wong GK, Li J, Li X, Deyholos MK, Cronk QC (2013) Phylogenetic pinpointing of a paleopolyploidy event within the flax genus (Linum) using transcriptomics. Ann Bot 113(5):753–761

Tejklová E (2002) Curly stem-an induced mutation in flax (Linum usitatissimum L.). Czech J Genet Plant Breed 38(3/4):125–128

Till BJ, Burtner C, Comai L, Henikoff S (2004) Mismatch cleavage by single-strand specific nucleases. Nucleic Acids Res 32(8):2632–2641. https://doi.org/10.1093/nar/gkh599

Triques K, Sturbois B, Gallais S, Dalmais M, Chauvin S, Clepet C, Aubourg S, Rameau C, Caboche M, Bendahmane A (2007 Sep) Characterization of Arabidopsis thaliana mismatch specific endonucleases: application to mutation discovery by TILLING in pea. Plant J 51(6):1116–1125

Tsai H, Howell T, Nitcher R, Missirian V, Watson B, Ngo KJ, Lieberman M, Fass J, Uauy C, Tran RK, Khan AA, Filkov V, Tai TH, Dubcovsky J, Comai L (2011) Discovery of rare mutations in populations: TILLING by sequencing. Plant Physiol 156:1257–1268

Vereshchagin AG (1973) The variability of individual flax seeds in fatty acid composition. Biokhimiya 38:573–582

Vrinten P, Hu Z, Munchinsky MA, Rowland G, Qiu X (2005) Two FAD3 desaturase genes control the level of linolenic acid in flax seed. Plant Physiol 139(1):79–87

Wang Z, Hobson N, Galindo L, Zhu S, Shi D, McDill J et al (2012a) The genome of flax (Linum usitatissimum) assembled de novo from short shotgun sequence reads. Plant J 72(3):461–473

Wang TL, Uauy C, Robson F, Till B (2012b) TILLING in extremis. Plant Biotechnol J 10(7):761–772

Wróbel-Kwiatkowska M, Starzycki M, Zebrowski J, Oszmianski J, Szopa J (2007) Lignin deficiency in transgenic flax resulted in plants with improved mechanical properties. J Biotechnol 128:919–934

You FM, Xiao J, Li P, Yao Z, Jia G, He L, ...Cloutier S (2018) Chromosome-scale pseudomolecules refined by optical, physical and genetic maps in flax. The Plant Journal, 95(2), 371–384

Zhang N, Deyholos MK (2016) RNASeq analysis of the shoot apex of flax (Linum usitatissimum) to identify phloem fiber specification genes. Front Plant Sci 7:950

Chapter 11
Linum Lignan and Associated Biochemical Pathways in Human Health and Plant Defense

Syed G. A. Moinuddin, John R. Cort ⓘ, Clyde A. Smith ⓘ, Christophe Hano ⓘ, Laurence B. Davin ⓘ, and Norman G. Lewis ⓘ

Flax (*Linum usitatissimum* L.) is a plant species with a long and rich history of use, whether for its fiber, oil, or its health-protecting lignans. Its seed is the source of both the oil and the important health-protecting lignans. For example, the seed coat-derived lignan, secoisolariciresinol diglucoside (SDG, Fig. 11.1), reportedly suppresses development of atherosclerosis (Prasad 2009), reduces incidence rates of type 1 diabetes, and delays development of type 2 diabetes progression in rats (Prasad and Dhar 2016). Flax stems are also highly valued for their fibers in, for example, linen production and paper products, such as bank notes.

The lignans are an abundant class of natural products of immense structural variety and complexity *in planta* (Vassão et al. 2010), many of which are optically active. They are typically dimeric (C_6C_3) phenylpropanoids, although higher oligomeric forms are well known throughout the plant kingdom (Vassão et al. 2010). Flax seed coat lignans are largely SDG derived, although the seed coat tissue itself does not accumulate significant levels of free SDG as such.

Lignan preparations containing ester-linked SDG – among other constituents – can be readily solubilized through an aqueous EtOH extraction of its flax seed hulls

S. G. A. Moinuddin · L. B. Davin · N. G. Lewis (✉)
Institute of Biological Chemistry, Washington State University, Pullman, WA, USA
e-mail: lewisn@wsu.edu

J. R. Cort
Institute of Biological Chemistry, Washington State University, Pullman, WA, USA

Earth and Biological Sciences Directorate, Pacific Northwest National Laboratory, Richland, WA, USA

C. A. Smith
Stanford Synchrotron Radiation Lightsource, Stanford University, Menlo Park, CA, USA

C. Hano
Laboratoire de Biologie des Ligneux et des Grandes Cultures, INRA USC1328, Université d'Orléans, Chartres, France

© Springer Nature Switzerland AG 2019
C. A. Cullis (ed.), *Genetics and Genomics of Linum*, Plant Genetics and Genomics: Crops and Models 23, https://doi.org/10.1007/978-3-030-23964-0_11

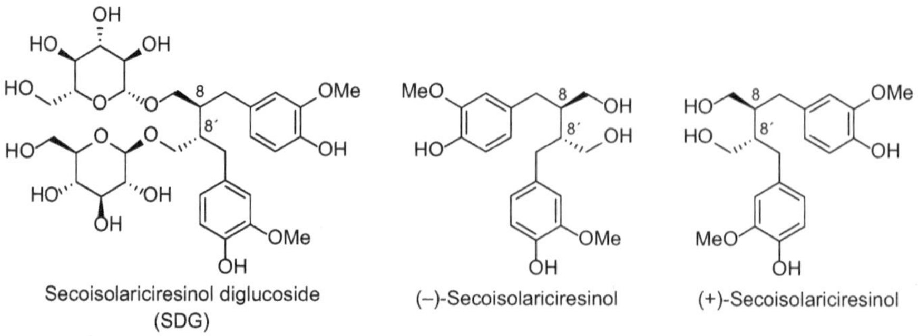

Fig. 11.1 Secoisolariciresinol diglucoside (SDG) and (-)- and (+)-secoisolariciresinols

at room temperature. The SDG component can also be facilely released by either treatment of the flax seed coat or the above aqueous EtOH solubilized lignan enriched isolate thereof, with dilute alkali at room temperature, i.e., through saponification of ester linkages (discussed below) (Johnsson et al. 2000; Ford et al. 2001). The resulting SDG, when deglycosylated by a β-glucuronidase, affords the aglycone secoisolariciresinol, which exists as a mixture of two enantiomers in a circa 99:1 ratio of the (+)- and (−)-antipodes (Fig. 11.1) (Ford et al. 2001).

Other lignans, namely pinoresinol, pinoresinol diglucoside, isolariciresinol, and matairesinol (Meagher et al. 1999; Qiu et al. 1999; Ramsay et al. 2017), have also been obtained in small amount from flax seed following alkali (base) treatment. Previous accounts of divanillyl tetrahydrofuran (anhydrosecoisolariciresinol, Fig. 11.2) in flax seed, as a natural product, were later found to be an isolation artefact (Meagher et al. 1999).

Additionally, p-coumaric, ferulic, and sinapic acids (Kozlowska et al. 1983; Ford et al. 1999), together with glucosides, p-coumaric and caffeic glucosides, named linocinnamarin and linocaffein, respectively (Klosterman and Muggli 1959), as well as ferulic acid glucoside (Kozlowska et al. 1983; Ford et al. 1999), and herbacetin 3,8-O-diglucoside (Qiu et al. 1999), were isolated following alkali (base) treatment (Fig. 11.2). 3-Hydroxy-3-methylglutaric acid (HMGA) had also previously been obtained through alkaline methanolysis, as its methyl ester (Klosterman and Smith 1954).

Flax leaf and stem tissues by contrast contain lignans of the dibenzylbutyrolactone type, as exemplified by (−)-yatein (Fig. 11.2) (Hemmati et al. 2010). Unlike the seed coat lignans, (−)-yatein and other lignans exist in their free form in these tissue types.

11.1 Initial Forays into Lignan Biosynthetic Pathway Discovery

At the beginning of our investigations into probing lignan biosynthetic pathways, there was a complete paucity of scientific knowledge on same, i.e., including lack of knowledge of the entry points into the biochemical pathways to afford various

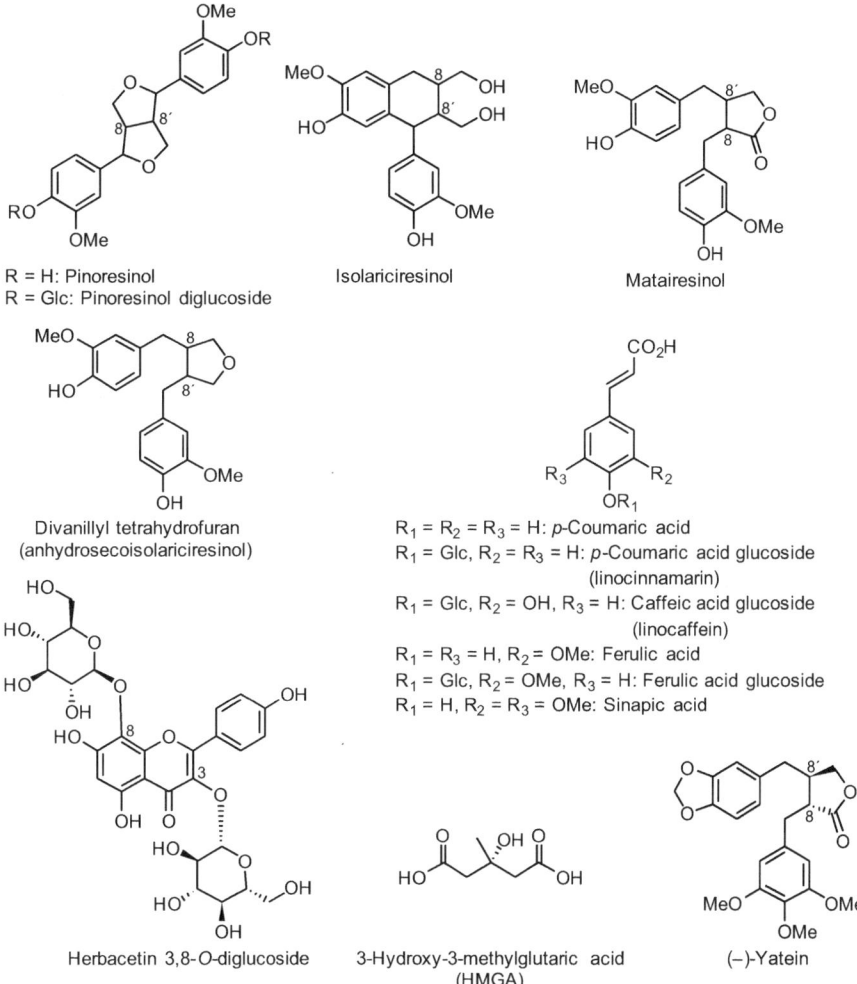

R = H: Pinoresinol
R = Glc: Pinoresinol diglucoside

Isolariciresinol

Matairesinol

Divanillyl tetrahydrofuran
(anhydrosecoisolariciresinol)

$R_1 = R_2 = R_3 = H$: *p*-Coumaric acid
$R_1 = Glc$, $R_2 = R_3 = H$: *p*-Coumaric acid glucoside
(linocinnamarin)
$R_1 = Glc$, $R_2 = OH$, $R_3 = H$: Caffeic acid glucoside
(linocaffein)
$R_1 = R_3 = H$, $R_2 = OMe$: Ferulic acid
$R_1 = Glc$, $R_2 = OMe$, $R_3 = H$: Ferulic acid glucoside
$R_1 = H$, $R_2 = R_3 = OMe$: Sinapic acid

Herbacetin 3,8-O-diglucoside

3-Hydroxy-3-methylglutaric acid
(HMGA)

(–)-Yatein

Fig. 11.2 Known flax seed constituents and the divanillyl tetrahydrofuran artifact. (–)-Yatein is present in the leaves/stems

lignan skeleta through phenylpropanoid (C_6C_3) monomer coupling, as well as in identification of the proteins and genes involved. This was also the case for the post-coupling transformations which were unknown at the beginning of our studies.

In particular, we considered it likely to be very instructive to establish how optical activity was frequently conferred in the lignan products found in vivo, as no protein or enzyme was known at that time to be able to produce optically active lignans (C_6C_3 dimers) from potential C_6C_3 monomeric precursors through stereoselective coupling. Instead, all attempts to enzymatically produce various lignans in vitro from C_6C_3 monomeric products by enzymatic (peroxidase, laccase) means engendered one electron oxidative coupling and only gave mixtures of racemic products (Freudenberg 1965).

Our earlier studies of lignan biosynthesis using *Forsythia* cell free extracts gave an indication that a stereoselective coupling entry point to some of the lignans might be occurring through the initial formation of (+)-pinoresinol from the achiral coniferyl alcohol (Umezawa et al. 1990; Davin et al. 1992). Further investigations led to the discovery of dirigent proteins, DPs (a name that we coined from the Latin *dirigere*, to guide or align). These studies established that the DP isolated from *Forsythia* was able to control the outcome of stereoselective coupling of coniferyl alcohol derived substrate to afford (+)-pinoresinol, as long as one electron oxidative capacity was also provided (Davin et al. 1997) (Fig. 11.3). The gene encoding this (+)-pinoresinol-forming DP was subsequently obtained (Gang et al. 1999), with the corresponding functional recombinant protein characterized in terms of kinetic properties (Halls and Lewis 2002; Halls et al. 2004). The native (+)-pinoresinol-forming DP was provisionally envisaged to be a trimer of circa 78 kDa (Davin et al. 1997).

Interestingly, when the *Forsythia* (+)-pinoresinol-forming DP was first isolated, its final purification step involved cation exchange chromatography separation from three other major proteins in the mixture (Davin et al. 1992), one of which was a one electron oxidase, a laccase. It was also purified and the presumed encoding gene cloned (Lewis et al. 1999). However, whether these additional proteins were part of an overall protein complex responsible for (+)-pinoresinol formation was not investigated at that time.

Our lignan biochemical pathway studies in *Forsythia* also resulted in the discovery of two downstream enantiospecific enzymes, pinoresinol-lariciresinol reductase (PLR) and secoisolariciresinol dehydrogenase (SDH) (Dinkova-Kostova et al. 1996; Xia et al. 2001), respectively. The *Forsythia* PLR catalyzed the sequential NADPH-dependent enantiospecific reduction of (+)-pinoresinol into initially (+)-lariciresinol and then (−)-secoisolariciresinol, whereas SDH catalyzed the NAD-dependent dehydrogenation of (−)-secoisolariciresinol into the dibenzylbutyrolactone lignan (−)-matairesinol (Fig. 11.4).

As rationalized below, the two chiral carbons or chiral centers at the C7 and C7′ positions in (+)-pinoresinol, as well as the one at C7′ in (+)-lariciresinol, are such as to establish the overall optical activity of both of these molecular species, i.e., to give a positive optical rotation with plane polarized light. On the other hand, the two chiral carbons at C8 and C8′ in (−)-secoisolariciresinol and (−)-matairesinol result in the opposite (now negative) optical rotation. Since, in their cases, the chirality at

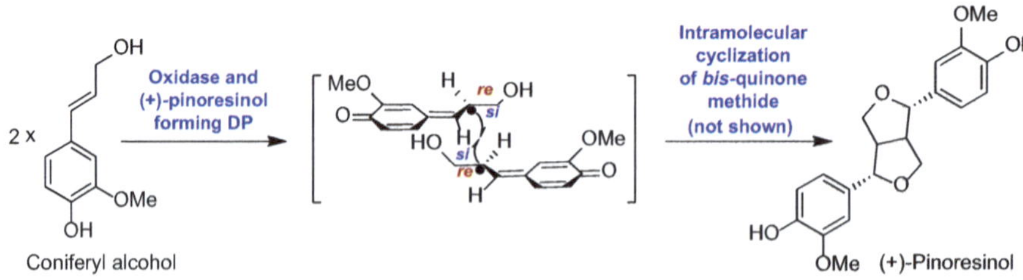

Fig. 11.3 *Forsythia* (+)-pinoresinol-forming dirigent protein (DP) stipulates formation of (+)-pinoresinol when incubated in the presence of coniferyl alcohol and an oxidase

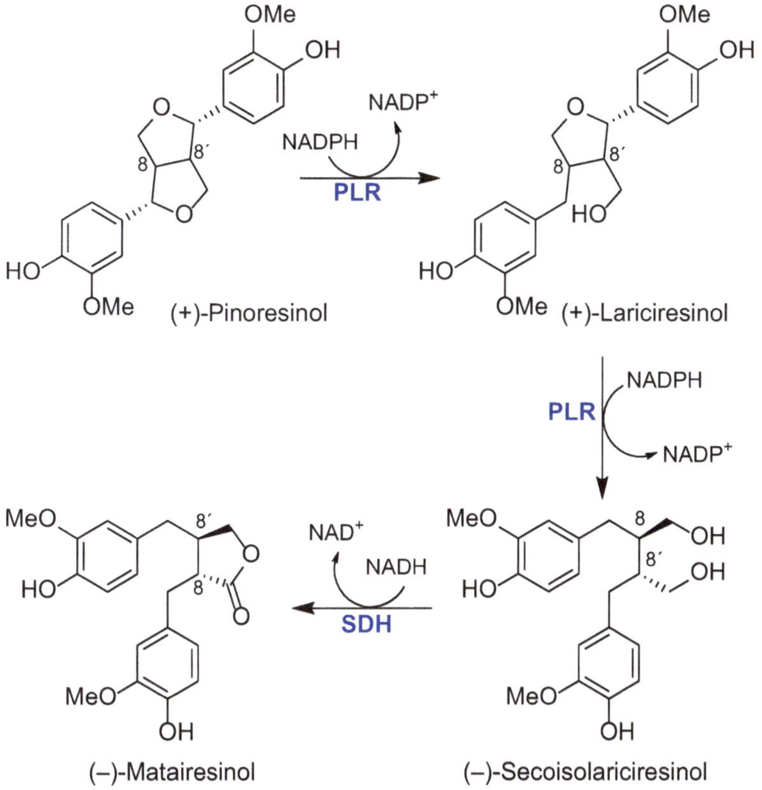

Fig. 11.4 Biochemical pathway from (+)-pinoresinol to (−)-matairesinol, via (+)-lariciresinol and (−)-secoisolariciresinol in *Forsythia* sp. PLR, pinoresinol lariciresinol reductase; SDH, secoisolariciresinol dehydrogenase

C7 and C7′ is absent, this reveals the greater influence on optical activity of the chiral centers at C7 and C7′ that results in (+)-pinoresinol and (+)-lariciresinol.

The corresponding genes encoding both PLR and SDH were individually cloned and the recombinant proteins expressed, with both PLR and SDH then characterized biochemically. Taken together, these studies thus demonstrated the overall pathway to (−)-secoisolariciresinol and (−)-matairesinol in *Forsythia* from coniferyl alcohol.

11.2   Flax Seed Coat Lignan Formation

To better understand the chemical nature of the SDG-containing component(s) in maturing flax seed tissue, an in-depth analysis of the natural products detected was carried out. This led to isolation of several new compounds, including 3-hydroxy-3-methylglutaric acid (HMGA) esters of SDG (Fig. 11.5), as well as *p*-coumaric and

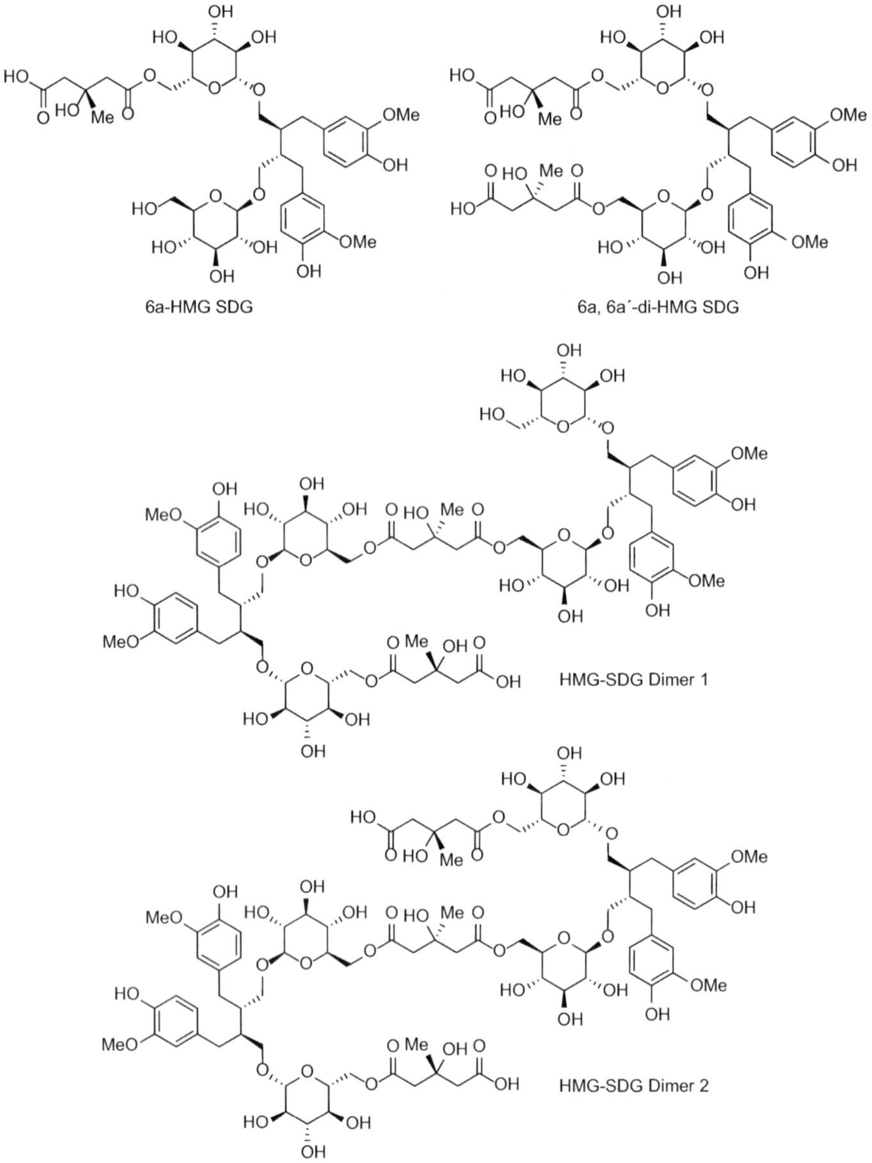

Fig. 11.5 6a-HMG SDG, 6a, 6a′-di-HMG SDG, HMGA, HMG-SDG dimer 1, and HMG-SDG dimer 2 from maturing whole flax seed tissue (Ford et al. 2001)

ferulic acid glucosides (Fig. 11.2) (Ford et al. 1999; Ford et al. 2001). Our discovery of the HMG-SDG esters, however, began to explain why SDG was not detected as such in the mature flax seed tissue, as it was covalently modified through ester linkages to HMGA and perhaps other moieties (Ford et al. 2001).

As previously indicated, the secoisolariciresinol released from the alkali-liberated SDG from whole flax seed tissue consisted of a circa 99:1 mixture of (+)- and (−)-secoisolariciresinols (Ford et al. 2001). This suggested the predominance of (−)-pinoresinol-forming DP activity, in contrast to the (+)-pinoresinol-forming DP described above, i.e., in order to account for the ~99% predominance of (+)-secoisolariciresinol formation. To account for the much smaller level (~1%) of (−)-secoisolariciresinol present in the SDG, one possibility was that it was derived from a much lower level of a (+)-pinoresinol-forming DP activity.

Preliminary assays in our laboratory resulted in detection of a (−)-pinoresinol-forming DP in the whole seed tissue. However, attempts to clone the encoding gene were unsuccessful. Instead, only a (+)-pinoresinol-forming DP flax gene homolog of the *Forsythia* (+) pinoresinol-forming DP was obtained, this presumably being a consequence of the primers employed at that time (Ford 2001).

Thus, to obtain the gene encoding the putative (−)-pinoresinol-forming DP in developing flax seed, a different strategy was clearly required. In this regard, independent studies by a team led by a previous Lewis lab researcher using *Arabidopsis* reported the presence of a (−)-pinoresinol reductase (PR) in *Arabidopsis* roots (Nakatsubo et al. 2008) that enantiospecifically converted (−)-pinoresinol into (−)-lariciresinol. That report strongly suggested to us that *Arabidopsis* must also have a gene or genes encoding one or more (−)-pinoresinol-forming DPs. This possibility was investigated, as the *Arabidopsis* genome had been sequenced earlier (Arabidopsis Genome Initiative 2000).

From our earlier studies (Kim et al. 2002), it was known that *Arabidopsis* had a DP multi-gene family of 25 members. Two of these, AtDIR5 and AtDIR6, were potentially pinoresinol-forming DP candidates as they had highest similarity (at the amino acid level) to the *Forsythia* (+)-pinoresinol-forming DP (61 and 65%, respectively). In addition, using a β-glucuronidase reporter gene strategy, *AtDIR6* was shown to be strongly expressed in *Arabidopsis* roots (Kim et al. 2012).

To determine whether *AtDIR5* and *AtDIR6* genes were encoding (−)-pinoresinol-forming DPs, both were individually cloned into pART17 and pMT/V5-His-TOPO expression vectors, respectively, for heterologous expression in *Solanum peruvianum* and *Drosophila* Schneider2 cells. Both recombinant DPs were established to be (−)-pinoresinol-forming DPs (Fig. 11.6) (Vassão et al. 2010; Kim et al. 2012), a finding that was further confirmed with in vivo studies (Kim et al. 2012). Our *Arabidopsis* study, together with the previous studies, thus clearly demonstrated the existence of two distinct modes of DP engendered stereoselective coupling to afford either (+)- or (−)-pinoresinol enantiomers. Another report of a (−)-pinoresinol-forming DP was also published (Pickel et al. 2010), an independent confirmation of our findings above.

Bioinformatics analysis of the known (+)-and (−)-pinoresinol-forming DPs next identified 14 residues differentially conserved between them. Ten of these were within a 49-amino-acid region of the DPs (see Fig. 11.7; region is shown by the green rectangle), whereas the others were in the N- and C-terminal regions. Thus, 49 amino acids of the region highlighted by the green rectangle (Asn-98 to Pro-146 in the (+)-pinoresinol-forming DP, ScDIR, Fig. 11.7) were swapped with those

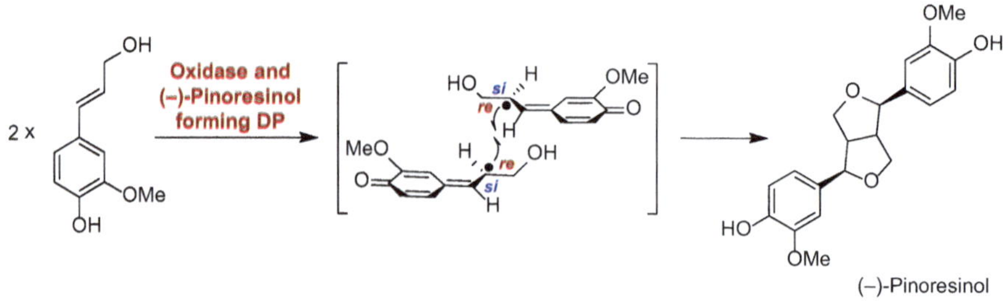

Fig. 11.6 *Arabidopsis* (–)-pinoresinol forming DP

```
ScDIR    ---MEGRKLIITIPLLLFFIAFFSVPPAAFGRKVTLPRKRMPQPCMNLVF
AtDIR6   MAFLVEKQLFKALFSFFLLVLLFSDTVLSFRKTIDQKK-----PCKHFSF

ScDIR    YFHDILYNGKNAANATSAIVGSPAWGNRTILAGQSNFGDMVVFDDPITLD
AtDIR6   YFHDILYDGDNVANATSAIVSPPG-------LGNFKFGKFVFFDGPITMD

ScDIR    NNLHSPPVGRAQGFYFYDRKDVFTAWLGFSFVFNNSDYRGSINFAGADPL
AtDIR6   KNYLSKPVARAQGFYFYDMKMDFNSWFSYTLVFNSTEHKGTLNIMGADLM

ScDIR    LIKTRDISVIGGTGDFFMARGIATLMTDAFEGEVYFRLRTDIKLYECY
AtDIR6   MEPTRDLSVVGGTGDFFMARGIATFVTDLFQGAKYFRVKMDIKLYECY
```

Fig. 11.7 Sequence alignments of (+)-pinoresinol (ScDIR) and (–)-pinoresinol (AtDIR6) forming dirigent proteins. The center region, used for swapping, is shown as a green rectangle (Kim et al. 2012)

in the (–)-pinoresinol-forming DP, AtDIR6 (Lys-90 to Leu-138). The resulting chimeric protein did not engender predominant formation of (+)-pinoresinol under the conditions employed. Instead, (–)-pinoresinol formation preferentially occurred (Fig. 11.8). This suggested that these 49 amino acid residues were involved in control of the two different stereoselectivities (Kim et al. 2012).

These studies had thus allowed us to discover and identify the nature of the genes that encoded the distinct (+)- and (–)-pinoresinol-forming DPs and the key regions in the DPs dictating different stereoselectivities.

Sequencing of the flax genome (Wang et al. 2012) next enabled us to readily investigate the types of DPs in flax and, in particular, those potentially stipulating the entry stereoselective coupling steps to the lignans present in its seed coat tissue. Although there were 44 different DPs annotated as such in the flax genome, only six of these were likely to be pinoresinol-forming DPs based on sequence similarity and identity. Two of these, trivially named LuDIR5 and LuDIR6, were ultimately shown to be responsible for (–)-pinoresinol formation in vitro, whereas another, LuDIR1, enabled production of the (+)-antipode (Dalisay et al. 2015). Based on sequence homology with other (+)- and (–)-pinoresinol forming DPs, it was deduced that LuDIR2–LuDIR4 could also potentially engender formation of (+)-pinoresinol, although this was not investigated. Nevertheless, our studies established there were indeed distinct (+)- and (–)-pinoresinol-forming DPs in flax.

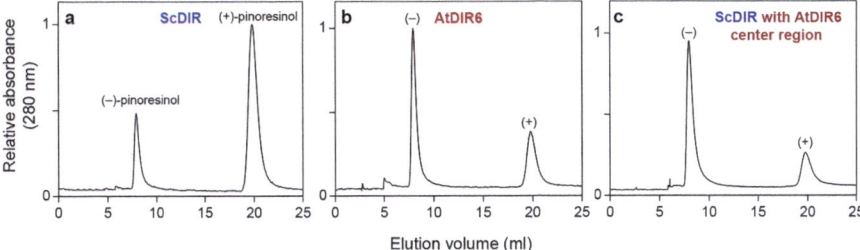

Fig. 11.8 Chiral chromatographic analyses of DP-engendered pinoresinol formation: Pinoresinol enantiomeric compositions obtained in assays using achiral coniferyl alcohol, an oxidase and a DP (Note: assay conditions were not optimized in terms of optimizing stereoselectivity). (+)-Pinoresinol-forming DP (ScDIR) engenders preferential formation of (+)-pinoresinol under the conditions employed (**a**), whereas (–)-pinoresinol-forming DP (AtDIR6) predominantly produces (–)-pinoresinol (**b**). Domain swapping to a chimeric protein changes the (+)-pinoresinol ScDIR into a (–)-pinoresinol-producing DP (**c**). (Adapted from Kim et al. (2012))

A gene encoding a PLR homolog, namely, PLR_Lu from flax seed, had also previously been found able to catalyze the conversion of (–)-pinoresinol into (–)-lariciresinol but not to (+)-secoisolariciresinol (Ford 2001; Teoh et al. 2003). This PLR homolog's biochemical activity was thus the same as that shown later by Nakatsubo et al. (2008) in *Arabidopsis* for which they named it as pinoresinol reductase (AtPrR). Another PLR homolog, LuPLR1, however, catalyzed formation of (+)-secoisolariciresinol from (–)-pinoresinol (von Heimendahl et al. 2005), whereas the homolog, LuPLR2, engendered formation of (–)-secoisolariciresinol from (+)-pinoresinol (Hemmati et al. 2010) as in *Forsythia* species (Dinkova-Kostova et al. 1996).

Interestingly, the LuPLR2 transcripts were only detected in stems and leaves, i.e., tissues that accumulate the (–)-secoisolariciresinol derived (–)-yatein, whereas transcripts of both LuPLR1 and LuPLR2 were detected in roots and seeds (Hemmati et al. 2010). RT-PCR analyses also established that LuPLR1 expression occurred in the seed coats at all five developmental stages studied (i.e., 10, 16, 20, 24, and 35 days after flowering) (Hano et al. 2006). Additionally, using a GUS reporter gene strategy, *LuPLR1* expression was found to be mainly localized to the seed coat, but was not detectable in the embryo (Fig. 11.9a), whereas expression of *LuPLR2* was highest in young leaves and to a lesser extent in old leaves and stems (Fig. 11.9b) (Corbin et al. 2017). These data were thus in agreement for preferential (+)-secoisolariciresinol formation in the seed coat and conversely of (–)-secoisolariciresinol formation leading to (–)-yatein in leaves and stems.

As indicated earlier, flax seed reportedly also contains small amounts of matairesinol and isolariciresinol. Their formation as bona fide natural products would require an enzyme catalyzing intramolecular ring closure to afford isolariciresinol and an enzymatic dehydrogenation and lactone ring formation to produce matairesinol.

As regards matairesinol formation, this is known to be catalyzed in *Forsythia* (FiSDH) and *Podophyllum* (PpSDH) species by the enantiospecific secoisolariciresinol dehydrogenase (SDH) (Fig. 11.4). However, to date there are no reports currently of a functional SDH in flax. A BLAST search of the *L. usitatissimum* genome

a
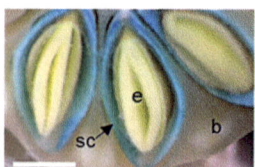
b

Fig. 11.9 (a) Histochemical X-Gluc visualization of the β-glucuronidase (GUS) activity in boll cross-section and seeds of stably transformed *Linum usitatissimum* transgenic plants. Blue staining indicates the localization of the GUS activity. Bars represent 1.5 mm. b, boll; sc, seed coat; and e, embryo. This figure is representative of five independent experiments performed under the same conditions. (**b**) Histochemical localization of GUS expression driven by the *LuPLR2* gene promoter showing the spatial *LuPLR2* gene expression in flax leaves. (Adapted from Hano et al. (2006), Renouard et al. (2012) and Corbin et al. (2017))

at Phytozome (Goodstein et al. 2012) was carried out with FiSDH as query. Of all of the genes retrieved, the first sixteen hits were selected with score values between ~300 and 230 and e-value $>e^{-74}$, with identities ranging from ~50–68% and ~48–50% as compared to FiSDH and PpSDH, respectively (Table 11.1). Although none of these were confirmed as SDH proper biochemically, this identity level is similar to that between FiSDH and PpSDH (~48%).

Additionally, gene expression profiling of the 16 putative SDHs was collected from RNA-seq data from eight different flax organs and compared to the accumulation profile of (−)-yatein (Fig. 11.10). In previous studies (Hemmati et al. 2010; Corbin et al. 2017), the leaves were established to be the main accumulation site of (−)-yatein. Six of the putative SDHs were actively expressed in leaves, with Lus10021312, Lus10021318, Lus10021314, and Lus10021313 only expressed in leaves, and thus could constitute good candidates for SDH proper. Lus10035258 was also highly detected in flowers and seedlings, with these tissues constituting secondary accumulation sites for (−)-yatein. However, high constitutive expression of Lus10021320 was detected in all tested organs, this not being consistent with (−)-yatein accumulation profile.

A glucosyltransferase (GT; UGT74S1) from flax also reportedly converted secoisolariciresinol into SDG (Ghose et al. 2014). Kinetic parameters, determined using 80 μg of His-tag purified UGT74S1, gave a k_{cat} value of 0.89 sec^{-1}, with estimated apparent K_m values for secoisolariciresinol and UDP-Glc of 79 and 1188 μM, respectively. An ethyl methane sulfonate (EMS) mutagenized flax population was also screened for UGT74S1 mutants, yielding 13 homozygous missense mutations and two homozygous nonsense mutations (Fofana et al. 2017). These mutations in the UGT74S1 coding regions altered SDG profiles in flax seed, with SDG not being detected in seeds of homozygous nonsense mutants from M3 and M4 generations. These data would suggest that this glucosyltransferase was required for SDG formation.

The specific enzymes responsible for subsequent downstream conversions of SDG into 6a-HMG SDG, 6a, 6a′-di-HMG SDG, and additional congeners have not

Table 11.1 Identity matrix (in %) of flax SDH protein homologs

	FiSDH	PpSDH	Lus10021312	Lus10021318	Lus10016991	Lus10021314	Lus10021313	Lus10005262	Lus10016995	Lus10014227	Lus10021319	Lus10014228	Lus10034626	Lus10029368	Lus10021320	Lus10016997	Lus10035258
FiSDH																	
PpSDH	48.0																
Lus10021312	51.1	49.3															
Lus10021318	50.0	49.8	92.2														
Lus10016991	50.2	49.8	91.5	93.1													
Lus10021314	50.4	49.6	91.9	92.7	94.3												
Lus10021313	50.4	48.9	96.0	92.3	92.5	91.7											
Lus10005262	52.2	50.0	91.6	90.8	88.7	89.4	90.9										
Lus10016995	48.9	48.3	90.7	95.6	93.1	90.5	90.8	88.6									
Lus10014227	52.0	49.1	64.3	64.9	64.2	63.7	64.0	63.6	64.6								
Lus10021319	49.8	49.5	52.0	52.0	52.3	51.8	51.8	51.8	52.8	55.6							
Lus10014228	52.7	48.0	60.7	60.9	60.2	60.8	60.3	61.0	59.8	83.4	55.6						
Lus10034626	51.2	48.5	54.0	53.8	55.0	53.6	54.8	55.5	53.8	56.3	51.1	54.0					
Lus10029368	52.3	50.4	48.9	49.4	48.7	49.2	49.3	50.2	48.7	51.3	52.1	51.3	57.0				
Lus10021320	68.0	48.6	50.9	51.9	51.8	51.3	51.3	51.3	50.7	52.4	50.4	53.5	50.0	51.5			
Lus10016997	67.6	48.6	50.9	51.9	51.8	51.3	51.3	51.3	50.7	52.0	49.6	53.1	49.6	51.2	96.8		
Lus10035258	50.8	48.9	53.2	53.1	54.2	52.9	54.0	54.8	53.4	55.5	52.3	53.2	95.5	57.4	50.0	49.6	
Lus10014226	53.5	48.2	61.3	62.8	63.1	63.3	61.0	61.7	62.8	83.4	55.0	77.5	53.8	50.8	53.9	54.3	52.7

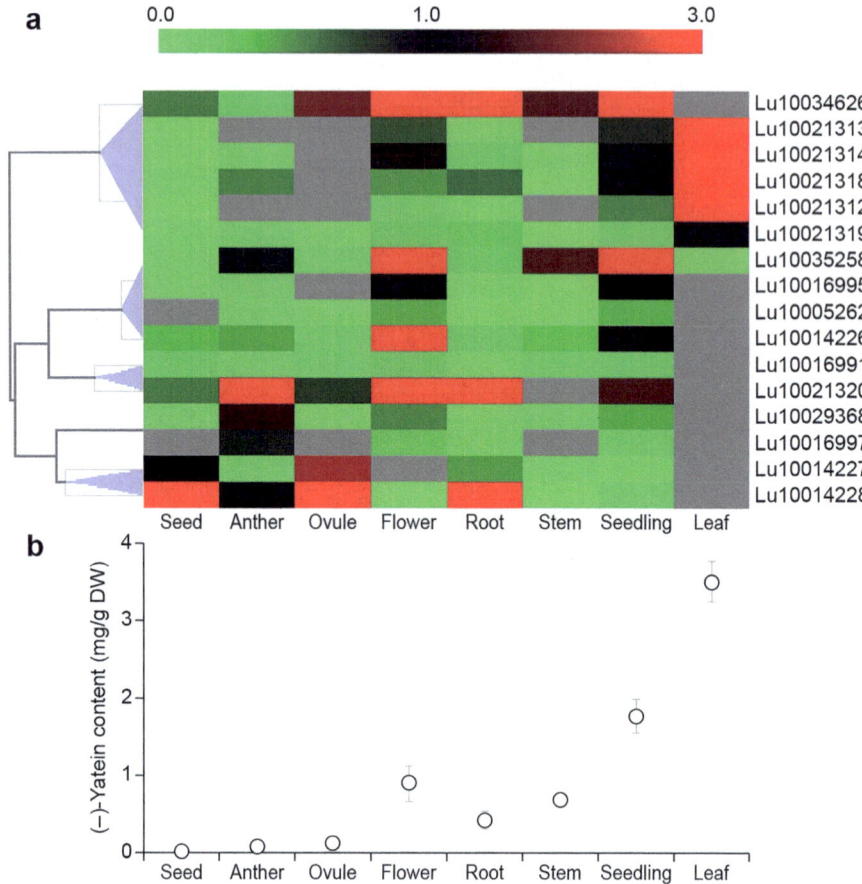

Fig. 11.10 Correlation of SDH RNAseq data with (–)-yatein amounts in different tissue types. (**a**) Clustering analysis of RNAseq data obtained for putative SDHs from flax. Available flax putative *SDH* transcriptomic data from different flax organs was normalized and converted in log form for a clustering classification realized by a hierarchical clustering analysis (HCA), based on the complete linkage Pearson uncentered correlation method performed with MeV. (**b**) (–)-Yatein accumulation in these organs. Data are expressed as the mean of four independent experiments ± standard deviation of the mean. (Adapted from Corbin et al. (2017) and Drouet et al. (unpublished results))

yet been described. On the other hand, HMGA biosynthesis is very well-known, being a product derived from acetyl CoA (Rudney 1957).

The biosynthetic formation of herbacetin 3,8-*O*-diglucoside in vivo can be envisaged to occur through hydroxylation of kaempferol at C8, this presumably being catalyzed by a yet to be described cytochrome P450; the kaempferol biosynthetic pathway has been extensively reported (Forkmann and Heller 1999). However, to our knowledge, the corresponding glucosyltransferases to afford not only herbacetin diglucoside, but also the hydroxycinnamic acid glucosides, have not been reported.

11.3   Lignan Pathway Structural Biology Studies

The discovery of DPs, PLRs, and SDHs and their facile expression in recombinant form provided a means to carry out much needed structural biology studies.

11.3.1   (+)- and (−)-Pinoresinol-Forming DPs

The first DP to have its 3D structure (4REV) solved was that of the (+)-pinoresinol-forming DP from pea (*Pisum sativum*), DRR206, obtained at 1.95 Å resolution (Kim et al. 2015). Determination of the structure proved that this DP exists as a tightly packed trimer, with each monomer having an eight-stranded β-barrel topology (Fig. 11.11a). Analysis suggested that each monomer in the trimer has a prominent deep pocket at one end of the barrel, surrounded by flexible loops. Interestingly, a Dali search of the Protein Data Bank (PDB) identified allene oxide cyclases (AOC) as having the same trimeric β-barrel structure as DRR206 despite their low sequence similarity (~11%). The closest AOC from *Physcomitrella patens* (PpAOC2;PDB code 4H69) can be superimposed with 2–3 Å *rmsd* over the entire barrel structure of one monomer (Fig. 11.11b). We have proposed that the pocket near the top of each barrel, oriented toward the outside of the trimer and lined with hydrophobic residues, is provisionally the substrate binding site for (+)-pinoresinol formation (Fig. 11.11c). However, because of the spatial distribution of each putative binding site in each monomer, intermolecular coupling cannot occur between the distinct binding sites in the trimer per se. The volume of the pocket suggested that two monolignol-derived substrates could bind in a single pocket.

In a related manner, the corresponding structure of the (−)-pinoresinol-forming DP was subsequently reported at 1.4 Å resolution, this further confirming the

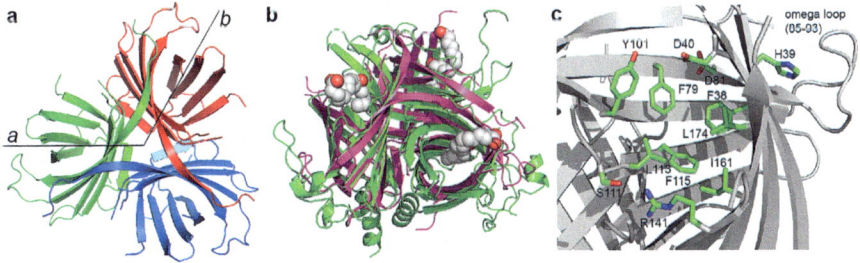

Fig. 11.11 (**a**) (+)-Pinoresinol-forming DP DRR206 trimer viewed down the threefold *c*-axis of the H32 space group, with *a*- and *b*-axes as indicated. (**b**) DRR206 (magenta) superposition with *P. patens* AOC (4H69, green, w/ligand, spheres), with *rmsd* ~ 2.1 Å over 102 residues. (**c**) Several residues lining putative active site pocket are differentially conserved in (+)- or (−)-pinoresinol-forming DPs, suggesting a role in determining substrate orientation. (Adapted from Kim et al. (2015))

trimeric constitution in vivo (Gasper et al. 2016). In contrast to DRR206, this structure had well-resolved loops surrounding the putative substrate binding pocket and, like DRR206, modeling suggested that two substrates could occupy the pocket.

Around 54% of the residues in the two pinoresinol-forming dirigent proteins (DRR206 and AtDIR6) are identical. Each is around 70% identical to the corresponding (+)- or (−)-pinoresinol-forming DP homologues in flax: LuDIR1 – LuDIR4 (+), and LuDIR5 and LuDIR6 (−). This high level of sequence similarity is sufficient for construction of accurate homology models. While only the activities of LuDIR1, LuDIR5 and LuDIR6 have been experimentally verified, the additional sequences listed are highly similar (>90% identity) to one or the other as indicated.

11.3.2 Pinoresinol-Lariciresinol Reductase (PLR) and PLR Homologs

The apo-PLR dimer structure (PLR_Tp1 from *Thuja plicata*, PDB code 1QYD) was solved at 2.5 Å resolution, this being responsible for the enantiospecific NADP(H) conversion of (−)-pinoresinol into (+)-secoisolariciresinol (Fig. 11.12) (Min et al. 2003). In that study, site-directed mutagenesis and substrate modeling studies provided evidence that Lys[138] was responsible for the catalytic activity, this being abolished when Lys[138] was mutated with Ala. In this catalysis, the 4-*pro*-R hydride from NADP(H) is also specifically abstracted.

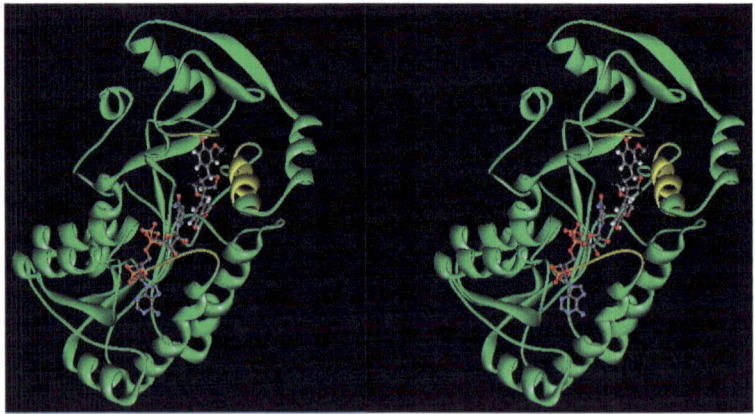

Fig. 11.12 Stereo view of ribbon diagram of PLR_Tp1. (Reproduced from Min et al. (2003) with permission)

Fig 11.13 Arrangement of the tetrameric SDH from *Podophyllum peltatum*. (Adapted from Youn et al. (2005))

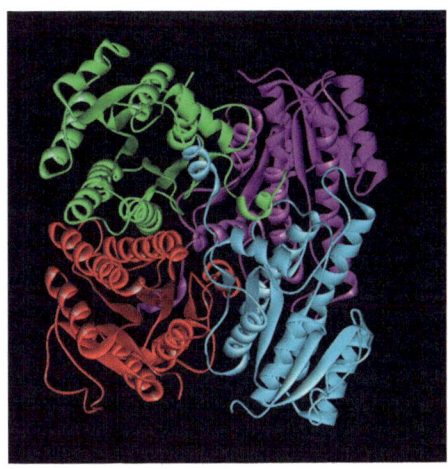

11.3.3   Secoisolariciresinol Dehydrogenase (SDH)

An apo-SDH structure in tetrameric form was determined at 1.6 Å resolution, using *Podophyllum peltatum* SDH (PDB code 2BGK, Fig. 11.13) (Youn et al. 2005). This enzyme catalyzes the enantiospecific conversion of (−)-secoisolariciresinol into (−)-matairesinol. As a short chain dehydrogenase (SDR), it has the well-known highly conserved catalytic triad (in this case, Ser[153], Tyr[167], and Lys[171]) with Tyr[167] in the SDH serving as general catalytic base.

Interestingly, analysis of the potential flax SDH sequences to that of the SDH above shows presence of the conserved glycine-rich motif, GXXGXG (Fig. 11.14), known to participate in binding of the pyrophosphate group of NAD⁺, as well as Asp[47] (numbering system as for PpSDH) found in SDRs that preferentially bind NAD(H). Flax SDH homologs also have the conserved catalytic triad containing Tyr, Lys, and Ser as above. Other amino acids involved in cofactor/substrate binding arc also present (red labeled residues in Fig. 11.14).

11.4   Metabolite Imaging and Other Localization Studies in Developing Flax Seed In Situ

MALDI mass spectrometry imaging (MSI) was deemed instructive to study localization of the lignan metabolites in developing (maturing) flax seed tissues in situ (Fig. 11.15a-d) (Dalisay et al. 2015). Of the different stages of flax seed maturation examined, however, no metabolites corresponding to either pinoresinol, lariciresinol, or secoisolariciresinol were detected under the conditions employed, perhaps reflecting their rapid turnover.

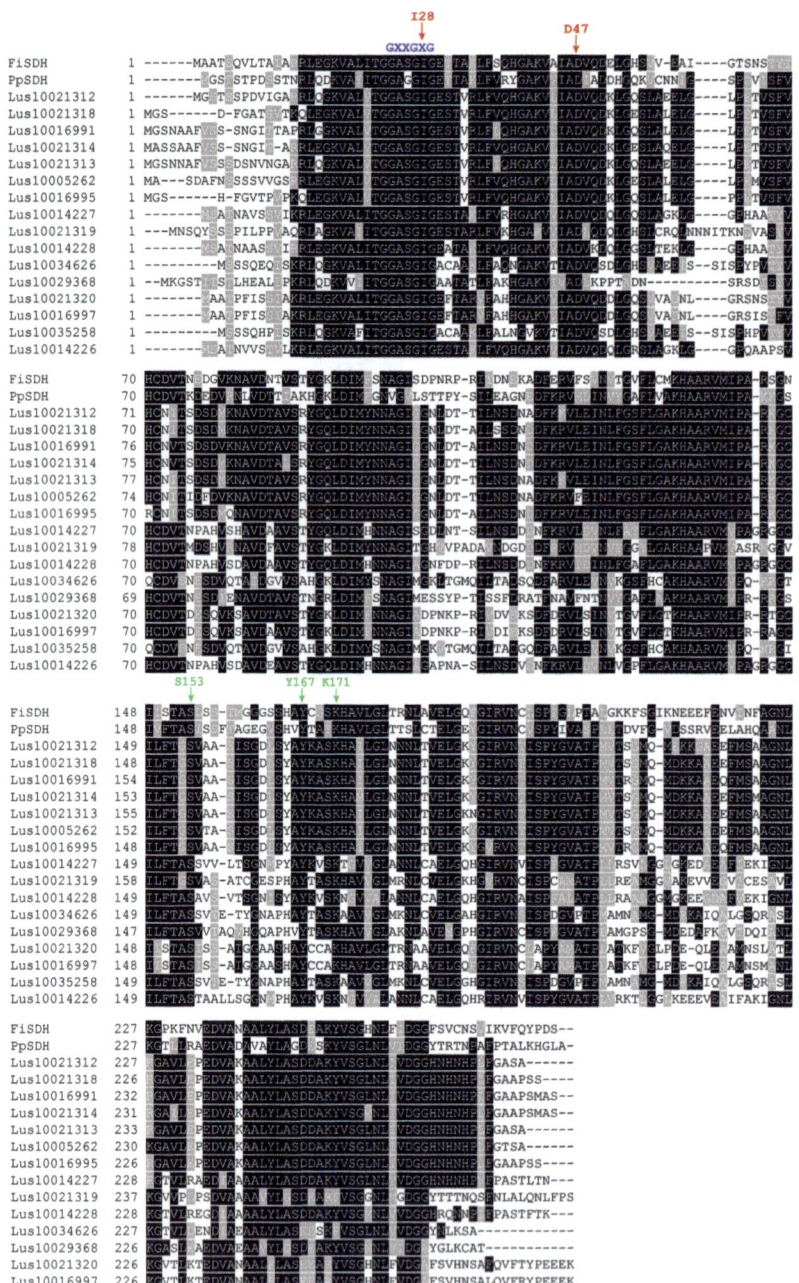

Fig. 11.14 Alignment of 16 flax SDH homologs to FiSDH and PpSDH. GXXGXG: conserved glycine-rich motif is shown in bold blue. Green arrows indicate the catalytic triad (Ser[153], Tyr[167], and Lys[171]). Ile[28] and Asp[47] involved in NAD binding are shown with red arrows

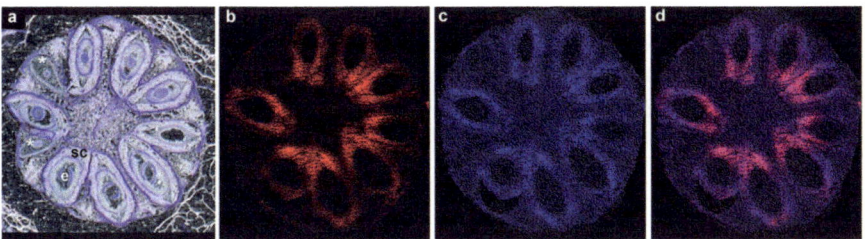

Fig. 11.15 MALDI mass spectrometry images of flax capsules. (**a**) Optical image of capsule (7 DAF) cross-section stained with toluidine blue O. (**b–d**) Positive ion [M+K]⁺ MALDI MS images of 7 DAF flax capsule: SDG (**b**), 6a-HMG SDG (**c**). Merged MALDI MS images of SDG (red) and 6a-HMG SDG (blue) (**d**). (Adapted from Dalisay et al. (2015). Abbreviations: e, embryo; sc, seed coat; * indicates aborted seeds)

On the other hand, both SDG (Fig. 11.15b) and the 6a-HMG SDG homolog (Fig. 11.15c) were readily detected in the seed coat tissue at an early developmental stage (6 and 7 days after flowering, DAF). Molecular ions corresponding to 6a, 6a'-di-HMG SDG were not detected though, as the ion intensity above m/z 900 was very low under these conditions. Accordingly, how additional steps in SDG covalent modification were potentially occurring post 6a-HMG SDG has not yet been determined in situ.

Importantly, using the OpenMSI tool, it was established that both SDG and 6a-HMG SDG metabolites were co-localized in the seed coat and not the embryo, as shown in Fig. 11.15d.

It should also be emphasized that unambiguous identification and localization of the above SDG and the 6a-HMG SDG dimer required meeting several quality criteria. These were 1) HRMS accurate mass values were within acceptable limits in comparison to the authentic standards; 2) MS/MS fragmentation patterns of the localized metabolites and the authentic standards were in agreement with each other; and 3) ion mobility analyses were in good agreement for both authentic standards and localized metabolites in situ.

While the data collection per sample for metabolite imaging can take up to a day or two to collect, the information obtained is very high throughput as regards the OpenMSI capability. It allows for an interactive analysis in real time, whereby exploration of numerous ion occurrences and their localization can be carried out to examine whether, for example, additional metabolites of any metabolic pathway of interest can be detected.

In this respect, the cyanogenic glucosides linamarin, lotaustralin, linustatin, and neolinustatin (Fig. 11.16) were readily detected and localized (Dalisay et al. 2015). This ability to search for numerous metabolites simultaneously is an extraordinarily useful capability for localization of such metabolites in situ, provided that the conditions employed permit the molecular species to be able to "fly" and fragment during the MALDI TOF MS/MS and Ion Mobility analyses.

Additionally, histochemical GUS staining was performed on developing seeds of transgenic flax plants, harboring the *LuPLR1* gene promoter fused to the *uidA* gene. This staining revealed strong GUS activity localized in the parenchymatous cells of the outer integument. No GUS activity was detected in either the inner integument or the embryo regardless of developmental stage (Fig. 11.17). These data are also consistent with the SDG being produced in the seed coat.

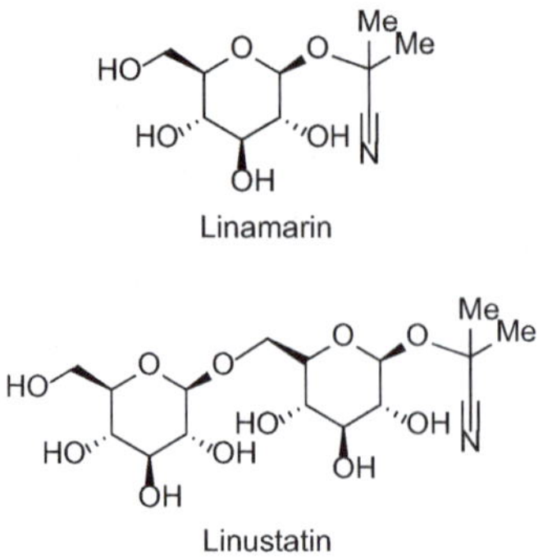

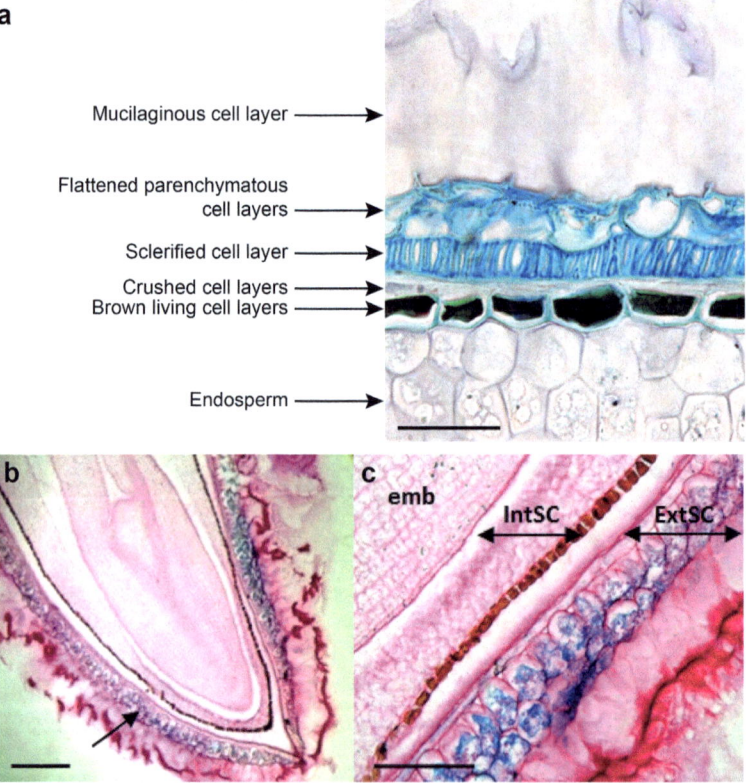

Fig. 11.16 Cyanogenic glucosides, linamarin, lotaustralin, linustatin, and neolinustatin

Fig. 11.17 GUS-staining visualization of *LuPLR1* gene promoter expression. (**a**) Micrograph of a semi-thin section (5 μm) of mature flax seed (cv. Barbara) stained with toluidine blue. Bar represents 50 μm. (**b, c**) Histochemical studies of the localization of *LuPLR1* promoter-driven *GUS* gene expression in transgenic flax immature seed (20 DAF). (**b**) Semi-thin longitudinal section (5 μm). Bars represent 250 μm. Arrow indicates *GUS* staining. (**c**) Magnification of the semi-thin longitudinal sections in (**b**). Bars represent 50 μm. Emb, embryo; IntSC, internal seed coat (inner integument); ExtSC, external seed coat (outer integument). (Adapted from Fang et al. (2016))

11.5 SDG-Containing Lignans in Developing Flax Seed Tissue and Their Constitution In Vivo

As indicated earlier, SDG liberation from either flax seed coat tissue or from an aqueous ethanol extract of said tissue by mild alkali treatment is indicative of ester-linkage cleavage in the molecular species (or mixtures) containing covalently linked SDG present in situ. This observation though places severe constraints as to how SDG can be covalently linked to other constituents in the flax seed coat tissue, as SDG only has free hydroxyl or phenolic groups available for esterification. Accordingly, SDG can only be ester-linked to another moiety containing a carboxylic acid group, such as with the previously noted HMGA (Fig. 11.2). Covalent linkage of SDG to the dicarboxylic acid HMGA does, however, provide the possibility of forming higher-molecular-weight natural products or pathway intermediates. These can build up from, for example, two SDG moieties linked together to three HMGA molecules as shown in HMG-SDG Dimer 2 in Fig. 11.5, with the remaining free carboxylic acid moiety in the esterified HMGA being available for further covalent modification.

Since the SDG-ester linked moieties in flax seed were readily aqueous EtOH solubilized, this perhaps suggested that their molecular weight distribution range was not very high (i.e., oligomeric rather than polymeric). Accordingly, to investigate this question, partial purification of the SDG-containing components from other constituents in the flax seed hull aqueous EtOH extract by Amberlite XAD-16 column chromatography was carried out by adapting the procedure of Kosinska et al. (2011a). The SDG-containing fractions so obtained were then combined prior to gel permeation chromatography (GPC).

A HiLoad® 16/600 Superdex® 30 pg column was then used for the GPC fractionation, with the SDG-containing sample applied and eluted from the column using EtOH:H_2O, to afford 120 individual fractions. A second GPC column, an analytical TSK gel G2000 SWXL (Tosoh) GPC column, eluted with CH_3CN:H_2O containing 0.1% CF_3CO_2H, was next used to investigate the molecular weight profiles of the obtained individual fractions by adapting the method of Kosinska et al. (2011b). This was calibrated aforehand with thyroglobulin (669 kDa) and bovine albumin (66 kDa) – both being eluted at the void volume (V_0) – together with other standards, namely, cytochrome c (12.4 kDa), apoprotein (6.5 kDa), cyclodextrin (1.7 kDa), rifampicin (823 Da), taxifolin hydrate (304 Da), and gallic acid (188 Da). While these were used for calibration purposes, the shape and size of these standards, as well as the SDG-containing components themselves, are not necessarily directly comparable from a molecular weight perspective.

Nevertheless, Fig. 11.18a shows the GPC chromatographic behavior of the eight standards for column calibration purposes and Fig. 11.18b that of the crude SDG-containing sample prior to HiLoad® 16/600 Superdex® 30 GPC fractionation. With the column calibrated as described, the crude SDG-containing "macromolecular" extract was eluted as essentially a monomodal polydisperse fraction. Based on the calibration curve, the MWD (molecular weight distribution) of the sample was estimated to largely range from 6.5 to 0.8 kDa.

Selected fractions obtained from the HiLoad® 16/600 Superdex® 30 GPC step were then individually applied to an analytical TSK gel G2000 SWXL (Tosoh) GPC column and eluted under the same conditions as for the standards (Moinuddin et al., manuscript in finalization). Analysis of these selected fractions containing SDG-derived components demonstrated that they were of decreasing molecular size (Fig. 11.18b). Under these conditions, there was no significant level of non-covalent association observed within each fractions 67–106 re-examined, i.e.,

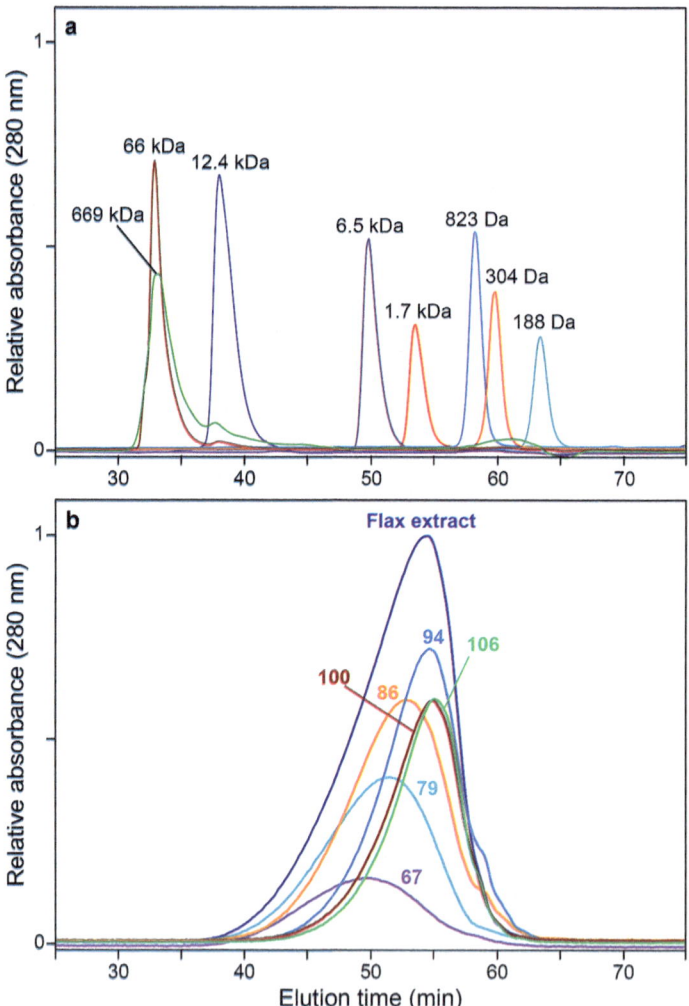

Fig. 11.18 (**a**) Separation of molecular weight standards: thyroglobulin (669 kDa), bovine albumin (66 kDa), cytochrome c (12.4 kDa), apoprotein (6.5 kDa), cyclodextrin (1.7 kDa), rifampicin (823 Da), taxifolin hydrate (304 Da), and gallic acid (188 Da). (**b**) Gel permeation chromatography of flax seed hull lignan extract and of selected fractions (67, 79, 86, 94, 100, and 106) obtained from HiLoad® 16/600 Superdex® 30 GPC

they were unable individually to regenerate or restore the original monomodal MWD profile.

From the above Superdex GPC fractionation, some 20 different fractions (between fractions 67 and 106) were next selected, with each individually subjected to mild alkali treatment, in order to establish the molecular species composition of the liberated compounds, with each being quantified on a molar basis using either quantitative HPLC or GC. Figure 11.19 illustrates the estimated changes in alkali released molar component proportions that occurred in the GPC separation, with the eight detected components being SDG, HMGA, herbacetin diglucoside, *p*-coumaric acid and its glucoside, ferulic acid and its glucoside, as well as caffeic acid glucoside. These data clearly showed that the proportion of SDG steadily increased as the presumed molecular weight average of each fraction increased. While the structures of these eight individual components released upon alkaline hydrolysis have been determined in earlier studies, a previous attempt to identify the precise chemical structure of the SDG-containing "macromolecule" preparations yielded limited clarification at that time (Struijs et al. 2009). This uncertainty of the(ir) precise chemical structure(s), as here, remains an important goal to solve.

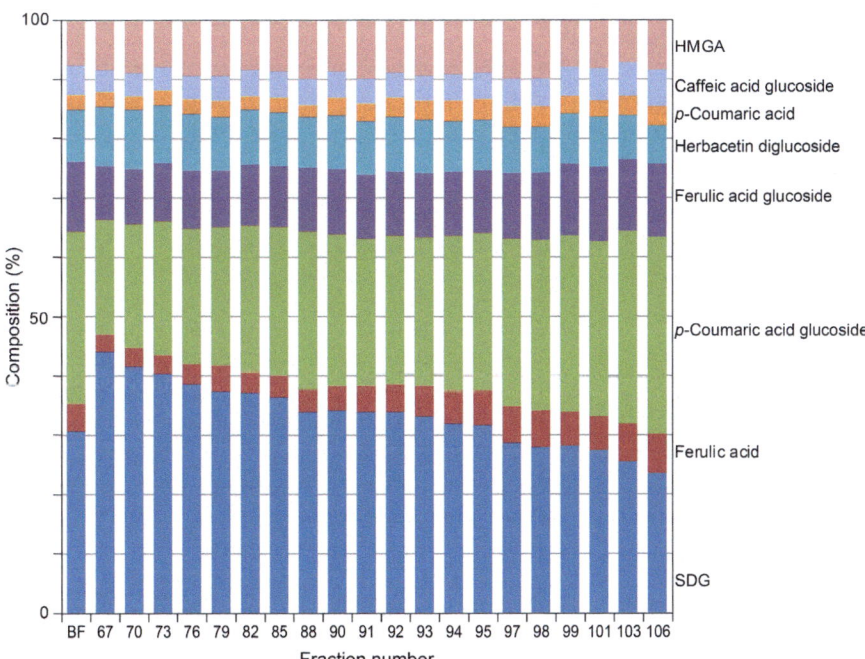

Fig. 11.19 Estimated molar proportions of flax seed components released by alkaline hydrolysis of SDG-containing components before fractionation (BF) and of different fractions after GPC. SDG, secoisolariciresinol diglucoside; HMGA, 3-hydroxy-3-methylglutaric acid

11.6 Lignan Formation in Flax Stem and Leaf Tissues

Flax leaf and stem tissues contain the lignan (−)-yatein, this being considered to be a plant defense response biosynthesized either constitutively or inducibly. Its formation derives from (+)-pinoresinol via (−)-matairesinol as depicted in Fig. 11.4. The intermediates between matairesinol and yatein are still, however, unknown/unclear as is the order of enzymatic reactions. Its formation though requires hydroxylation at C5, O-methylation at C4 and C5, and methylenedioxy bridge formation at C3′-C4′.

In *Anthriscus sylvestris*, from feeding experiments using [13]C- or [2]H-labeled precursors, (−)-yatein is considered to be formed from matairesinol through a biosynthetic sequence including thujaplicatin, 5-methylthujaplicatin, and 4,5-dimethylthujaplicatin (Sakakibara et al. 2003).

A cDNA isolated from an *A. sylvestris* cDNA library was also obtained, whose recombinant protein was shown to encode an O-methyltransferase (OMT) involved in methylation of thujaplicatin to 5-methylthujaplicatin in yatein biosynthesis (Ragamustari et al. 2013). Several orthologs of this OMT, as well as CYP81Q (yatein synthase), were also identified in the flax genome (unpublished results).

This metabolic pathway apparently differs from that proposed by Hendrawati et al. (2011) in the same plant species, as well as that of Marques et al. (2014) and Lau and Sattely (2015) where (−)-pluviatolide is reported to be an intermediate in podophyllotoxin biosynthesis.

To date, it is thus unclear if yatein biosynthesis from matairesinol involves either a single route or a plant specific route resulting from convergent evolution, or whether it exists as a metabolic grid.

11.7 The Flax 44-Membered DP Multi-gene Family

It is of considerable interest that the emergence of DPs in the plant kingdom appears to be coincident with the transition of aquatic plants to a terrestrial environment.

As for many other plant species, flax contains a large multi-gene family that encodes DPs, of which there are a total of 44 members in all (Corbin et al. 2018). These are shown in Fig. 11.20, as well as the different clades they have been assigned to. However, other than the (+)- and (−)-pinoresinol-forming DPs, no other physiological or biochemical functions are currently known for the others.

This lack of knowledge though provides a wonderful and exciting opportunity to systematically establish in future the functions and physiological roles of the remainder of the DPs in flax that are of unknown biochemical function.

In this regard, a few other DP homologs have been reported in other plant species with biochemical functions unrelated to that of lignan biosynthesis. These include stereoselective formation of (+)-gossypol from its achiral precursor, hemi-gossypol,

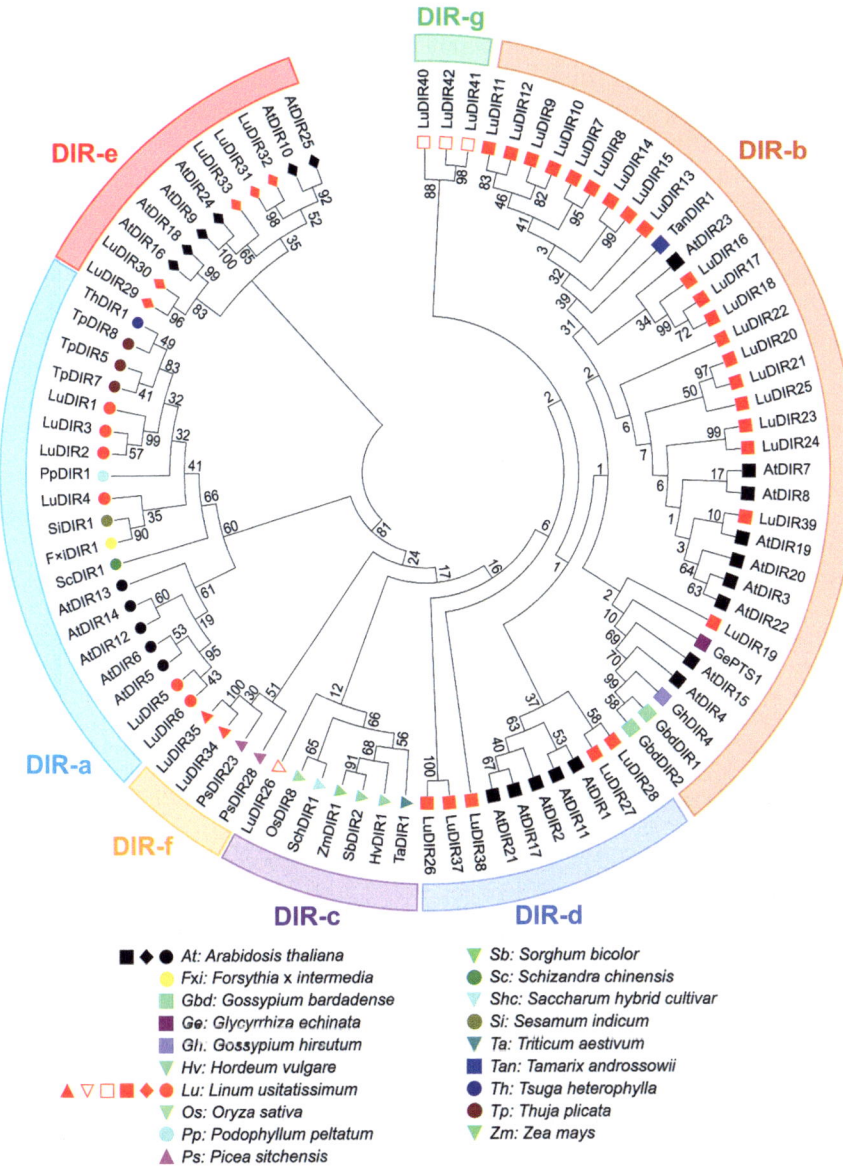

Fig. 11.20 Phylogenic tree of flax DIRs. (Reproduced with permission from Corbin et al. (2018))

in cotton (Liu et al. 2008; Effenberger et al. 2015); enantiospecific formation of the pterocarpan medicarpin in liquorice from its precursor, 7,2'-dihydroxy-4'-methoxyisoflavanol (Uchida et al. 2017); and controlled deposition of Casparian band lignin in *Arabidopsis* (Hosmani et al. 2013).

As regards work studying biopolymeric lignin formation in the *Arabidopsis*

Casparian band tissue, the presence of a supramolecular protein complex was proposed as being responsible for its controlled lignin deposition (Hosmani et al. 2013). This proposed complex consisted of a scaffold protein (CASP), different DPs, and oxidases (laccase/peroxidase) and was viewed as being required for its constitutive lignification. Significantly, when these specific DPs were knocked out, uniform lignin deposition did not occur.

Whether the mixed biochemical pathway to SDG lignan and non-lignan components requires a supramolecular protein complex, such as including a (membrane) scaffold, oxidase(s), and DPs, is unknown at this time but worthy of inquiry in the future.

On the other hand, the vast bulk of DP genes, and their encoding proteins, in the plant kingdom have no known biochemical and/or physiological function at present. Yet, in terms of investigating DP functions, flax appears to be one of the best plant candidates to do this. This is because the roles of its 44-membered DP family can be systematically explored using all of the major technologies currently available at the gene, protein/enzyme, and metabolomics level.

Acknowledgments The authors thank the Chemical Sciences, Geosciences and Biosciences Division, DOE Office of Basic Energy Sciences (DE-FG-0397ER20259) and the National Science Foundation (MCB-1052557) for support, as well as the Arthur M. and Katie Eisig-Tode Foundation. MALDI MS-based imaging analysis was performed on an instrument acquired through a Major Research Instrumentation grant (DBI-1229749) from the National Science Foundation. A portion of the research was performed using the Environmental Molecular Sciences Laboratory (EMSL), a DOE Office of Science User Facility sponsored by the Office of Biological and Environmental Research and located at Pacific Northwest National Laboratory (PNNL) in Richland, WA. Use of the Stanford Synchrotron Radiation Lightsource is supported by the DOE Office of Basic Energy Sciences under Contract No. DE-AC02-76SF00515. The SSRL Structural Molecular Biology Program is supported by the DOE Office of Biological and Environmental Research and by the National Institutes of Health, National Institute of General Medical Sciences (including P41GM103393).

References

Arabidopsis Genome Initiative (2000) Analysis of the genome sequence of the flowering plant *Arabidopsis thaliana*. Nature 408:796–815

Corbin C, Drouet S, Mateljak I, Markulin L, Decourtil C, Renouard S, Lopez T, Doussot J, Lamblin F, Auguin D, Lainé E, Fuss E, Hano C (2017) Functional characterization of the *pinoresinol-lariciresinol reductase-2* gene reveals its roles in yatein biosynthesis and flax defense response. Planta 246:405–420

Corbin C, Drouet S, Markulin L, Auguin D, Lainé E, Davin LB, Cort JR, Lewis NG, Hano C (2018) A genome-wide analysis of the flax (*Linum usitatissimum* L.) dirigent protein family: from gene identification and evolution to differential regulation. Plant Mol Biol 97:73–101

Dalisay DS, Kim K-W, Lee C, Yang H, Rübel O, Bowen BP, Davin LB, Lewis NG (2015) Dirigent protein-mediated lignan and cyanogenic glucoside formation in flax seed: integrated omics and MALDI mass spectrometry imaging. J Nat Prod 78:1231–1242

Davin LB, Bedgar DL, Katayama T, Lewis NG (1992) On the stereoselective synthesis of (+)-pinoresinol in *Forsythia suspensa* from its achiral precursor, coniferyl alcohol. Phytochemistry 31:3869–3874

Davin LB, Wang H-W, Crowell AL, Bedgar DL, Martin DM, Sarkanen S, Lewis NG (1997) Stereoselective bimolecular phenoxy radical coupling by an auxiliary (dirigent) protein without an active center. Science 275:362–366

Dinkova-Kostova AT, Gang DR, Davin LB, Bedgar DL, Chu A, Lewis NG (1996) (+)-Pinoresinol/(+)-lariciresinol reductase from *Forsythia intermedia*: Protein purification, cDNA cloning, heterologous expression and comparison to isoflavone reductase. J Biol Chem 271:29473–29482

Effenberger I, Zhang B, Li L, Wang Q, Liu Y, Klaiber I, Pfannstiel J, Wang Q, Schaller A (2015) Dirigent proteins from cotton (*Gossypium* sp.) for the atropselective synthesis of gossypol. Angew Chem Int Ed 54:14660–14663

Fang J, Ramsay A, Renouard S, Hano C, Lamblin F, Chabbert B, Mesnard F, Schneider B (2016) Laser microdissection and spatiotemporal pinoresinol-lariciresinol reductase gene expression assign the cell layer-specific accumulation of secoisolariciresinol diglucoside in flaxseed coats. Front Plant Sci 7:1743

Fofana B, Ghose K, Somalraju A, McCallum J, Main D, Deyholos MK, Rowland GG, Cloutier S (2017) Induced mutagenesis in *UGT74S1* gene leads to stable new flax lines with altered secoisolariciresinol diglucoside (SDG) profiles. Front Plant Sci 8:1638

Ford JD (2001) Cancer chemopreventive flax seed lignans: delineating the metabolic pathway(s) to the SDG-HMG ester-linked polymers. PhD thesis, Washington State University, p 246

Ford JD, Davin LB, Lewis NG (1999) Plant lignans and health: cancer chemoprevention and biotechnological opportunities. In: Hemingway RW, Gross GG, Yoshida T (eds) Plant polyphenols 2: chemistry, biology, pharmacology, ecology. Kluwer Academic/Plenum Publishers, New York, pp 675–694

Ford JD, Huang K-S, Wang H-B, Davin LB, Lewis NG (2001) Biosynthetic pathway to the cancer chemopreventive secoisolariciresinol diglucoside-hydroxymethyl glutaryl ester-linked lignan oligomers in flax (*Linum usitatissimum*) seed. J Nat Prod 64:1388–1397

Forkmann G, Heller W (1999) Biosynthesis of flavonoids. In: Barton Sir DHR, Nakanishi K, Meth-Cohn O (eds) Comprehensive natural products chemistry, vol 1. Elsevier, Oxford, pp 713–748

Freudenberg K (1965) Lignin: its constitution and formation from *p*-hydroxycinnamyl alcohols. Science 148:595–600

Gang DR, Costa MA, Fujita M, Dinkova-Kostova AT, Wang H-B, Burlat V, Martin W, Sarkanen S, Davin LB, Lewis NG (1999) Regiochemical control of monolignol radical coupling: a new paradigm for lignin and lignan biosynthesis. Chem Biol 6:143–151

Gasper R, Effenberger I, Kolesinski P, Terlecka B, Hofmann E, Schaller A (2016) Dirigent protein mode of action revealed by the crystal structure of AtDIR6. Plant Physiol 172:2165–2175

Ghose K, Selvaraj K, McCallum J, Kirby CW, Sweeney-Nixon M, Cloutier SJ, Deyholos M, Datla R, Fofana B (2014) Identification and functional characterization of a flax UDP-glycosyltransferase glucosylating secoisolariciresinol (SECO) into secoisolariciresinol monoglucoside (SMG) and diglucoside (SDG). BMC Plant Biol 14:82

Goodstein DM, Shu S, Howson R, Neupane R, Hayes RD, Fazo J, Mitros T, Dirks W, Hellsten U, Putnam N, Rokhsar DS (2012) Phytozome: a comparative platform for green plant genomics. Nucleic Acids Res 40:D1178–D1186

Halls SC, Lewis NG (2002) Secondary and quaternary structures of the (+)-pinoresinol forming dirigent protein. Biochemistry 41:9455–9461

Halls SC, Davin LB, Kramer DM, Lewis NG (2004) Kinetic study of coniferyl alcohol radical binding to the (+)-pinoresinol forming dirigent protein. Biochemistry 43:2587–2595

Hano C, Martin I, Fliniaux O, Legrand B, Gutierrez L, Arroo RRJ, Mesnard F, Lamblin F, Lainé E (2006) *Pinoresinol-lariciresinol reductase* gene expression and secoisolariciresinol diglucoside accumulation in developing flax (*Linum usitatissimum*) seeds. Planta 224:1291–1301

Hemmati S, von Heimendahl CBI, Klaes M, Alfermann AW, Schmidt TJ, Fuss E (2010) Pinoresinol-lariciresinol reductases with opposite enantiospecificity determine the enantiomeric composition of lignans in the different organs of *Linum usitatissimum* L. Planta Med 76:928–934

Hendrawati O, Woerdenbag HJ, Michiels PJA, Aantjes HG, van Dam A, Kayser O (2011) Identification of lignans and related compounds in *Anthriscus sylvestris* by LC-ESI-MS/MS and LC-SPE-NMR. Phytochemistry 72:2172–2179

Hosmani PS, Kamiya T, Danku J, Naseer S, Geldner N, Guerinot ML, Salt DE (2013) Dirigent domain-containing protein is part of the machinery required for formation of the lignin-based Casparian strip in the root. Proc Natl Acad Sci U S A 110:14498–14503

Johnsson P, Kamal-Eldin A, Lundgren LN, Åman P (2000) HPLC method for analysis of secoisolariciresinol diglucoside in flaxseeds. J Agric Food Chem 48:5216–5219

Kim MK, Jeon J-H, Davin LB, Lewis NG (2002) Monolignol radical-radical coupling networks in western red cedar and *Arabidopsis* and their evolutionary implications. Phytochemistry 61:311–322

Kim K-W, Moinuddin SGA, Atwell KM, Costa MA, Davin LB, Lewis NG (2012) Opposite stereoselectivities of dirigent proteins in *Arabidopsis* and *Schizandra* species. J Biol Chem 287:33957–33972

Kim K-W, Smith CA, Daily MD, Cort JR, Davin LB, Lewis NG (2015) Trimeric structure of (+)-pinoresinol forming dirigent protein at 1.95 Å resolution with three isolated active sites. J Biol Chem 290:1308–1318

Klosterman HJ, Muggli RZ (1959) The glucosides of flaxseed. II. Linocaffein. J Am Chem Soc 81:2188–2191

Klosterman HJ, Smith F (1954) The isolation of β-hydroxy-β-methylglutaric acid from the seed of flax (*Linum usitatissimum*). J Am Chem Soc 76:1229–1230

Kosinska A, Penkacik K, Wiczkowski W, Amarowicz R (2011a) Presence of caffeic acid in flaxseed lignan macromolecule. Plant Foods Hum Nutr 66:270–274

Kosinska A, Urbalewicz A, Penkacik K, Karamac M, Amarowicz R (2011b) SE-HPLC-DAD analysis of flaxseed lignan macromolecule and its hydrolysates. Pol J Food Nutr Sci 61:263–271

Kozlowska H, Zandernowski R, Sosulski FW (1983) Phenolic acids in oilseed flours. Nahrung 27:449–453

Lau W, Sattely ES (2015) Six enzymes from mayapple that complete the biosynthetic pathway to the etoposide aglycone. Science 349:1224–1228

Lewis NG, Davin LB, Sarkanen S (1999) The nature and function of lignins. In: Barton Sir DHR, Nakanishi K, Meth-Cohn O (eds) Comprehensive natural products chemistry, vol 3. Elsevier, Oxford, pp 617–745

Liu J, Stipanovic RD, Bell AA, Puckhaber LS, Magill CW (2008) Stereoselective coupling of hemigossypol to form (+)-gossypol in moco cotton is mediated by a dirigent protein. Phytochemistry 69:3038–3042

Marques JV, Dalisay DS, Yang H, Lee C, Davin LB, Lewis NG (2014) A multi-omics strategy resolves the elusive nature of alkaloids in *Podophyllum* species. Mol BioSyst 10:2838–2849

Meagher LP, Beecher GR, Flanagan VP, Li BW (1999) Isolation and characterization of the lignans, isolariciresinol and pinoresinol, in flaxseed meal. J Agric Food Chem 47:3173–3180

Min T, Kasahara H, Bedgar DL, Youn B, Lawrence PK, Gang DR, Halls SC, Park H, Hilsenbeck JL, Davin LB, Lewis NG, Kang C (2003) Crystal structures of pinoresinol-lariciresinol and phenylcoumaran benzylic ether reductases and their relationship to isoflavone reductases. J Biol Chem 278:50714–50723

Nakatsubo T, Mizutani M, Suzuki S, Hattori T, Umezawa T (2008) Characterization of *Arabidopsis thaliana* pinoresinol reductase, a new type of enzyme involved in lignan biosynthesis. J Biol Chem 283:15550–15557

Pickel B, Constantin M-A, Pfannstiel J, Conrad J, Beifuss U, Schaller A (2010) An enantiocomplementary dirigent protein for the enantioselective laccase-catalyzed oxidative coupling of phenols. Angew Chem Int Ed 49:202–204

Prasad K (2009) Flaxseed and cardiovascular health. J Cardiovasc Pharmacol 54:369–377

Prasad K, Dhar A (2016) Flaxseed and diabetes. Curr Pharm Des 22:141–144

Qiu S-X, Lu Z-Z, Luyengi L, Lee SK, Pezzuto JM, Farnsworth NR, Thompson LU, Fong HHS (1999) Isolation and characterization of flaxseed (*Linum usitatissimum*) constituents. Pharm Biol 37:1–7

Ragamustari SK, Nakatsubo T, Hattori T, Ono E, Kitamura Y, Suzuki S, Yamamura M, Umezawa T (2013) A novel *O*-methyltransferase involved in the first methylation step of yatein biosynthesis from matairesinol in *Anthriscus sylvestris*. Plant Biotechnol 30:375–384

Ramsay A, Fliniaux O, Quéro A, Molinié R, Demailly H, Hano C, Paetz C, Roscher A, Grand E, Kovensky J, Schneider B, Mesnard F (2017) Kinetics of the incorporation of the main phenolic compounds into the lignan macromolecule during flaxseed development. Food Chem 217:1–8

Renouard S, Corbin C, Lopez T, Montguillon J, Gutierrez L, Lamblin F, Lainé E, Hano C (2012) Abscisic acid regulates pinoresinol-lariciresinol reductase gene expression and secoisolariciresinol accumulation in developing flax (*Linum usitatissimum* L.) seeds. Planta 235:85–98

Rudney H (1957) The biosynthesis of β-hydroxy-β-methylglutaric acid. J Biol Chem 227:363–377

Sakakibara N, Suzuki S, Umezawa T, Shimada M (2003) Biosynthesis of yatein in *Anthriscus sylvestris*. Org Biomol Chem 1:2474–2485

Struijs K, Vincken J-P, Doeswijk TG, Voragen AGJ, Gruppen H (2009) The chain length of lignan macromolecule from flaxseed hulls is determined by the incorporation of coumaric acid glucosides and ferulic acid glucosides. Phytochemistry 70:262–269

Teoh KH, Ford JD, Kim M-R, Davin LB, Lewis NG (2003) Delineating the metabolic pathway(s) to secoisolariciresinol diglucoside hydroxymethyl glutarate oligomers in flaxseed (*Linum usitatissimum*). In: Thompson LU, Cunnane SC (eds) Flaxseed in human nutrition, 2nd edn. AOCS Press, Champaign, pp 41–62

Uchida K, Akashi T, Aoki T (2017) The missing link in leguminous pterocarpan biosynthesis is a dirigent domain-containing protein with isoflavanol dehydratase activity. Plant Cell Physiol 58:398–408

Umezawa T, Davin LB, Yamamoto E, Kingston DGI, Lewis NG (1990) Lignan biosynthesis in *Forsythia* species. J Chem Soc Chem Commun:1405–1408

Vassão DG, Kim K-W, Davin LB, Lewis NG (2010) Lignans (neolignans) and allyl/propenyl phenols: biogenesis, structural biology, and biological/human health considerations. In: Mander LN, Liu H-W (eds) Comprehensive natural products II chemistry and biology, Structural diversity I, vol 1. Elsevier, Oxford, pp 815–928

von Heimendahl CBI, Schäfer KM, Eklund P, Sjöholm R, Schmidt TJ, Fuss E (2005) Pinoresinol-lariciresinol reductases with different stereospecificity from *Linum album* and *Linum usitatissimum*. Phytochemistry 66:1254–1263

Wang Z, Hobson N, Galindo L, Zhu S, Shi D, McDill J, Yang L, Hawkins S, Neutelings G, Datla R, Lambert G, Galbraith DW, Grassa CJ, Geraldes A, Cronk QC, Cullis C, Dash PK, Kumar PA, Cloutier S, Sharpe AG, Wong GK-S, Wang J, Deyholos MK (2012) The genome of flax (*Linum usitatissimum*) assembled *de novo* from short shotgun sequence reads. Plant J 72:461–473

Xia Z-Q, Costa MA, Pélissier HC, Davin LB, Lewis NG (2001) Secoisolariciresinol dehydrogenase purification, cloning and functional expression: Implications for human health protection. J Biol Chem 276:12614–12623

Youn B, Moinuddin SGA, Davin LB, Lewis NG, Kang C (2005) Crystal structures of apoform, and binary/ternary complexes of *Podophyllum* secoisolariciresinol dehydrogenase, an enzyme involved in formation of health-protecting and plant defense lignans. J Biol Chem 280:12917–12926

Chapter 12
Flax Transformation via Floral-Dipping

Nasmah Bastaki and Christopher A. Cullis

12.1 Introduction

Flax (*Linum usitatissimum*) is an important crop grown widely for its fibers and oils. Transformation of the flax genome is possible with techniques such as wounding, *Agrobacterium* infection, and co-cultivation in tissue culture, applying biolistic particles or ultrasound sonication followed by regeneration (Zhan et al. 1988; Dong and McHughen 1993; Mlynárová et al. 1994; Beranová et al. 2008). However, these techniques have disadvantages, including the proclivity to many mutational events and an extended time to obtain the transgenic lines. Some of these methods can also be expensive and require skilled and efficient manipulation of the instruments, resulting in the low recovery of seedlings. Most importantly, these techniques often result in low transformation rates (Beranová et aal. 2008; Dong and McHughen 1993). *Agrobacterium*-mediated plant transformation via floral-dip is a simple and efficient approach to generate transgenic plants. It has been routinely and successfully used for many plant species such as *Arabidopsis thaliana* (Clough and Bent 1998; Bent 2006), *Medicago truncatula* (Trieu et al. 2000), tomato (Yasmeen et al. 2009), wheat (Zale et al. 2009), and maize (Mu et al. 2012). However, it has not been considered a viable technique for high-throughput flax transformation due to several factors, such as the low numbers of flowers produced by flax, the limited number of seeds obtained from each flower, the large seed size, and the thick coat, which could also be problematic for such genetic transformation process. The normal selection scheme for the floral-dip technique requires germinating transformed

N. Bastaki
Kuwait University, Saft, Kuwait
e-mail: nasmah.bastaki@ku.edu

C. A. Cullis (✉)
Department of Biology, Case Western Reserve University, Cleveland, OH, USA
e-mail: cac5@case.edu

© Springer Nature Switzerland AG 2019
C. A. Cullis (ed.), *Genetics and Genomics of Linum*, Plant Genetics and
Genomics: Crops and Models 23, https://doi.org/10.1007/978-3-030-23964-0_12

seeds on media containing an antibiotic, with transformed progenies distinguished based on their ability to germinate and stay green, while non-transformed progenies either do not germinate or germinate but bleach out quickly and die. However, it has been noted that wild-type flax tends to escape high concentration of antibiotic selections, producing false-positive results and making the selection of T1 progenies based on antibiotic resistance more difficult (Zhan et al. 1988; Dong and McHughen 1993), and when a high concentration of antibiotic was added to the selection medium, the rate of observed transformation dropped dramatically (Mlynárová et al. 1994).

In spite of these potential obstacles, the floral-dip method with *Agrobacterium* was used to transform the two different flax varieties, Stormont cirus (PL), a fiber variety described elsewhere in this volume (Chap. 6), and Bethune, an oilseed variety use to develop the assembled flax genome (Chap. 4). The rationale for using these two varieties was that since Stormont cirus had been shown to reorganize its genome in response to environmental stress (Cullis 2005; Chen et al. 2005), the transformation rate may be higher in Stormont cirus than the stable oilseed variety Bethune. This hypothesis was also supported in previous literature, since differences have been observed in the transformation rates among varieties/ecotypes/genotypes of the same species. For example, in the ecotypes of *Arabidopsis thaliana*, the first and best model in which transformation using *Agrobacterium* and floral-dipping was accomplished, the floral-dip transformation method worked well for the majority of ecotypes, such as Col-0, Ws-0, Nd-0, and No-0, but its efficiency is reduced when using the Ler-0 ecotype (Clough and Bent 1998; Zhang et al. 2006). Similar observations pertain to transformation of tomato (Qiu et al. 2007) and maize (Ishida et al. 2007). The reason for such differential transformation rates has not been definitively determined.

Following transformation, the screening for the identification of positive transformants was simply done by growing T1 plants on soil and screening their leaves soon after they germinated for vector and target, bypassing the use of antibiotic selection. In the direct PCR testing of leaves, and using the appropriate T-DNA primers, positive transformants can be rapidly selected, although a priori, the number of seeds that would need to be screened could not be determined. At a transformation rate of 1%, hundreds of seeds would need to be screened, and if the transformation rate was lower, then it might not be practical to identify any transformants. In spite of these potential problems in identifying transformants, the floral-dip technique was applied to the two selected flax varieties.

12.2 Transformation Protocol

A total of 10 parents (T0) of the inbred flax variety Stormont cirus and a total of 25 parents (T0) of Bethune were sown and grown to flowering stage under long daylight (14 hours light and 10 hours dark) and used for the floral-dip treatment (Bastaki and Cullis 2014).

The plants were ready for transformation when the buds were visible and had just formed in the inflorescence. The leaves around the buds were removed to expose them to the *Agrobacterium* cells. For each treated (T0) plant, the main branch and one or two of the side branches were used for the dipping, and another side branch was left untreated and to serve as a control non-dipped branch. Successful transformation depended on the use of the best flower stage, which was determined by using a range of different flower stages (Bastaki and Cullis 2014).

12.2.1   Floral-Dipping

The details of the method of floral-dipping transformation have been reported elsewhere (Bastaki and Cullis 2014). Since one of the expected difficulties in applying the technique to flax was the small number of seeds per boll and the recalcitrance of using the antibiotic selection, a rigorous branch and tracking protocol was followed. Figure 12.1 illustrates the process of individual flower tracking, seed collection, and labeling that were done for every branch very precisely.

12.2.2   Choice of Selection/Screening Method

In the floral-dip method, selection of positive transformants normally involves a selectable agent/ antibiotic, which is applied directly to the media. First, transformed seed are germinated on a growth medium containing an antibiotic. The result of the process is transformed progenies which are distinguished based on their ability to germinate and stay green, while non-transformed progenies either do not germinate or germinate but bleach out quickly and die. This method of screening was first used for screening *Arabidopsis thaliana*, following floral-dipping (Das and Joshi 2011; Clough and Bent 1998; Bent 2006; Davis et al. 2009; Mara et al. 2010) but later was the norm in many other plant species (Agarwal et al. 2009; Horsch et al. 1985; Yasmeen et al. 2009; Mu et al. 2012).

However, it has been noted that wild-type flax tends to escape a high concentration of antibiotic selections, producing false-positive results and making the selection of positive transformants based on antibiotic resistance more difficult (Zhan et al. 1988; Dong and McHughen 1993). Also, when a high concentration of antibiotic was added to the selection medium, the rate of observed transformation dropped dramatically (Mlynárová et al. 1994).

In an attempt to confirm previous findings about wild-type flax escaping high selections, wild-type flax seeds were germinated in MS media with increasing concentrations of kanamycin. The escape of wild-type flax from high concentrations of kanamycin, up to 2 mg/ml, is shown in Fig. 12.2, confirming previous findings in the literature (Zhan et al. 1988; Dong and McHughen 1993; Mlynárová et al. 1994). Also, the data shown in Fig. 12.2 confirms that such an antibiotic selection would

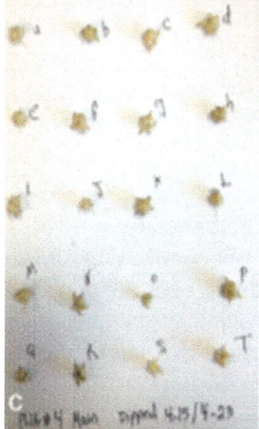
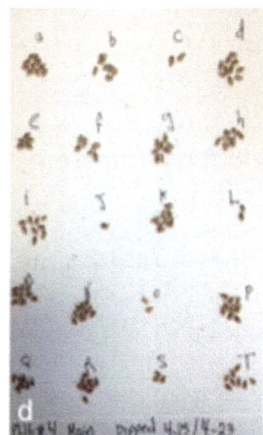
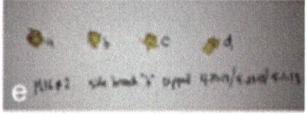
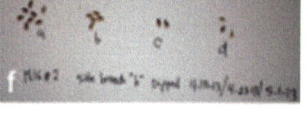

Fig. 12.1 The process of flower tracking and seed collections, from the T0 treated plants. (**a**) An example of the whole plant with the main branch (the tallest branch in the center) and the side branches. (**b**) Different branches from the same plant, control non-dipped, main dipped, and side dipped. (**c, d**) An example of the seeds collected from individual flowers (labeled a-t) from a main branch. (**e, f**) An example of the seeds collected from individual flowers (labeled a-d) from a side branch. Note the differences in the number of flowers obtained from the main and side branches, also the variation between the number of seeds in each flower

not work to identify positive transformants following the floral-dipping. Currently, the antibiotic or herbicide selection are the major methods of identifying transformed plants. In a few cases, the modified trait can be observed in the T1 generations without using a selection agent (Yasmeen et al. 2009).

An alternative to antibiotic selection would be the direct testing of leaf DNA by PCR to identify those individuals that had incorporated the vector or gene of interest.

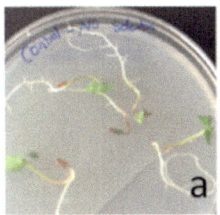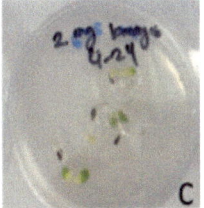

Fig. 12.2 Antibiotic escape, a problem for T1 selection, is overcome by direct PCR screening. (**a**) Wild-type flax seeds germinated on MS plant media without antibiotic. (**b**) Wild-type flax seeds germinated on MS plant media+ increasing concentrations of kanamycin (200 µg/ml, 600 µg/ml, 1 mg/ml). (**c**) Wild-type flax seeds germinated on MS plant media with 2 mg/ml kanamycin

This method was chosen to attempt to identify positive T1 transformants following PL and Bethune transformation, in spite of the theoretical drawbacks of expected low transformation rates.

12.2.3   Identification of Positive Transformants with Direct PCR

The T1 seeds were either sown in soil or in Murashige and Skoog basal salt medium (MS medium). The seedlings were kept under long daylight (14 hours light and 10 hours of dark); they usually germinated in about 4–6 days (Fig. 12.3). Approximately 10–14 days after germination, when true leaves develop, direct PCR was performed on leaf punches of these leaves to test for positive transformants.

Positive transformants were identified based on correct size of PCR amplifications followed by sequencing of the pCR product. Positive transformants were grown to maturity, and the stability of the transgene was subsequently confirmed, either by direct PCR or by isolating DNA from the upper leaves of the putative transformed plants and confirming the presence of the transgene.

12.2.4   Selection of Positive (T1) Transformants with Direct PCR from Leaf DNA

Since the transformation rate was expected to be low, one seed per flower for every branch on each of the T0 plants was chosen for testing, following transformation. The number of seeds per flower varied from one seed to a maximum of ten seeds per flower/capsule, while the main branch always had more flowers than side branches. A collection of primers that would identify regions of the vector, the transgene, or the junctions between the two were designed to screen the T1 plants. For Stormont

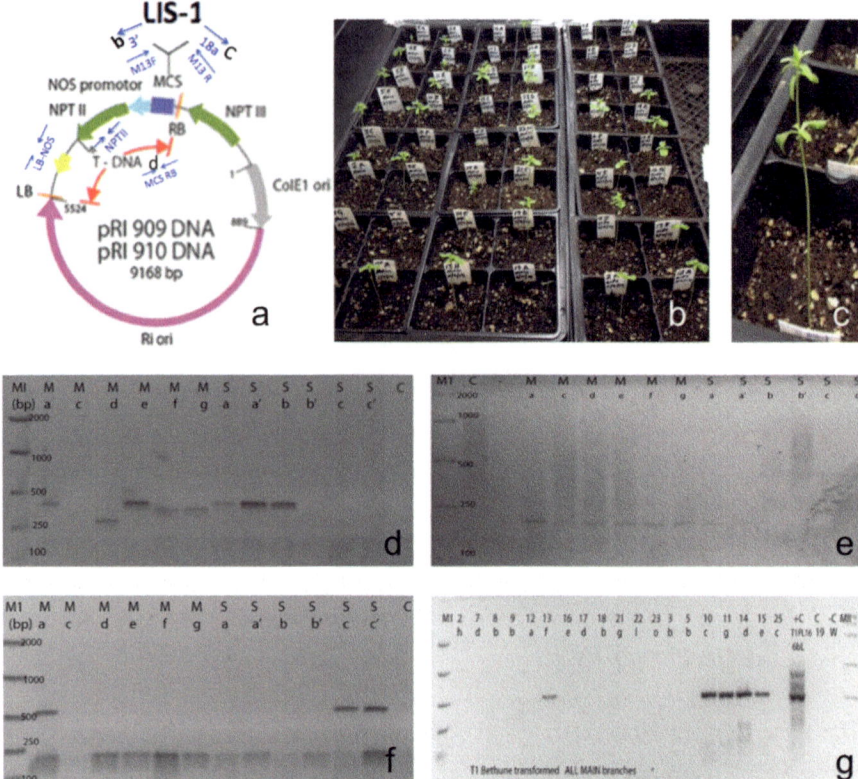

Fig. 12.3 Selection of T1 positive transformants with PCR. (**a**) Diagram of the plant binary vector+ the cloned LIS-1 insert. Blue arrows indicate the position of PCR primers used in the direct PCR screening. (Modified from Takara). (**b**) In the T1 seedlings at early stage, each seedling is precisely labeled to know the origin of the seed. (**c**) Close up in from (**b**). (**d–f**) PCR from transformed PL. (**d**) PCR with primers M13F+3′. (**e**) PCR with primers M13R+18a. (**f**) PCR with primers right border (RB) and multiple cloning site (MCS). Each lane represents a T1 from individual flowers collected from C, control branch (non-dipped); M, main branch flower a–g; S, side branch flower b–c′, from transformed T1 PL. (**g**) PCR with primers NPTII (kanamycin) from transformed T1 Bethune. Letters refer to a single seed from specific flower, numbers 2–25: refers to the parent T0 from which the flowers were collected. C: non-dipped branch

cirus, 10 pants were used for the floral-dipping, resulting in approximately 160 flowers from all the branches. Of these a total of 137 seedlings were successfully germinated and tested with direct PCR on their leaf DNA using the appropriate primers (Table 12.1 and Fig. 12.3a). When positive transformants were identified, these plants were transplanted to soil and grown to maturity.

For Bethune transformation, 25 parent (T0) plants were used for the floral-dipping resulting in about 400 flowers. From these treated flowers, 1 seed per flower for 130 flowers was tested to generate a comparable population to the 137 seedlings tested from PL resulting in 103 seedlings.

Table 12.1 Some of the primers used for the direct PCR testing

Forward primer sequence source	Sequence 5'-3'	Reverse primer sequence source	Sequence 5'-3'	Annealing temp. (°C)	Extension time (sec)	Exp. size (bp)	Another name for this pair
M13F (T-DNA)	CTGCAAGGCGATTAAGTTGG	3' (LIS-1 insert)	GAGGATGGAAGATGAAGAAGG	57°	40	450	–
18a (LIS-1 insert)	CATAAATTCAGTCCTATCGAC	M13R (T-DNA)	TAAAGCCTGGGGTGCCTAAT	57°	40	540	–
MCS (T-DNA)	TGGTCATAGCTGTTTCCTGTG	RB (T-DNA)	TTTAAACTGAAGGCGGGAAA	60°	20	200	PriC
NPTII (T-DNA)	GCGATACCGTAAAGCACGAG	NPTII (T-DNA)	GCTCGACGTTGTCACTGAAG	65°	40	500	Kanamycin

Figure 12.3d–g shows representations of some of the primers used for the screening in the T1 of PL and Bethune. Figure 12.3d uses primers M13F+3′ to amplify the junction between the multiple cloning site of the vector and LIS-1 insert. Figure 12.3e uses primers M13R+18a to amplify the other junction between the multiple cloning site of the vector and LIS-1 insert. Figure 12.3f uses primers PriC to amplify the region between the right border (RB) and multiple cloning site (MCS) of the vector. Figure 12.3e uses primers NPTII to amplify the kanamycin transgene.

Some of these bands were sequenced, such as bands obtained from amplifications with kanamycin primers and PriC primers, to confirm their identities and to confirm that these amplified bands are indeed an exact match to the predicted pieces. These results confirm that *Agrobacterium*-mediated plant transformation via floral-dipping is an applicable method for making transgenic flax and can be monitored by direct PCR rather than by selection.

12.2.5 Differences in the Transformation Rates

The transformation rate was predicted to be different between main and side branches and it was predicted to be higher in Stormont cirus (PL) than Bethune, since PL is the plastic (See Chap. 6), responsive line, and Bethune is not. The beak-down of the total number of T1 progenies, collected from main and side branches, from all T0 of PL and Bethune, were put together in Tables 12.2a, 12.2b, and 12.3.

Table 12.2a shows the results of PL transformation. There were 99 plants screened from main branches, and out of those, 52 plants were positive. There were 38 plants screened from side branches, and out of those, 17 were positive. Therefore, for the transformation of PL, it appears that there is no significant difference in the transformation rate between main and side branches. In PL transformation, the total number of screened plants from all branches is 137, and out of those, 69 are positive transformants, giving a PL transformation rate via floral-dip of 50.36%.

Table 12.2b shows the results of Bethune transformation. There were 67 plants screened from main branches, and out of those, 8 plants were positive. There were 36 plants screened from side branches, and out of those, 6 were positive. From Bethune transformation, it appears that there is also no significant difference in the transformation rate between main and side branches; both transformed well. For the Bethune transformation, the total number of screened plants from all branches was 103, and out of those, 14 were positive transformants. Therefore, Bethune transformation rate via floral-dip is 13.6%.

Table 12.2a Summary for transformation rate of PL and Bethune. Overall transformation rate of PL

Total number of T1 plants screened	Total number of positive transformants obtained	Overall transformation rate of PL
137	69	50.36%

Table 12.2b Summary for transformation rate of PL and Bethune. Overall transformation rate of Bethune

Total number of T1 plants screened	Total number of positive transformants obtained	Overall transformation rate of Bethune
103	14	13.6%

Table 12.3 Summary for transformation rate of Bethune. Breakdown of positive transformants of main and side branches from each T1 plant (Table and most of the PCR were generated by Jennifer Piechowski)

#T0 Treated parent	# T1 Plants screened from Main branch	# T1 Positive transformants obtained from main branch	# T1 Plants screened from side branches	#T1 Positive transformants obtained from side branches	Total #T1 positive transformants
#1	1	0	1	0	0
#2	2	0	2	0	0
#3	2	0	0	0	0
#4	0	0	3	1	1
#5	3	0	0	0	0
#6	2	1	–	–	1
#7	5	0	1	0	0
#8	2	0	2	1	1
#9	2	0	1	0	0
#9'	2	0	0	0	0
#10	5	1	0	0	1
#11	6	2	0	0	2
#12	2	1	2	0	1
#13	5	1	5	2	3
#14	5	1	0	0	1
#15	6	1	0	0	1
#16	2	0	2	0	0
#17	1	0	5	0	0
#18	2	0	5	0	0
#19	1	0	0	0	0
#20	–	–	–	–	–
#21	2	0	2	1	1
#22	2	0	2	0	0
#23	2	0	2	0	0
#24	1	0	1	1	1
#25	4	0	0	0	0
Total	**67**	**8**	**36**	**6**	**14**

The results from flax floral-dip in both PL and Bethune confirm that floral-dip works very well in flax. The transformation rate was unexpectedly high in PL, with 50.36% of the seeds from treated plants being positive transformants. This is a higher transformation rate than those reported for *Arabidopsis thaliana* (Bent 2006) and other plant species (Yasmeen et al. 2009; Mu et al. 2012), using floral-dip transformation. It is also the highest reported rate for any method of flax transformation. In Bethune, the transformation rate is still high for floral-dip method (13.6%); however, it is much lower than PL, which is consistent with the expectation that since PL is plastic and responsive line, it is responding to the transformation much better than Bethune, which is known to be nonresponsive and stable.

12.2.6   Whole or Subset Seeds of the Flower Are Transformed?

Previous studies of plant transformation via *Agrobacterium* and floral-dipping using *Arabidopsis thaliana* have shown that T1 transformants are typically hemizygous, that is, carrying T-DNA at only one of two alleles of a given locus. If transformation events occurred prior to or early in floral development of the T0 plant, it is expected to give rise to identically transformed male and female gametophytes, which upon self-fertilization could produce a significant number of homozygous T1 plants for the T-DNA insertion. The rarity or total absence of these homozygous lines indicates that the transformation events usually occur late in floral development, during germline development after divergence of male and female gametophyte cell lineages or soon after fertilization of the T1 embryo (Clough and Bent 1998; Horsch et al. 1985; Bechtold et al. 2000; Feldmann 1991; Desfeux et al. 2000). Although the cell type that is transformed and the timing of transformation still remained unknown (Desfeux et al. 2000), it is still possible to investigate the frequency rate of transformation per individual flower.

The seed tracking and labeling approaches that were applied during the floral-dipping procedure, combined with the simple anatomy of flax branching, permitted the determination of the frequency of transformation per individual flower in PL and Bethune. The expectation was that since PL is a plastic and responsive line, it will respond better to the process of dipping, and PL individual flowers will transform at a higher frequency rate than Bethune individual flowers; the hypothesis is that PL individual flowers will then produce more transformants than Bethune individual flowers.

The parental branches of PL and Bethune received the same dipping treatment. Three to four seeds were selected from some of previously identified positive transformed bolls from the T1 of PL and Bethune (Fig. 12.4a). The seeds were germinated on soil and tested by direct PCR from leaf DNA using kanamycin transgene (Fig. 12.4b).

Table 12.4a and 12.4b illustrates the number of progenies tested from each flower from PL (Table 12.4a) and Bethune (Table 12.4b). The results from T1 of PL individual flowers suggest that the frequency of transformation in individual flowers

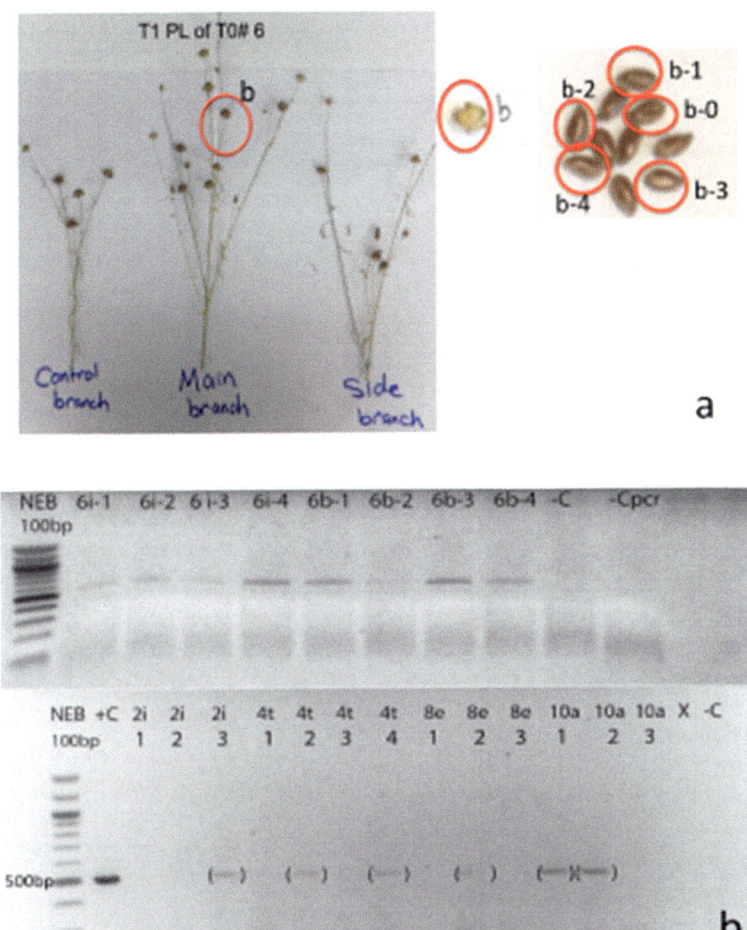

Fig. 12.4 PCR to show transformation rate from seeds of the same flower. (**a**) Seeds from T1 of PL main branch flower "b"; flower "b" has ten seeds; one of them (b-0) was previously conformed to be positive. Therefore, four more seeds (b-1, b-2, b-3, b-4) were selected and screened with kanamycin. The same strategy of labeling was used for the other flowers from PL and Bethune. (**b**) PCR of kanamycin primers from seeds of the same flower of PL

from PL is higher than Bethune. Furthermore, in some cases in PL, all flowers produced 100% transformants, such as T1 PL# 6 main branch flowers (b) and (i), T1 PL#7 main branch flower (c), and T1 PL# 9 side branch flower (f). However, for Bethune, the frequency of transformation per individual flower was lower; moreover, none of the screened flowers from Bethune provided 100% transformants.

Table 12.4a Breakdown of the individual flower tested from PL, showing the number of progenies tested (from same flower) and the total number of positive transformants, obtained from each flower

T1 PL flower	# Progenies screened from this flower	# Positive transformants from this flower
#2 main branch, flower (i)	4	2
#3 side branch "b", flower (a)	4	3
#4 main branch, flower (t)	5	3
#6 main branch, flower (i)	5	5
#6 main branch, flower (b)	5	5
#7 main branch, flower (b)	5	4
#7 main branch, flower (c)	3	3
#8 main branch, flower (g)	4	2
#9 side branch, flower (f)	4	4
#10 main branch, flower (a)	4	2

Table 12.4b Breakdown of the individual flower tested from Bethune, showing the number of progenies tested (from same flower) and the total number of positive transformants, obtained from each flower

T1 Bethune flower	# Progenies screened from this flower	# Positive transformants from this flower
#7 main branch, flower (d)	4	2
#10 main branch, flower (c)	3	1
#11 main branch, flower (g)	4	2
#13 side branch, flower (a)	4	3
#13 main branch, flower (f)	4	2
#14 main branch, flower (d)	3	1
#15 main branch, flower (e)	4	2
#17 side branch, flower (a)	4	1
#18 side branch, flower (e)	4	0
#25 main branch, flower (c)	3	2

These results suggest that not only is PL responding better to the transformation process but also PL individual flowers produce more transformed progenies than Bethune individual flowers.

12.2.7 Phenotype of the T1 Generation as They Were Maturing

Figure 12.5 shows pictures of the screened 137 T1 from the 10 T0 PL plants, as they were maturing over time. The plant pictures in Fig. 12.5a were taken approximately 2 months post-germination; each pot contains T1 progenies that came from either

Fig. 12.5 Phenotypic differences in the T1 of transformed PL. (**a**) In the 137 (T1) plants of PL, which were screened by direct PCR, 69 were tested positive for the transgene. The picture was taken approximately 2 months post-germination. (**b**) T1 from plant #2. The left pot contains T1 plants from side branch, and the right pot contains T1 plants from main branch. Note the phenotypic differences in the height of the plants and the shape and size of the leaves. (**c**) T1 from plants #3, 2, and 9. Each pot contains T1s from either main or side branches. Pictures were taken approximately 6 months post-germination. While T1 of #3 collected from side branches was shorter, T1 of #2 collected from main branches was taller. However, T1s of main and side branches from #9 don't look much different than each other

main or side branches from the same T0-treated plant. In some plants, there were phenotypic differences between the T1 progenies taken from main and side branches of the same T0 plant (Fig. 12.5b) in which the T1 progenies taken from the main branch look different than the T1 progenies taken from side branch, even though both were from the same PL T0 plant# 2. The phenotypic differences include differences in the height of the plants, curving of the leaves, size of the leaves, and the internode (distance) between the leaves on the stem of the plant.

As time proceeds, some of these plants still had these differences (Fig. 12.5c). The first picture on the left was taken approximately 6 months post-germination, and it shows that the T1 progenies of PL#3 and the phenotypic differences between the T1 progenies from main and side branches were very noticeable compared to the middle and right picture of the same figure in which T1 progenies of PL#2 and 9 don't look much different than each other.

Similar differences were also observed in some of the T1 progenies of Bethune.

12.2.8 Stability of the Transgene with Time in the T1 Generation

To determine if the transgene is stable over time, the topmost leaves on each positive matured T1 transformants were collected and rechecked for the presence of the transgene 7–9 months later. Figure 12.6a, b are examples from mature T1 of PL transformants 9 months post-initial screening, whereas Fig. 12.6c, d are examples from mature T1 of Bethune, 7 months post-initial screening. The arrows point to the position of the topmost leaves, and the red circle points to the mature T2 flower.

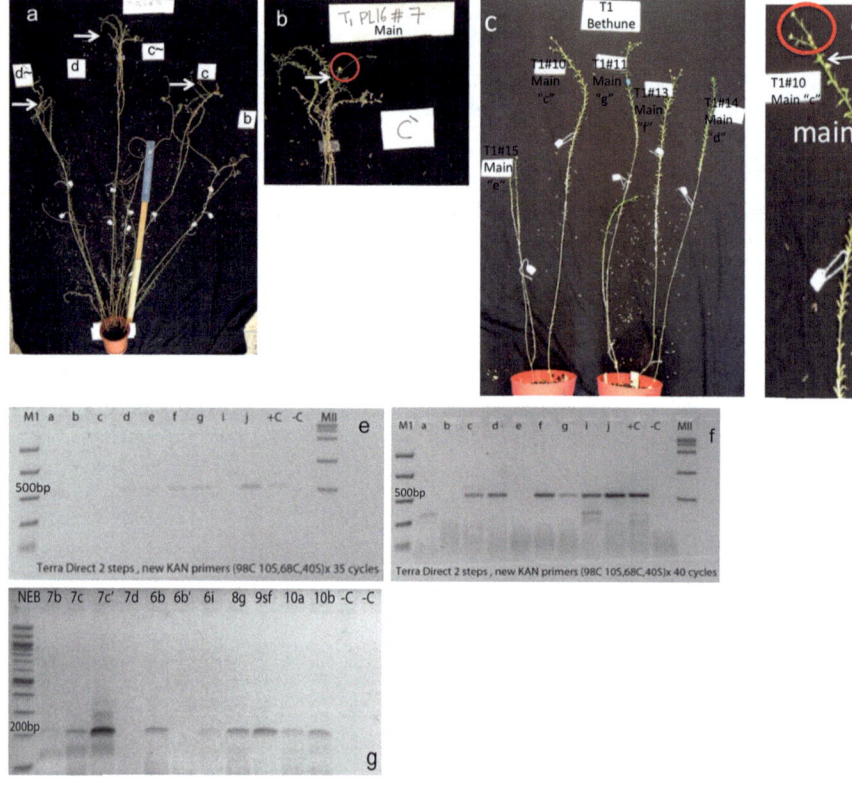

Fig. 12.6 Stability of the transgene with time. (**a**) Mature T1 transformed PL 9 months post-initial selection. (**b**) Zoom in from (**a**), the arrow points to the position of the leaves, surrounding the T2 flower, which was tested again to confirm the stability of the transgene. (**c**) Mature T1 transformed Bethune, 7 months post-initial selection. (**d**) Zoom in from (**c**). (**e**) PCR of the kanamycin transgene from T1 PL 9 months post-initial selection with direct PCR. PCR with 35 cycles (PCR by Morgan MacBeth). (**f**) Same PCR as in (**e**) but done at 40 cycles (PCR by Morgan MacBeth). (**g**) PCR of some positive transformants from PL with primers PriC of T-DNA, 9 months post-initial screening. Letters refer to single seed from specific flowers of the main branch (labeled a–j from single T0 parent. –C is non-dipped control branch. White arrows: point to the position of the leaves tested. Red circle: shows the T2 flowers to be collected for further testing

The stability of the transgene was checked with direct PCR using primers PriC of T-DNA primers (Fig. 12.6g) and kanamycin primers (Fig. 12.6f and e). Figures 12.6e and g show an example of PCR from T1 of PL#1 progenies (a–j), done at 35 and 40 cycles, respectively, while Fig. 12.6g shows an example of PCR from selected T1 progenies of PL#6, 7, 8, 9, and 10.

These results confirm that the transgene is stable as the transgenic plants matured and was expected to be transmitted to the next generation (T2).

12.2.9 Inheritance of the Transgene to the T2 Generation

Matured T2 flowers were collected from T1-positive transformants. Collection of T2 flowers were precisely done with labeling depending on the branching anatomy of each plant. Some plants had many branches with matured T2 flowers on each branch, other plants had only one matured main branch and no side branches, while a few plants did not develop T2 flowers from the main branch and had T2 flowers only from side branches. All of the selected T2 flowers were collected from branches that were confirmed to be positive for the transgene to test whether that transgene was transmitted to the T2 generation.

The stability and inheritance pattern of the transgene in the T2 generation was tested using one T2 flower from ten T1-positive transformants from PL and one T2 flower from a the positive transformant from Bethune. All the seeds inside each T2 flower, which varied from three to nine seeds per flower, were sown.

The inheritance of the transgene was confirmed using the kanamycin transgene primers and direct PCR. Figure 12.7 shows an example of PCR amplification of the

Fig. 12.7 Terra direct PCR from T2 generations using kanamycin transgene primers, from selected T2 flowers of PL. (**a**) T2 progenies (of T1 PL#9 main branch flower f) from side branch. (**b**) T2 progenies (T1 PL#9 main branch flower f) from main branch. (**c**) T2 progenies (of T1 PL#8 main branch flower g) from main branch. NOTE: progenies are taken from single T2 flower. -C: non-dipped branch control

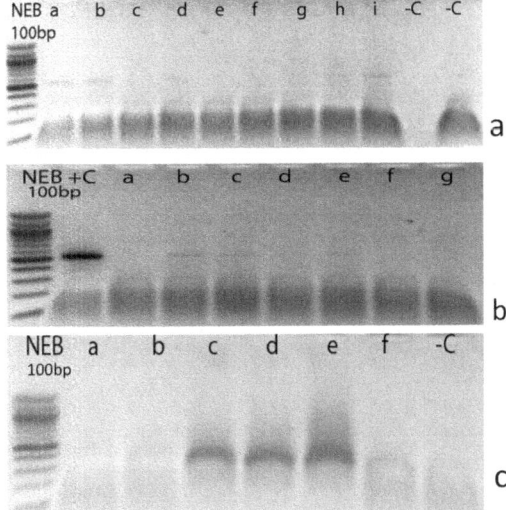

kanamycin transgene from selected T2 plant. Tables 12.5a and 12.5b show the breakdown of the total number of T2 progenies screened from each T2 flower and the number of positive transformants, obtained from each T2 flower.

These results confirm that transformation of flax via *Agrobacterium* and floral inherited to subsequent generations.

Table 12.5a Stability of the transgene in the T2 generation. Breakdown of the number of progenies screened for main and side branches (from single T2 flower) and the number of positives from them

Origin of T2 flower screened	# Of T2 progenies screened from main branch (all from single flower)	# Positive for transgene	# T2 progenies screened from side branch (all from single flower)	# Positive for transgene
T1 PL# 1 main branch, flower (g)	8	4	None	N/A
T1 PL# 3 side branch "b," flower (a')	4	2	4	4
T1 PL# 6 main branch, flower (i)	6	0	8	0
T1 PL# 6 main branch, flower (b)	None	N/A	8	6
T1 PL# 7 main branch, flower (b')	None	N/A	8	4
T1 PL# 7 main branch, flower (c)	None	N/A	4	0
T1 PL# 7 main branch, flower (c')	7	6	7	1
T1 PL# 8 main branch, flower (g)	6	3	None	N/A
T1 PL# 9 side branch, flower(f)	7	4	8	6
T1 PL# 10 side branch, flower (a)	3	2	6	0
T1 Bethune# 10 main branch, flower (c)	9	6	None	N/A

None, not available T2 flower to screen; N/A, not applicable

Table 12.5b Stability of the transgene in the T2 generation. Total number of screened and positive progenies

Total progenies screened from main branches	Total # positive for transgene from main branches	Total progenies screened from side branches	Total # positive for transgene from side branches	Total # of progenies from all branches	Total # of positive for transgene from all branches
50	27	53	21	103	48

12.3 Discussion

In some plant species, such as flax (*Linum usitatissimum*), high-throughput successful plant transformation has been limited. Previously, transformation in flax has been achieved by wounding and co-cultivation with *Agrobacterium* infection, by applying biolistic particles or using ultrasound sonication, followed by regeneration, a process that is both long and prone to being accompanied by many mutational events. Moreover, the selection process of these techniques requires the use of antibiotic selectable markers such as kanamycin. However, it has been noted in the literature that this method of selection produces many false positives, as flax tends to escape high concentrations of antibiotics (Zhan et al. 1988; Dong and McHughen 1993; Mlynárová et al. 1994). Another disadvantage of the previous techniques in flax transformation has been the low transformation rates (Beranová et al. 2008; Dong and McHughen 1993).

To overcome the antibiotic escape problem, direct PCR testing of DNA from T1 leaves was chosen instead of selection. The simple anatomy of flax to track specific flowers at the time of treatment made this strategy possible to perform. This tracking system allowed selection of seeds from specific flowers. Positive transformants were simply identified by testing DNA obtained from leaves using the fast and efficient method of direct PCR.

Fiber flax has few branches (one main branch and few side branches) and few flowers, producing approximately 100 seeds per plants, while oilseed flax is bred for a branching habit and seed yield, so it is generally shorter with many more branches and seeds. However, overall the growth habit makes it possible to track individual flowers and to select specific seeds during screening process. This tracking strategy was not used in other plants with more complex branching anatomy such as *Arabidopsis thaliana*, which has many branches and many flowers, which makes it difficult to distinguish dipped and non-dipped flowers on the same plant. Therefore, large numbers of seeds, approximately 20,000 seeds per plant, need to be screened to identify positive transformants (Mara et al. 2010; Bernhardt et al. 2012).

These data demonstrated that the floral-dip method can work very efficiently in flax and resulted in a surprisingly very high transformation rate (50.36% for PL and 13.6% for Bethune). This is higher than those previously observed for *Arabidopsis thaliana*, which was reported to be 0.1–1%, and also higher than other plant species (Mu et al. 2012; Yasmeen et al. 2009). Moreover, flax transformation via floral-dipping gave a much higher transformation rate than the other available methods of flax transformation (Zhan et al. 1988; Dong and McHughen 1993).

The results show that the transgene was stably integrated in the genome as the T1 transformants matured and set T2 flowers. The transgene was, also, successfully transmitted to the T2 generation. There was no significant difference in transformation rate between T1 flowers collected from main and side branches of the T0 plants. However, the results confirm the expectation that the responsive line (plastic), PL, is responding to the transformation much better than the stable and nonresponsive line, Bethune, since it was transformed at much higher rate.

Flax transformation via floral-dipping is simple to perform, inexpensive, and results in high transformation rates. However, the critical steps in the procedure include the selection of the best flower stage and the best surfactant concentration, so that the *Agrobacterium* can penetrate into the plant cells without killing the flower organs. If an early bud stage is used (with high Silwet-77 concentration of more than 0.05%), the flower will neither develop nor set seeds. If the late bud stage is used, a much lower rate of transformation is observed. Similar results were obtained with *Arabidopsis* floral-dip transformation (Clough and Bent 1998; Bent 2006). The optimum floral stage was middle bud stage with a second dipping at the late bud stage. The most important factors for high transformation rates were found to be using healthy plants at the correct flower stages and using the best Silwet-77 concentration. It was noticeable that two dipping intervals works somehow better than a single dipping, even though the single dipping also works.

A major limitation of this technique is the low number of flowers produced by flax, the limited number of seeds obtained from each flower, and the long life cycle of flax. It takes 6–8 weeks from seed sowing to have the primary buds ready for the first dipping and an additional 8–10 weeks post-dipping to get to the T1 generation. In total, a range of 5–6 months is needed to obtain the T1 generation. Unlike other plant species, which flower anytime of the year, some flax varieties flower better at specific times of the year.

In summary, the results of floral-dip with two different flax varieties, the fiber flax, Stormont cirus (responsive and plastic), and the oil flax, Bethune (stable and non-responsive), show that *Agrobacterium*-mediated plant transformation via floral-dip is an applicable and efficient method for flax transformation, as it results in a much higher transformation rate than those previously reported for *Arabidopsis thaliana* and other plant species using this method. It is also the highest reported transformation rate for flax. Therefore, flax transformation via floral-dipping can be used to replace the previously used techniques for flax transformation. It is also an applicable method to transform any related species of flax, a genus of approximately 200 species.

The modifications of the floral-dip method in this protocol will be applicable for use with any other plant species and not limited to flax. The direct PCR screening of T1 leaf DNA is an efficient way to overcome the problem of antibiotic resistance escape that often produces many false positives. It can also overcome the problem of having an antibiotic gene present in the transformed plant since it would no longer be necessary to have this resistance present on the T-DNA. This change might make transgenic flax more acceptable to the general public. The simple seed tracking technique employed in this protocol can be applied to any other plant species with branching anatomy similar to flax.

Acknowledgments We acknowledge the assistance of Morgan MacBeth and Jennifer Piechowski in the screening of T1 and T2 plants.

References

Agarwal S, Loar S, Steber C, Zale J. (2009) Floral transformation of wheat. Methods Mol Biol 478:105–113. https://doi.org/10.1007/978-1-59745-379-0_6

Bastaki NK, Cullis CA (2014) Floral-dip transformation of flax (Linum usitatissimum) to generate transgenic progenies with a high transformation rate. JoVE (94):e52189–e52189

Bechtold N, Jaudeau B, Jolivet S et al (2000) The maternal chromosome set is the target of the T-DNA in the in planta transformation of Arabidopsis thaliana. Genetics 155(4):1875–1887

Bent A (2006) Arabidopsis thaliana floral dip transformation method. Methods Mol Biol 343:87–103. https://doi.org/10.1385/1-59745-130-4:87

Beranová M, Rakouský S, Vávrová Z, Skalický T (2008) Sonication assisted Agrobacterium-mediated transformation enhances the transformation efficiency in flax (Linum usitatissimum L.). Plant Cell Tiss Org Cult 94(3):253–259

Bernhardt K, Vigelius SK, Wiese J, Linka N, Andreas PM (2012) An overview of T-DNA binary vectors, floral dip and screening for homozygous lines. J Endocytobiosis Cell Res 19:28

Chen Y, Schneeberger RG, Cullis CA (Jul 2005) A site-specific insertion sequence in flax genotrophs induced by environment. New Phytol 167(1):171–180

Clough SJ, Bent AF (1998) Floral dip: a simplified method for Agrobacterium-mediated transformation of Arabidopsis thaliana. Plant J 16(6):735–743

Cullis CA (Jan 2005) Mechanisms and control of rapid genomic changes in flax. Ann Bot 95(1):201–206

Das P, Joshi NC (2011) Minor modifications in obtainable Arabidopsis floral dip method enhances transformation efficiency and production of homozygous transgenic lines harboring a single copy of transgene. Adv Biosci Biotechnol 2(02):59

Davis AM, Hall A, Millar AJ, Darrah C, Davis SJ (2009) Protocol: streamlined sub-protocols for floral-dip transformation and selection of transformants in Arabidopsis thaliana. Plant Methods 5(1):3

Desfeux C, Clough SJ, Bent AF (2000) Female reproductive tissues are the primary target of Agrobacterium-mediated transformation by the Arabidopsis floral-dip method. Plant Physiol 123(3):895–904

Dong J-Z, McHughen A (1993) An improved procedure for production of transgenic flax plants using Agrobacterium tumefaciens. Plant Sci 88(1):61–71

Feldmann KA (1991) T-DNA insertion mutagenesis in Arabidopsis: mutational spectrum. Plant J 1(1):71–82

Horsch RB, Fry JE, Hoffmann NL, Eichholtz D, SGa R, Fraley RT (1985) A simple and general method for transferring genes into plants. Science 227:1229–1231

Ishida Y, Hiei Y, Komari T (2007) Agrobacterium-mediated transformation of maize. Nat Protoc 2(7):1614–1621

Mara C, Grigorova B, Liu Z (2010) Floral dip transformation of Arabidopsis thaliana to examine pTSO2::beta-glucuronidase reporter gene expression. JoVE (40) pii: 1952. https://doi.org/10.3791/1952

Mlynárová Ľ, Bauer M, Nap J-P, Preťová A (1994) High efficiency Agrobacterium-mediated gene transfer to flax. Plant Cell Rep 13(5):282–285

Mu GC, Chang N, Xiang K, Sheng Y, Zhang Z, Pan G (2012) Genetic transformation of maize female inflorescence following floral dip method mediated by agrobacterium. Biotechnology 11(3):178–183

Qiu D, Diretto G, Tavarza R, Giuliano G (2007) Improved protocol for Agrobacterium mediated transformation of tomato and production of transgenic plants containing carotenoid biosynthetic gene CsZCD. Sci Hortic 112(2):172–175

Trieu AT, Burleigh SH, Kardailsky IV et al (2000) Transformation of Medicago truncatula via infiltration of seedlings or flowering plants with Agrobacterium. Plant J 22(6):531–541

Zhan X-c, Jones DA, Kerr A (1988) Regeneration of flax plants transformed by Agrobacterium rhizogenes. Plant Mol Biol 11(5):551–559

Yasmeen A, Mirza B, Inayatullah S et al (2009) In planta transformation of tomato. Plant Mol Biol Report 27(1):20–28

Zale JM, Agarwal S, Loar S, Steber CM (2009) Evidence for stable transformation of wheat by floral dip in Agrobacterium tumefaciens. Plant Cell Rep 28(6):903–913

Zhang X, Henriques R, Lin S-S, Niu Q-W, Chua N-H (2006) Agrobacterium-mediated transformation of Arabidopsis thaliana using the floral dip method. Nat Protoc 1(2):641–646

Chapter 13
Disease Resistance Genes in Flax

Christopher A. Cullis

13.1 Introduction

Flax has long been a model for identifying and understanding the interaction between a host plant and disease organisms. The interaction between flax and the rust fungus (*Melampsora lini*) was the basis for the proposal by Flor (1956, 1954, 1941, 1971) that there was a gene-for-gene interaction between the plant and the fungus that controlled whether the particular fungal isolate would cause disease or would be resisted by the plant. Thus, it was proposed that there was a pair of interacting genes, the resistance (R) gene in the plant and a complementary Avr gene in the pathogen, both of which had to be present for the resistant phenotype to be exhibited. This identification of the gene-for-gene relationship between the plant and fungus in flax was an essential basis for the management of diseases in many important crop plants. In particular, the subsequent isolation of the disease resistance genes in flax, model systems such as Arabidopsis, as well as for important crop plants was facilitated by the knowledge of these gene interactions. The characterization of the interactions between flax and its rust pathogens has been advanced by the isolation of the relevant genes from both the host and the pathogen. The assembled genome sequences for the plant and pathogens, allied with the extensive genetic data for the resistances, will also inform the genetic mechanisms underlying quantitative trait loci (QTLs) for resistance.

C. A. Cullis (✉)
Department of Biology, Case Western Reserve University, Cleveland, OH, USA
e-mail: cac5@case.edu

© Springer Nature Switzerland AG 2019
C. A. Cullis (ed.), *Genetics and Genomics of Linum*, Plant Genetics and
Genomics: Crops and Models 23, https://doi.org/10.1007/978-3-030-23964-0_13

13.2   Gene-for-Gene Model for Disease Resistance

The long history of the understanding of the genetic relationship between the flax and its rust fungus extends back to Henry (1930) who first reported the genetic control of rust resistance in flax. However, the definitive work in developing the gene-for-gene hypothesis, that is, a gene in the host plant and a gene in the pathogen that interact to either give resistance (through the hypersensitive response) or result in disease, was reported by Flor (1956). The gene-for-gene relationship between pathogenicity in flax rust and resistance in the host is of a highly specific complementary type. Thus, if the pathogen has an avirulence gene (Avr gene) that is complementary to the plant resistance gene (R gene), an incompatible interaction occurs in which recognition of the rust infection gives rise to localized cell death (the hypersensitive response, HR). However, if a compatible interaction occurs, that is, one in which the avirulence gene in the pathogen is not recognized by the resistance gene present in the plant, the result is disease development. The outcome of the reactions between the complementary genes results in either the hypersensitive response, that is, resistance, or a development of disease, as shown in Table 13.1. Therefore, the response of a plant variety to a race of the rust depends on both the resistance gene in the plant and the gene for pathogenicity in the fungus. The flax genes responsible of rust resistance were developed empirically. A series of differentials for each of the rust resistance genes was developed through crosses between the variety Bison and varieties with different rust resistances followed by line selection, resulting in inbred lines that were each differentials for a specific single rust resistance gene (Flor 1956).

13.3   Flax Rust Resistance Genes

There are at least 31 rust resistance genes known in flax placed into 5 groups, the L, M, N, P, and K groups. The number of 34 resistance genes has been reported (Islam and Mayo 1990, Islam and Shepherd 1991) which included 2 new groups, the D and Q genes (Kutuzova and Kulikova 1989). However, little more is known about the possible specificities and organization of these two groups, and they will not be considered further here. The number of genes/alleles in each of these groups varies. There are 13 within the L group, 7 in the M, 3 in the N, and 6 in the P,

Table 13.1 Reaction between plants and fungal races depending on the gene complements

Plant genotype	Pathogen genotype	
	Avr_a	Avr
R_a	Resistant (hypersensitive response)	Sensitive
$r_a r_a$	Sensitive	Sensitive

R_a, plant resistance gene present either heterozygous or homozygous; Avr_a, the pathogen avirulence gene complementary to R_a; avr, any other or no avirulence gene present in the pathogen

while 2 genes have been identified for the K locus. All the rust resistance genes isolated encode an intracellular Toll interleukin 1 receptor–nucleotide binding site–leucine-rich repeat (TIR–NBS–LRR) class of proteins (Anderson et al. 1997; Dodds et al. 2001a, b; Lawrence et al. 2010, 1995).

One feature of the plant genes that control the specific response to pathogens is that they are often arranged in groups, that is, all the genes for each group are closely linked, but the groups themselves mostly segregated independently, although there is linkage between the N and P groups. The genes within each these groups are not present in the same organization, namely, the genes in the L and P loci are allelic, whereas the genes from the M and N loci form a multigene family. The initial isolation of the resistance genes was achieved through tagging with transposable elements, either the Ac element from maize or the endogenous dLUTE element (Ellis et al. 1997).

13.4   The Flax L Rust Resistance Genes

The L6 gene was the first of the flax rust resistance genes to be cloned. This was achieved by identifying the gene in an Ac tagging experiment (Lawrence et al. 1995). This L6 gene was also the site of the identification of an endogenous flax transposable element, dLute (Luck et al. 1998). The sequence of this first flax rust resistance gene was shown to represent a Toll-interleukin-1 receptor–nucleotide binding site–leucine-rich repeat (TIR–NBS–LRR) disease resistance protein. The important specificity determinants of the L genes are in the LRR region of the C terminal part of the genes. Transcription of the L6 gene results in at least four transcript classes as a result of alternative splicing with the most abundant transcript containing the NBS and LRR regions which encodes the functional protein. The three other transcripts lack most of the C-terminal region and do not appear to function in the rust resistance system (Ayliffe et al. 1999). The L6 and L11 proteins differ by 32 amino acid polymorphisms, all in the LRR domain. This difference would appear to be sufficient for the differential recognition of the cognate Avr genes in the rust fungus.

The LRR region defines the specificity of the allele, but other structural elements of the R genes can affect the response of particular genes to the relevant avirulent rust strain. The L6 and L7 proteins differ by only ten amino acid changes, all in the TIR region, and show similar recognition specificities. However, the L7 allele mediates a weaker resistance response than the L6 allele. Other L alleles, such as L5, L10, and L2, contain similar polymorphisms to L7 in this region of the TIR domain, but their rust resistance activity is similar to that of L6. Therefore, it is likely that other amino acid differences in the sequences of L5, L10, and L2 compensate for the inhibitory TIR polymorphisms in L7 and that L7 is substantially different from these other three L alleles in these compensatory regions (Ravensdale et al. 2011).

A comparison of two of the L alleles, the L6 and the weaker L7, identified two polymorphic regions in the TIR and the NBS domains that regulate both effector ligand-dependent and ligand-independent cell death signaling as well as nucleotide binding to the receptor. This suggests that a negative functional interaction between the TIR and NB domains holds L7 in an inactive/ADP-bound state more tightly than L6, hence decreasing its capacity to adopt the active/ATP-bound state and explaining its weaker activity in planta. L6 and L7 variants with a more stable ADP-bound state failed to bind to AvrL567 in yeast two-hybrid assays, while binding was detected to the signaling active variants (Ravensdale et al. 2011).

13.5 The Flax M Rust Resistance Genes

In contrast to the L rust resistance genes, in which the different specificities are allelic, the M locus is comprised of separate rust resistance genes. Four of the seven genes are closely linked but separable by recombination (Islam and Shepherd 1991). However, these loci are complex, and each can contain up to 15 tandemly arranged paralogues. The L6 and M resistance genes share about 86% nucleotide identity (Ellis et al. 1997). Flax is an ancient tetraploid, and the molecular data on the relationship between the L and M loci indicate that they are likely to be homoeologous. If this is, in fact, the case, then the two loci must have evolved differently since the genome duplication, with one, the L locus, only containing a single allele at the locus, while the M locus evolved into a tandem array. However, for the M locus, although all the members of the cluster can be transcribed, only one appears to have a resistance gene function. Whether this is due to the remaining members not being resistance genes or the appropriate Avr complement has not been identified is not yet resolved (Ellis et al. 1995).

13.6 N Locus

The three rust resistance specificities at the N locus were cloned through the use of degenerate PCR targeted to the NBS region present in many plant resistance genes (Dodds et al. 2001a). As noted elsewhere, since there are many resistance gene analogs (RGAs) in the flax genome, the result was the amplification of numerous fragments. However, one of the RGA amplicons co-segregated with N locus, and the subsequent cloning of this locus identified four tandemly arranged paralogues termed Ngc-A to D. The Ngc-D gene was shown to be responsible for the N resistance gene specificity, but although the other three genes, Ngc-A, Ngc -B, and Ngc -C, are transcribed and encode apparently full-length proteins, no known resistance specificities are known for these three genes (Dodds et al. 2001a). These four sequences could be placed into three groups. The Ngc-A and Ngc-B groups each contain a single gene in the N locus haplotypes, while the Ngc-C and Ngc-D each contain two paralogues in each N locus haplotype. Sequence exchanges have only been observed within each group, and those occurring within the Ngc-C/Ngc-D group have contributed to generating significant variation (Dodds et al. 2001a), which is of biological importance since the Ngc-D gene is responsible for the N resistance gene specificity.

13.7   P Locus

There are at least six rust resistance specificities that map to the P locus, which are designated as P and P1 to P5, in flax (Dodds et al. 2001b). The first member to be isolated from this family, P2, was through transposon tagging with the maize Ac element that had also been used to isolate the L6 and M genes (Lawrence et al. 1995; Anderson et al. 1997). This P2 gene, similar to the L and M genes, encoded a protein with the TIR–NBS–LRR structure but differed from the other two loci in that it also included a C-terminal non-LRR domain (CNL domain) of 150 amino acids (Dodds et al. 2001b). The activity and specificity of the genes at this locus can be modified by variants in different regions of the gene. Thus, for example, a truncation of the CNL domain caused loss of function. The comparison between the P2 and P genes identified ten amino acid differences, of which only six were sufficient to alter the specificity from P2 to P (Dodds et al. 2001b). These six variable residues were all in the LRR β-sheet region and are possibly involved in the direct R-Avr interactions since they are exposed on a concave β-sheet surface and available for participation in R-Avr interactions (Dodds et al. 2001b). Therefore, the specificity for the P alleles may be determined by a shorter region of polymorphisms than that found necessary when the L6 and L11 alleles were compared.

13.8   Resistance Gene Analogs

As noted earlier, the rust resistance genes have a particular structure (TIR–NBS–LRR). This gene structure, or regions of similarity to parts of the structure or to other known resistance genes, can be used to interrogate the genome sequence to identify all the sequences that have a sufficient similarity to this overall structure. These sequences have been termed resistance gene analogs (RGAs), but the existence of a similar structure does not necessarily directly translate into a functional resistance activity. Again, as noted earlier, the M resistance loci can contain many tandemly arrayed paralogues which have not been shown to have resistance specificities. The RGAs present in the flax genome have been identified (You et al. 2018). The range of structures that have been identified for these RGAs in the flax genome and the number of representatives in each class, in addition to the TIR-NBS-LRR motif (88 representatives), include those with the NBS domain only (15 representatives), a coiled coil (CC) CC–NBS domain (3 representatives), a CC–NBS–LRR domain (21representatives), an NBS–LRR domain (32 representatives), a TIR–NBS domain (32 representatives), a TIR–unknown domain (43 representatives), a receptor-like protein kinase (RLK) (778 representatives), a receptor-like protein (RLP)

(106 representatives), and a transmembrane (TM) coiled-coil protein (TM–CC) (213 representatives) as well as 9 classified as other (You et al. 2018). Within the flax genome, these 1327 regions of RGA similarity have been identified (You et al. 2018) of which 348 are present in 87 clusters, the remainder being single motifs. These clusters contain between 3 and 19 motifs in a cluster, the cluster with 19 being on chromosome 8. The most frequent number of motifs in a cluster was 3 (51 clusters) followed by 4 (18 clusters), 5 (9 clusters), 6 (4 clusters) 8 (2 clusters), 9 (1 cluster), 10 (1 cluster), and 19 (1 cluster).

13.9 Avr Genes

Avr genes have been identified through genetic analysis. These genes generally did not occur in similar clusters as those observed for the plant R genes, although there was some clustering of subsets of the Avr genes. Thus, one cluster contained the AvrL3, AvrL4, and AvrL10; another the AvrL5, AvrL6, and AvrL7; a third the AvrM1 and AvrM4; and the fourth the AvrP, AvrP1, AvrP2, and AvrP3 genes (Ellis et al. 2007). As noted above, the cloning of the plant genes was done using both a transposon insertion and through map-based cloning of using RGAs. The rust Avr genes were cloned by mapping candidate genes; genomic clones were then isolated and tested for avirulence function in flax with different resistance profiles (Ellis et al. 1997). Four of the Avr families representing nine recognition specificities have been cloned (Barrett et al. 2009; Catanzariti et al. 2006). Unlike the plant R genes, there was no sequence similarity between these four Avr gene families (Ellis et al. 2007). The interactions between the gene products from the plant and pathogen have been determined since the relevant genes in both the host and pathogen have been isolated. These Avr genes encode small secreted proteins that interact directly with the plant R genes (Ellis et al. 2007; Dodds et al. 2004). The direct interaction between the R genes products and the Avr gene products was confirmed by yeast two-hybrid interactions (Dodds et al. 2006).

The complete genome of the rust fungus has been assembled and annotated for the presence of possible effector proteins ((Nemri et al. 2014). The annotation took advantage of the 4 cloned flax rust avirulence genes and 2 basidiomycete effectors, to identify 725 flax rust candidate effectors. Additional specific avirulence gene cloning was achieved through a high-density RADseq linkage map (Anderson et al. 2016).

13.10 Other Disease Resistance Needs in Flax

Three other diseases are of importance in flax, namely, pasmo (caused by *Septoria linicola*), powdery mildew (caused by *Oidium lini* Skoric), and fusarium wilt (caused by *Fusarium oxysporum* f. sp. *lini* Schlecht). The extensive collections of varieties and molecular resources provide the basis for identifying the genes and mechanisms underlying the resistance phenotypes and subsequently can be incorporated into breeding programs to produce more disease-resistant varieties.

13.11   Pasmo

Pasmo (*Septoria linicola*) is a fungal disease causing major losses in seed yield and quality and stem fiber quality in flax. Pasmo resistance (PR) is quantitative and has many loci associated with the resistance. Genome-wide association studies (He et al. 2018) identified 67 quantitative trait loci (QTL) with large effects (3–23%) that explained 32–64% of the total variation for PR and were mainly additive. Forty-five of these QTL spanned 85 resistance gene analogs including a large TIR–NBS–LRR gene cluster on chromosome 8. The accuracy of genomic prediction (GP) was evaluated using these large data sets to improve the PR breeding efficiency (He et al. 2019), and the results demonstrated the GP models based on marker information from all identified QTL were highly effective for PR prediction.

13.12   Powdery Mildew

Four loci that putatively provide resistance to Canadian strains of powdery mildew (PM) were identified in a segregating population derived from a cross of French fiber flax cultivars (Booker, Chap. 3, this volume). Previously, QTL analysis had identified three PM resistance QTL located on flax chromosomes 1, 7, and 9 (Asgarinia 2013). These QTL explained 97% of the phenotypic variation exhibiting a mainly dominant gene action, with the QTL on linkage group 9 explaining nearly 50% of the phenotypic variation and the other two loci explaining about the same amount each. The molecular resources available for flax will support the map-based cloning of the candidate genes underlying the QTL for PM resistance is possible using the molecular resources available, with the added ability to transform flax to directly confirm the resistance gene identification.

13.13   Fusarium Wilt

Two studies have characterized the response of sensitive and resistant flax lines to fusarium infection (Dmitriev et al. 2017; Galindo-Gonzalez and Deyholos 2016). In one (Dmitriev et al. 2017), the early events that occur in the roots within 2 days postinfection (dpi) were characterized using high-throughput Illumina sequencing, while in the second (Galindo-Gonzalez and Deyholos 2016), the changes in expression at 2, 4, 8, and 18 days postinfection (dpi) were determined in both resistant and susceptible flax genotypes. The greatest changes were observed in the samples18 dpi, where more than 1000 genes responded. These changes in expression were compiled into a model detailing the plant defense responses of flax following *F. oxysporum* inoculation and is shown in Fig. 13.1. Genes that were up- or downregulated in resistant cultivars under infection by *F. oxysporum*, but did not show a change in expression in

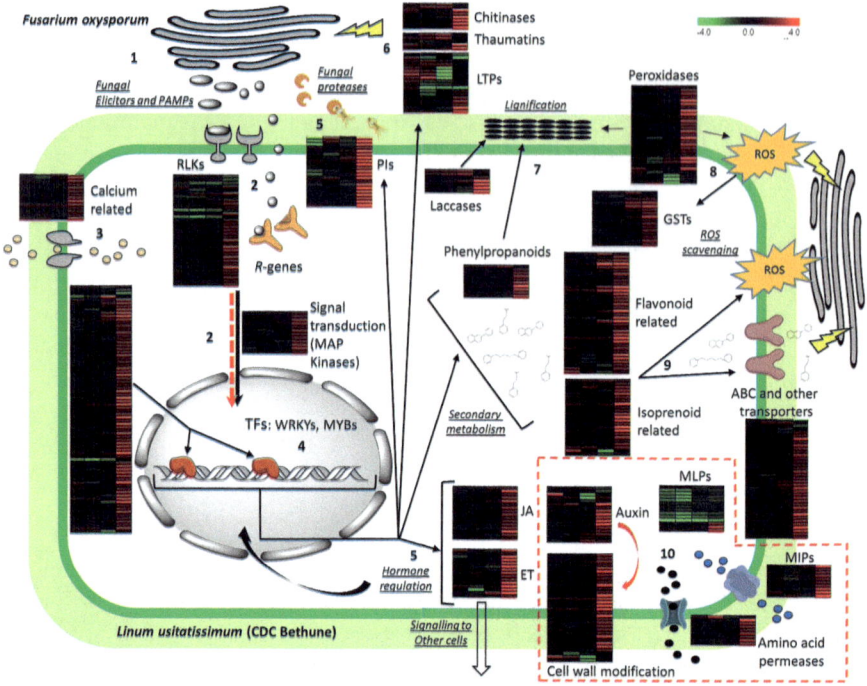

Fig. 13.1 Heatmaps for log2-fold gene expression changes at 2, 4, 8, and 18 DPI are shown besides each major gene group analyzed. The full deployment of plant defense is evidenced 18 DPI as seen in the fourth column of the heatmaps. (1) During fungal attack, *Fusarium oxysporum* liberates elicitors (pathogen-associated molecular patterns – PAMPs), effectors, and fungal proteases (which are also considered effectors) to facilitate infection. (2) Membrane receptors including receptor-like kinases (RLKs), an NBS-LRR (*R*-genes), interact with the PAMPs and effectors, respectively, causing downstream changes in phosphorylation of kinases (e.g., MAP kinases). (3) At the same time, an influx of calcium causes changes in calcium-binding proteins that are also involved in signal transduction. (4) Regulation of transcription factors results in activation of hormone-related, defense, and secondary metabolism genes. (5) Presence of jasmonate (JA), ethylene (ET) biosynthetic genes indicates further signaling to other cells and feedback loops to activate more defense genes. (6) Protease inhibitors (PIs) neutralize fungal proteases, while chitinases, thaumatins, and lipid transfer proteins (LTPs) act directly on the fungal cell wall or membrane. (7) Lignin precursors are created via phenylpropanoid metabolism and are polymerized into lignin by the action of laccases and peroxidases. (8) Peroxidases are also involved in the generation of reactive oxygen species (ROS) which are regulated by enzymes like glutathione *S*-transferases (GSTs). (9) Flavonoids and isoprenoids can act as antioxidants against ROS or be directly translocated outside the cell by ATP-binding cassette (ABC) transporters to impair fungal function and growth. (10) Some unexpected regulation was found in some specific transcripts of several gene groups: auxin-related genes, major latex proteins (MLPs), cell wall modification proteins, major intrinsic proteins (MIPs), and amino acid permeases; the potential manipulation of the host by the pathogen to regulate such genes is indicated by a red arrow that parallels signal transduction and by the gene groups surrounded by dashed red lines. (Reprinted from Galindo-Gonzalez and Deyholos 2016)

susceptible cultivars, were likely candidate genes for resistance. The protein beta-1-3-glucanase, which hydrolyzes beta-1,3-glucans of the cell wall in fungi, is the most well-known fungal-responsive protein in flax. This protein has a well-documented role in plant defense against pathogens, including increased resistance to *Fusarium* in transgenic flax lines containing the potato beta-1,3-glucanase gene and in plants overexpressing the beta-1,3-glucanase gene (Wojtasik et al. 2014; Balasubramanian et al. 2012). One of the important findings of these two transcriptomic documentations is the identification of transcripts that are differentially regulated upon infection but remain to be annotated. These novel transcripts may be important for their possible novel functions.

13.14 Summary

Flax has been the model for the development of the gene-for-gene hypothesis of the interaction between plants and some of their fungal pathogens. This hypothesis has been vital in the development of many disease resistance crops. The long history of use of flax in its two commercial forms, tall unbranched stems for fiber and bushy habit for oil yield, has also resulted in extensive collections of historical varieties and germplasm collections, both of the commercial *Linum usitatissimum* and other related species. These collections have been used to map the loci conditioning disease reactions. Similar resources have also been assembled for the pathogens. The molecular resources, in terms of genomic sequence assemblies, transcriptomic data for both genes and small RNAs, for both the plant and the pathogens, position the flax–flax pathogen interactions as a valuable resource for understanding the complex interactions and the coevolutionary forces shaping these interactions in disease development. The characterization of these interactions will be enhanced through the availability of transformation of both partners involved and the previous development of transposon tagging systems. The genomic assemblies and the identification of resistance gene analogs are important to determine the reservoir of potential disease resistance genes available and whether they have been employed within the evolutionary history of the arms race between plant and pathogen.

References

Anderson PA, Lawrence GJ, Morrish BC, Ayliffe MA, Finnegan EJ, Ellis JG (1997) Inactivation of the flax rust resistance gene M associated with loss of a repeated unit within the leucine-rich repeat coding region. Plant Cell 9:641–651

Anderson C, Khan MA, Catanzariti AM, Jack CA, Nemri A, Lawrence GJ, Upadhyaya NM, Hardham AR, Ellis JG, Dodds PN, Jones DA (2016) Genome analysis and avirulence gene cloning using a high-density RADseq linkage map of the flax rust fungus, Melampsora lini. BMC Genomics 17:667

Asgarinia P (2013) Mapping quantitative trait loci for powdery mildew resistance in flax (Linum usitatissimum L.). Crop Sci 53:2462–2472

Ayliffe MA, Frost DV, Finnegan EJ, Lawrence GJ, Anderson PA, Ellis JG (1999) Analysis of alternative transcripts of the flax L6 rust resistance gene. Plant J 17:287–292

Balasubramanian V, Vashisht D, Cletus J, Sakthivel N (2012) Plant beta-1,3-glucanases: their biological functions and transgenic expression against phytopathogenic fungi. Biotechnol Lett 34:1983–1990

Barrett LG, Thrall PH, Dodds PN, Van Der Merwe M, Linde CC, Lawrence GJ, Burdon JJ (2009) Diversity and evolution of effector loci in natural populations of the plant pathogen Melampsora lini. Mol Biol Evol 26:2499–2513

Catanzariti AM, Dodds PN, Lawrence GJ, Ayliffe MA, Ellis JG (2006) Haustorially expressed secreted proteins from flax rust are highly enriched for avirulence elicitors. Plant Cell 18:243–256

Dmitriev AA, Krasnov GS, Rozhmina TA, Novakovskiy RO, Snezhkina AV, Fedorova MS, Yurkevich OY, Muravenko OV, Bolsheva NL, Kudryavtseva AV, Melnikova NV (2017) Differential gene expression in response to Fusarium oxysporum infection in resistant and susceptible genotypes of flax (Linum usitatissimum L.). BMC Plant Biol 17:253

Dodds PN, Lawrence GJ, Ellis JG (2001a) Contrasting modes of evolution acting on the complex N locus for rust resistance in flax. Plant J 27:439–453

Dodds PN, Lawrence GJ, Ellis JG (2001b) Six amino acid changes confined to the leucine-rich repeat beta-strand/beta-turn motif determine the difference between the P and P2 rust resistance specificities in flax. Plant Cell 13:163–178

Dodds PN, Lawrence GJ, Catanzariti AM, Ayliffe MA, Ellis JG (2004) The Melampsora lini AvrL567 avirulence genes are expressed in haustoria and their products are recognized inside plant cells. Plant Cell 16:755–768

Dodds PN, Lawrence GJ, Catanzariti AM, Teh T, Wang CIA, Ayliffe MA, Kobe B, Ellis JG (2006) Direct protein interaction underlies gene-for-gene specificity and coevolution of the flax resistance genes and flax rust avirulence genes. Proc Natl Acad Sci U S A 103:8888–8893

Ellis JG, Lawrence GJ, Finnegan EJ, Anderson PA (1995) Contrasting complexity of 2 rust resistance loci in flax. Proc Natl Acad Sci U S A 92:4185–4188

Ellis J, Lawrence G, Ayliffe M, Anderson P, Collins N, Finnegan J, Frost D, Luck J, Pryor T (1997) Advances in the molecular genetic analysis of the flax-flax rust interaction. Annu Rev Phytopathol 35:271–291

Ellis JG, Dodds PN, Lawrence GJ (2007) Flax rust resistance gene specificity is based on direct resistance-avirulence protein interactions. Annu Rev Phytopathol 45:289–306

Flor HH (1941) Inheritance of rust reaction in a cross between the flax varieties Buda and JWS. J Agric Res 63:0369–0388

Flor HH (1954) The genetics of host-parasite interaction in flax rust. Phytopathology 44:488–488

Flor HH (1956) The complementary genic systems in flax and flax rust. Adv Genet 8:29–54

Flor HH (1971) Current status of the gene-for-gene concept. Annu Rev Phytopathol 9:275–296

Galindo-Gonzalez L, Deyholos MK (2016) RNA-seq transcriptome response of flax (Linum usitatissimum L.) to the pathogenic fungus Fusarium oxysporum f. sp. lini. Front Plant Sci 7:1766

He L, Xiao J, Rashid KY, Yao Z, Li P, Jia G, Wang X, Cloutier S, You FM (2018) Genome-wide association studies for pasmo resistance in flax (Linum usitatissimum L.). Front Plant Sci 9:1982

He L, Xiao J, Rashid KY, Jia G, Li P, Yao Z, Wang X, Cloutier S, You FM (2019) Evaluation of genomic prediction for Pasmo resistance in flax. Int J Mol Sci 20. pii: E359

Henry AW (1930) Inheritance of immunity from flax rust. Phytopathology, 20, 707–721.

Islam MR, Mayo GME (1990) A compendium on host genes in flax conferring resistance to flax rust. Plant Breed 104:89–100

Islam MR, Shepherd KW (1991) Present status of genetics of rust resistance in flax. Euphytica 55:255–267

Kutuzova SN, Kulikova AE (1989) Identification of the genes for resistance in varieties of the international set of differentiators of Melampsora lini (Pers.) Lév. Sbornik Nauchnykh Trudov po Prokladnoi Botanike, Genetike i Selektsii, 125: 65–69

Lawrence GJ, Finnegan EJ, Ayliffe MA, Ellis JG (1995) The L6 gene for flax rust resistance is related to the Arabidopsis bacterial-resistance gene Rps2 and the tobacco viral resistance gene-N. Plant Cell 7:1195–1206

Lawrence GJ, Anderson PA, Dodds PN, Ellis JG (2010) Relationships between rust resistance genes at the M locus in flax. Mol Plant Pathol 11:19–32

Luck JE, Lawrence GJ, Finnegan EJ, Jones DA, Ellis JG (1998) A flax transposon identified in two spontaneous mutant alleles of the L6 rust resistance gene. Plant J 16:365–369

Nemri A, Saunders DGO, Anderson C, Upadhyaya NM, Win J, Lawrence GJ, Jones DA, Kamoun S, Ellis JG, Dodds PN (2014) The genome sequence and effector complement of the flax rust pathogen Melampsora lini. Front Plant Sci 5:98

Ravensdale M, Nemri A, Thrall PH, Ellis JG, Dodds PN (2011) Co-evolutionary interactions between host resistance and pathogen effector genes in flax rust disease. Mol Plant Pathol 12:93–102

Wojtasik W, Kulma A, Boba A, Szopa J (2014) Oligonucleotide treatment causes flax beta-glucanase up-regulation via changes in gene-body methylation. BMC Plant Biol 14:261

You FM, Xiao J, LI P, Yao Z, Jia G, He L, Kumar S, Soto-Cerda B, Duguid SD, Booker HM, Rashid KY, Cloutier S (2018) Genome-wide association study and selection signatures detect genomic regions associated with seed yield and oil quality in flax. Int J Mol Sci 19. pii: E2303

Chapter 14
Origin and Induction of the Flax Genotrophs

Christopher A. Cullis

14.1 Introduction

Flax (*Linum usitatissimum* L.) is a self-pollinated annual species ($2n = 2x = 30$) belonging to the Linaceae family. It has been utilized for two different primary products, seed oil and stem fiber. It was domesticated in the Near East more than 7000 years ago. The two crops are phenotypically different with the fiber varieties being taller with suppressed side branching for improved fiber length and quality and the oilseed varieties being more branched for high seed yield. The agronomic practices also contribute to these phenotypic differences, with the fiber crop being grown at very high density, much higher than the oilseed crop. There have been a number of comments over the years on the quality of flaxseed. In the 1800s it was agreed that the best seed was "Riga" seed (Aiton 1834) and that it was important to return to seed produced in isolation from the fiber crop to have continued yield success. It was believed some localities produce poor seed, but it was generally believed that plants tended to grow out of maternal effects so that, at maturity, little difference would be observed between plants grown from what is perceived to be good seed or poor seed. In preliminary experiments at Aberystwyth by Durrant, flaxseed collected in a good harvest year gave plants nearly twice as large as from seed collected in the following poor year when grown together the year after, but there was no carry over to the subsequent years. These conclusions were supported for plants that were well-spaced and grown under good cultural conditions.

C. A. Cullis (✉)
Department of Biology, Case Western Reserve University, Cleveland, OH, USA
e-mail: cac5@case.edu

14.2   Flax Genotrophs

The effect of differing cultural conditions on the progeny was determined initially in the flax variety Stormont cirus (subsequently designated as Pl as a plastic or responsive variety), which was grown with all eight combinations of nitrogen, phosphorus, and potassium fertilizers. The progeny from the 16 different treatments (Durrant 1962) were then grown the following year again in all combinations of the fertilizers. There were large differences in plant weight observed due to some of the fertilizer regimes applied in the previous generation. The two extremes were termed large (L) and small (S) genotrophs based on their final plant weight at maturity (Fig. 14.1), and the nutrient conditions applied in the first generation were termed inducing conditions. The nutrient conditions not resulting in this type of stable inherited variation were termed non-inducing conditions, and the plants grown under such non-inducing conditions could be subsequently induced to produce the extreme genotrophs. These two extreme types bred true, and the differences have been maintained for more than 50 generations. The plant weight at maturity was dependent on the growth conditions so that with poorer growth at lower light levels, the difference was minimized. However, as also apparent in Fig. 14.1, the heights of the three lines were different, with the original line being taller than either of the two derived lines. This difference in height was maintained irrespective of the plant weight difference when the three lines were grown under non-inducing conditions.

As noted earlier, the growth of the fiber flax crop from seed consistently collected from the crop became variable. It is noteworthy that the plant height of this inbred seed became more variable. However, the genotrophs were stably uniform in their heights within a line and consistently different between the two lines (Fig. 14.2).

What was notable was that not all of the nutrient treatments resulted in stable heritable phenotypic changes. Therefore, there had to have been two different sets of variation, one that conditions the changes in phenotype and the other for the loss of ability to subsequently respond. Since the large and small genotrophs have different phenotypes, but both are subsequently stable, the loss of the ability to respond to the nutrient environment must be separate from the control of the phenotype.

Fig. 14.1 Growth of the original Stormont cirus variety and the two extreme genotrophs (L and S) in the first generation after induction (**b**) and following 53 generations of propagation under non-inducing conditions (**a**)

Fig. 14.2 Growth of the small (S) and large (L) genotrophs 15 generations after their original induction. Note the uniform height of the individuals of the two lines and the consistent difference between them although there is little difference in final plant weight

S   L

14.3 Variation Between the Genotrophs and the Original Pl Line

The differences in phenotypes and some molecular markers that have been described include plant height and weight, the number of hairs on the false septa of the seed capsules, the pattern of isozymes for peroxidase and acid phosphatase, the nuclear DNA amount, and the number of ribosomal RNA genes. The differences in plant height have been illustrated above. The inheritance of both plant weight and plant height in crosses between the genotrophs and Pl were consistent with these variations being conditioned as quantitative characters; there is no evidence that, singly or together, these variants were due to single genes with any dominance/recessive relationship.

14.3.1 Hairs on the False Septa of Seed Capsules

The original Pl line had hairy false septa, with about 60 hairs per septum. However, the L and S genotrophs differed for this character in that the L genotroph had hairless septa, while the S genotroph had hairy septa, as did the Pl line. In crosses between individuals with hairy and hairless septa, hairy septa were dominant, and in the F2, there was segregation into 4 classes having 0, 30, 40, and 60 hairs. This is consistent with a quantitative character conditioned by multiple genes. However, most interestingly, the change between Pl and L is from a dominant characteristic to a recessive characteristic after one generation (Durrant and Nicholas 1970). Since there was no evidence that Pl was heterozygous for this character, this was a change from a homozygous-dominant to a homozygous recessive form after growth for a single generation under inducing conditions. This change on induction from being homozygous for one form to being a homozygous for an alternative form is a recurring theme as both phenotypic and molecular markers are considered.

14.3.2   Isozyme Mobility Variation

The mobility of peroxidase and acid phosphatase isozymes in Pl and the genotrophs L and S has been examined (Cullis and Kolodynska 1975). A number of the
bands visible in the lines have altered mobility between L and S, while Pl could be
the same as either of the genotrophs or a combination of both. Crosses were made
between L and S, and the relative mobilities from the main stem homogenates
obtained for the F1 and F2 selfed or backcross generations. The data indicated that
the control over each isozyme's relative mobility was by a simple system with a
dominant allele in L and a recessive one in S (Fieldes et al. 1977; Tyson et al. 1978;
Cullis 1979). These persistent relative mobility shifts for peroxidases, acid phos-
phatases, and other glycoproteins are probably controlled by modifier loci. Again,
here the form with the faster mobility was dominant, while the form with the
slower mobility was recessive. Since Pl was homozygous, these isozyme mobility
shifts were a result of a change in control from homozygous dominant to homozygous recessive (Tyson et al. 1986).

14.3.3   Nuclear DNA Variation

The total nuclear DNA amount as determined by Feulgen staining was shown to be
different when comparisons between Pl, L, and S were made. The data showed that
L had the greatest amount of nuclear DNA, Pl an intermediate amount, while S had
a smaller value (Evans et al. 1966, Evans 1968a, b). One component of this nuclear
DNA variation is the fraction coding for the ribosomal RNAs. This set of genes for
the large (25S and 18S) and the 5S ribosomal RNA genes comprises up to 5% of the
total genome (Timmis and Ingle 1973, Cullis 1976, 1979). The variation in these
genes could account for 3–4% of the observed 15% difference in nuclear DNA
amount between the L and S genotrophs.

14.4   Site and Timing of the Induced Heritable Changes

Most of the characteristics so far described have been measured on progeny of
plants grown for a single generation under a range of nutrient environments. The
question that needed to be answered was when did the variation underpinning the
altered phenotypes arise – was it during vegetative growth, flowering, meiosis, or
postfertilization? However, these differential growth conditions have a direct effect
on the phenotype of the plants while they are growing, making it difficult, or impossible, to follow the induction of any possible changes in whole plant characteristics
in the treated generation. Therefore, characteristics that were not affected by the
environment but could be measured in different parts of the plant were needed.

The two that were amenable with the technology available at the time were, first, the nuclear DNA amount and, second, the ribosomal RNA gene number. Both of these characters were shown to change while the plants were growing under the inducing conditions, before the transition from vegetative to reproductive stages had been reached. Thus, the nuclear DNA in the apical meristem was determined for Pl plants growing under a variety of inducing conditions, which could result in L or S genotrophs in the next generation. It was shown that the nuclear DNA amount changed during the vegetative growth and the nuclear DNA content at flowering was the value transmitted to the next generation (Evans et al. 1966; Evans 1968a). These data indicate that one site of induced variation is in the nuclei of the apical meristem. However, this technique was not suitable to determine if changes in the nuclear DNA also occurred within differentiated tissue. The ribosomal RNA gene number variation was useful in determining the site and timing of the induction of variation. The ribosomal RNA gene number was determined along the stem of plants growing under inducing conditions (Cullis and Charlton 1981). It was shown that the change in the number of ribosomal genes occurred over a relatively short period of growth, again, as observed for the total nuclear DNA amount, prior to flowering, remained stable, and was then transmitted to the next generation. An important characteristic of the changes in the ribosomal RNA gene number was that the plant was chimeric, such that the cells laid down earlier were different from those from later stages of growth. However, even at maturity, the cells from the lower part of the stem still had the same number of ribosomal RNA genes as the parent plants but different from the stem cells at the upper parts of the plants. These data are consistent with the proposal that the genomic variations occur at or near the apical tip (probably within the apical meristem) and not in differentiated cells. If all cells could participate in such genomic restructuring, then the complete plant should be modified and chimeric individuals should not be observed. These data, and subsequent extensive studies for multiple regions of the flax genome, have conclusively shown that the variation does occur during vegetative growth under inducing conditions.

14.5   Is the Variation Induced Adaptive?

This is a difficult question to answer. One possible response is to look at the performance of the genotrophs (and the original line) grown in the nutrient regimes in which they were induced. If the relative plant growth (performance) is altered depending on the environment supplied, then it is possible that the response to the inducing environment could be considered adaptive. The growth of the Pl line is more difficult to interpret since it should be responding and therefore might change in its response during the growth/monitoring period. Although not definitive, there is evidence supporting the suggestion that the relative genotroph performance is linked to the original inducing environment. Shown in Fig. 14.3 are plants grown under conditions that induced the L and S genotrophs, respectively. It can be seen

Fig. 14.3 Growth of Pl and the L and S genotrophs under conditions that are non-inducing (control) and under the growth conditions under which L was induced (NPK). Note the improved performance of the L genotrophs under this high nutrient environment, compared to Pl, which was less vigorous, and S, which was largely unaffected

that the conditions for the induction of the L genotroph resulted in the L genotroph having the most vigorous growth and under these conditions, L was even taller than Pl. All three lines grew well under these conditions. Under non-inducing conditions, as previously shown, the Pl line was the tallest, and the relative heights of L and S were as expected, with L being taller and more branched than S.

What about growth of the individual plants under inducing conditions? Observations on individuals of the original line and the large and small genotrophs growing under such inducing conditions also support the suggestion that the meristems (apical and adventitious) might be the sites of variation. The original Pl line, but neither of the genotrophs, appears to respond while growing under inducing conditions. The Pl plant phenotype can alter dramatically under such conditions, with a slow-growing individual with small leaves changing its growth rate and form and then growing as if it was in an improved environment (although nothing in the growth environment had been altered) (Fig. 14.4). This type of variation has not been observed with the stable genotrophs. Figure 14.4a and a' show an individual where the primary apical meristem senesced and new growth occurred from an adventitious meristem. This type of outgrowth is not what normally occurs if the primary meristem is removed – sprouting occurs from the basal meristems preferentially. Figure 14.4b and b' also demonstrate the change in growth habit, but in this case, the modification appeared to occur in the primary apical meristem. Here the growth was initially slow with rapid senescence of the leaves. Over a short region,

a   a'   b   b'

Fig. 14.4 Growth of Pl under low nutrient conditions. The plants grew slowly with small leaves initially. Two different responses are shown. a and a' show new growth from an adventitious meristem. b and b' show the change from small rapidly senescing leaves to a vigorous shoot

the form changed with new leaves now appearing and having a persistence normally observed for plants in good growing environments.

An associated question is why should such a system exist? Flax is close to being completely inbred and therefore homozygous at all loci. Therefore, if there is any selection by the environment for particular combinations of alleles, then the population is likely to become homogeneous with little variation. Under these conditions any changes in the environment cannot be accommodated within the existing germplasm. Therefore, a mechanism by which nonlethal variability can arise in response to growth challenges would have an advantage. The mechanism is not one in which a specific gene could be targeted in response to a particular environmental stress. The process would be one in which regions of the genome can be modified resulting in potential phenotypic variability and the new combinations tested by the environment. When an improved growth was conditioned by a subset of the genomic modifications, then the cells that have this combination would be able to grow at an increased rate. If the genomic modifications took place within the apical meristem, then cells with the improved growth potential could take over the meristem and when flowers are subsequently produced transmit this altered genome to the next generation.

If these induced changes are, in fact, an evolutionary mechanism, then the L and S genotrophs are essentially dead ends since they have lost the ability to respond to subsequent changes in their growth environment. However, it is important that the focus on these two lines has been because of this stability – most environments do not result in stable, heritable lines but are capable of repeated responses. The ability to respond would result in the correlation between phenotype of a particular generation being associated with the environment of the immediately preceding generation, a phenomenon of maternal carryover effects. The combination of induced phenotypic/genomic variation coupled with, in rare cases, the loss of the ability to subsequently respond was essential for this phenomenon to be investigated, as a simple maternal carryover set of observations would not have been particularly novel. As noted earlier, the potential for appearance of such variation was heralded by the many observations on the instability of the fiber flax crop in spite of its normally selfed, inbred characteristics.

References

Aiton W (1834) On the cultivation of flax in Holland and Scotland. Q J Agric 4:159–183

Cullis CA (1976) Environmentally induced changes in ribosomal RNA cistron number in flax. Heredity 36:73–80

Cullis CA (1979) Quantitative variation in the ribosomal RNA genes in flax genotrophs. Heredity 42(2):237–246

Cullis CA, Charlton L (1981) The induction of ribosomal DNA changes in flax. Plant Sci Lett 20:213–217

Cullis CA, Kolodynska K (1975) Variation in the isozymes of flax (Linum usitatissimum) genotrophs. Biochem Genet 13:687–697

Durrant, A. 1962. Induction, reversion and epitrophism of flax genotrophs. Nature 196:1302–1304

Durrant A, Nicholas DB (1970) An unstable gene in flax. Heredity 25:513–527

Evans GM (1968a) Nuclear changes in flax. Heredity 23:25–38

Evans GM (1968b) Induced chromosomal changes in Linum. Heredity 23:301–310

Evans GM, Durrant A, Rees H (1966) Associated nuclear changes in the induction of flax genotrophs. Nature 212:697–699

Fieldes MA, Deal CL, Tyson H (1977) Preliminary characterization of peroxidase isozymes isolated from two flax genotrophs. Can J Bot 55:1464–1473

Timmis JN, Ingle J (1973) Environmentally induced changes in rRNA gene redundancy. Nat New Biol 244:235–236

Tyson H, Taylor SA, Fieldes MA (1978) Segregation of the environmentally induced relative mobility shifts in flax genotroph peroxidase isozymes. Heredity 40:281–290

Tyson H, Fieldes MA, Starobin J (1986) Genetic control of acid phosphatase R_m and its relation to control of peroxidase R_m in flax (Linum) genotrophs. Biochem Genet 24:369–383

Chapter 15
Flax Genome "Edits" in Response to the Growth Environment

Christopher A. Cullis and Margaret A. Cullis

15.1 Introduction

Particular varieties of flax have been shown to respond to certain growth environments resulting in heritable phenotypic variations (Durrant 1962; Cullis 1977, 2005). The stable lines derived from the original fiber flax variety Stormont cirus were initially termed genotrophs (Durrant 1962). These differences in form of the genotrophs were also accompanied by variation in the nuclear DNA content (Evans et al. 1966; Evans 1968), by quantitative variation in rDNA copy number (Cullis 1976), and by the appearance of insertions, deletions, and multiple single-nucleotide polymorphisms in various sequence families (Cullis 2017). The development of next-generation sequencing technologies along with the assembly of the flax genome (Wang et al. 2012) has permitted the larger-scale comparison of these lines to characterize the regions of the genome that had been modified. The Illumina sequencing of the fiber variety Stormont cirrus (Pl), and many related lines (genotrophs), has allowed the identification of sites of rapid genomic variation. Eleven DNA samples have been sequenced to varying levels of multiplicity, and the genomes have been aligned to the flax reference genome derived from the oilseed flax variety Bethune. Therefore, only differences that are part of the Bethune genome have been characterized but not the variation specific to the Stormont cirus genome.

The whole genome sequence data has mainly been derived from individuals derived from the original Stormont cirus variety (Fig. 15.1).

Only the sample from the *Agrobacterium*-transformed callus culture had not passed through at least a single meiotic generation. Therefore, during development the nuclear DNA changes occurred had to be determined by following specific loci

C. A. Cullis (✉) · M. A. Cullis
Department of Biology, Case Western Reserve University, Cleveland, OH, USA
e-mail: cac5@case.edu

© Springer Nature Switzerland AG 2019
C. A. Cullis (ed.), *Genetics and Genomics of Linum*, Plant Genetics and
Genomics: Crops and Models 23, https://doi.org/10.1007/978-3-030-23964-0_15

Flax Genotrophs Family Tree

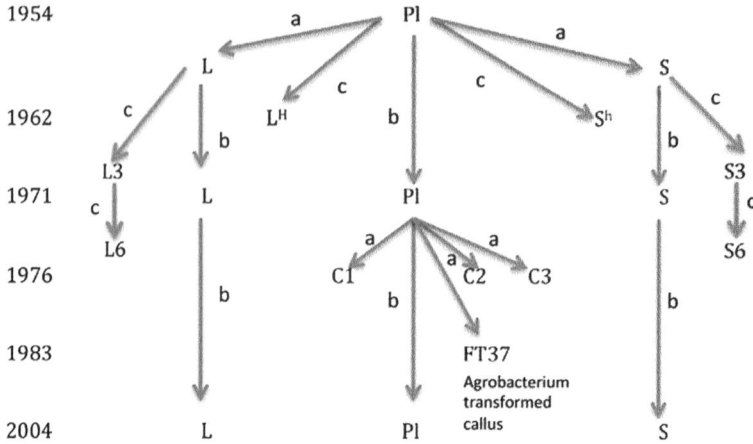

a – grown under nutrient inducing environment
b – grown under non-inducing environment
c – grown under different temperature regime

Fig. 15.1 The generational relationships between Stormont cirrus (Pl) and the derived lines that have had whole genome sequence data (highlighted)

during the growth of plants under conditions that were conducive for the changes to occur (termed inducing environments).

Not all flax varieties respond to the growth environment with heritable differences transmitted to subsequent generations. The control of the ability to respond to the growth environment has also been of interest. It was shown to be genetically controlled (Durrant and Timmis 1973) and the characterization of the responsiveness of F2 populations from a cross between a responsive and a stable line provided a platform for determining the possible genes controlling this property (Bickel et al. 2012).

15.2 Nuclear DNA Variation

The nuclear DNA variation associated with the heritable differences has been characterized both as a whole and for specific individual sequences. The initial variation was determined by Feulgen cytophotometry (Evans et al. 1966). Subsequently data using renaturation and hybridization (Cullis 1973, 1983; Cullis and Cleary 1986) has indicated that the nuclear DNA differences occurred in subsets of the genome. The whole genome sequencing of the individual lines indicated in Fig. 15.1 has enabled the identification of the sequences, and sequence families, involved in these rapid changes genome-wide. An initial, important check on the consistency of the

Table 15.1 The relative rDNA gene content of various lines as determined from hybridization or from relative abundance in next-generation sequencing data sets

Source of DNA	Relative rDNA amount by hybridization	Relative rDNA amount from whole genome sequence alignment
Pl	1.0	1.0
C1	0.8	0.78
C3	0.4	0.47
S	0.4	0.46
FT37	0.2	0.2

data among the many diverse experimental systems was made by parsing the aligned fragments and identifying regions where the difference in the number of reads across 1-kb blocks was greater than, or less than, the average coverage of a single copy region across the whole genome. Summing this data across the genome gave estimates of total variation between the lines that were consistent with the earlier data using estimated nuclear DNA contents. An alternative comparison of the data from different techniques, namely, from hybridization (Cullis 1976, 1979) and the number of sequence reads, was also available for the ribosomal RNA genes (rDNA) and found to be consistent between the two sets of data shown as shown in Table 15.1.

15.3 Comparisons of Pl and the Genotroph Genomes Through Whole Genome Sequencing

The lowering of the cost of whole genome sequencing coupled with the availability of an assembled flax genome has facilitated whole genome comparisons among the genotrophs. Sequencing data (from the Illumina platform as 100b paired-end reads) were aligned to the Bethune genome using Bowtie using the CyVerse bioinformatics environment. The alignments were characterized both by identifying single-nucleotide polymorphisms (SNPs) through the mpileup pipeline and through visualization using the Integrated Genome Viewer (http://software.broadinstitute.org/software/igv/). The differences identified between the genotrophs covered the spectrum of those observed between different accessions and included SNPs, insertions, and deletions. The nature of the analysis did not permit the identification of larger-scale chromosomal rearrangements.

The polymorphisms identified generally appear to fall into two classes – a genomic region identical to the reference genome Bethune or an alternative "locus." For those regions that have been characterized within the wild progenitor of flax, *Linum bienne*, both forms have been identified. The classification of the variable regions in the induced lines tended to identify two groups, one of which contained C_1, C_2, and S^h and the other S, C_3, Ft37, and S_6. The genomic organization within these two groups was conserved but different between the groups. It is important to stress that these data are again consistent with proposal that the variation is limited to specific regions of the genome.

15.4 Deletions in Genotroph Genomes

The simplest modifications to understand are those in which there is a deletion in one of the genotrophs compared with the original plastic line. Such variants could be due to the loss of a transposable element. Such a mobilization of transposons in stress environments has been observed in other systems. A region where this has occurred is shown in Fig. 15.2. Here a region where Pl, C1, C2, and Bethune are identical are shown, while C3, S, and FT37 all have a deletion in this region of 758bp. Since this is an alignment of reads across the genome, a movement of this region elsewhere in the genome would still be visible in the alignment. The lack of any reads across this region indicates that the sequences are missing from this genome. The structure at this point of the genome can be confirmed by amplifying across the region by the polymerase chain reaction (PCR) using a pair of primers within the conserved sequence flanking the deletion. Such a confirmation of the structure of the deletion is shown in Fig. 15.2a with the expected shorter fragment

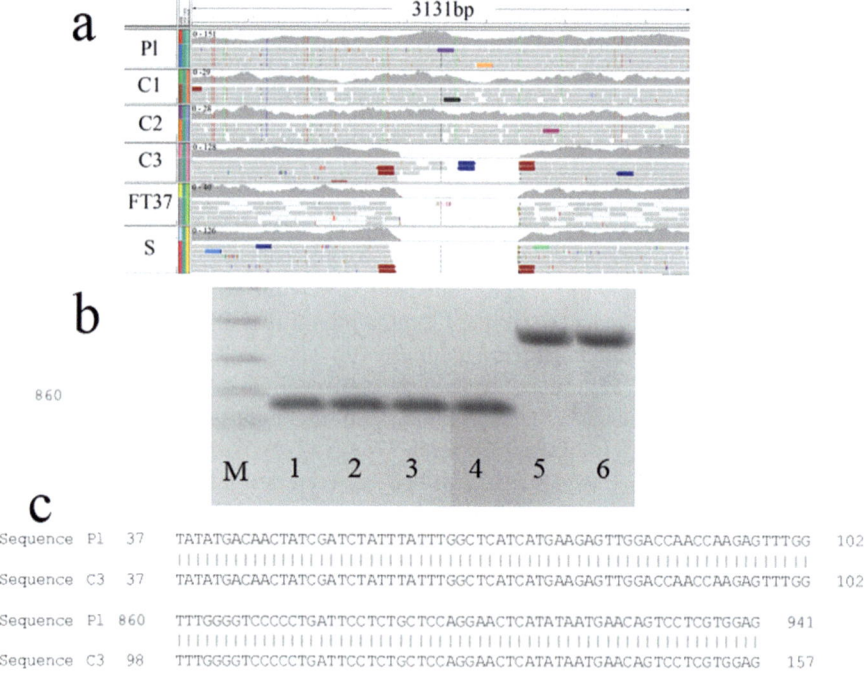

Fig. 15.2 (**a**) IGV trace for deletion in C3, FT 37, and S, while it is identical for Pl, C1, and C2. (**b**) The PCR amplification across the region using template DNAs extracted from Lane 1, S; Lane 2, C3; Lane 3, L6; Lane 4, FT37; Lane 5, Pl; and Lane 6, Bethune. Lane M is Bioline molecular weight marker I. (**c**) The long and short amplified fragments were sequenced and aligned. The gap in the alignment between bases 102 and 860 corresponds to the expected size difference from the whole genome shotgun alignment. Note the duplication of the sequence TTTGG in the long form flanking the deletion

being amplified from those individuals for which the alignment indicated no reads. Both length amplified fragments have been sequenced to confirm the structure at this point of the genome (Fig. 15.2b).

What are the ends of the deletion? The alignment of the sequencing of the PCR products (Fig. 15.2c) shows that the position of the deletion occurs at a short tandem repeat of TTTGG, there being two copies of this sequence in the long form but only one in the short sequence. The flanking of such deletions by direct repeats is common but not universal among such deletions. In addition, there is neither a consensus nucleotide sequence nor a standard length for these repeats.

The region deleted in this example is in fact part of a gene annotated in the assembly and gene finding compendium. One of the expected controls for such activities would be transposable elements which are known to be mobilized under stress. Although the sequence removed in this example does not have any similarity to known transposons, the direct repeat (TTTGG) flanking the deleted region could have been the target for a transposase. Since transposases normally have a restricted target site for excision, the plethora of sequences flanking such excisions would necessitate a suite of transposases all acting within the organism or a transposase that had a very wide range of target sequences. With these caveats, it is possible to conclude that the deletions occurring in response to the growth environment could be mediated by the increased activity of a series of transposases. Since the direct repeats are short, the access of such transposases must also be restricted so as not to completely destroy the genome.

15.5   New Insertions in the Genotroph Genomes

The flip side of deletions are insertions. Clearly one possible source would be the insertion of a transposon into a new site. If this was a replicative insertion, then a duplication would have occurred, which might be detected in the whole genome sequence data set by the increase in the number of reads of that transposon but not at the site of the insertion. The alignment of the sequence reads would show a dip as the reads no longer matched the reference sequence moving into the transposon sequence, followed by a rise as the reads exited the new insertion, but there would not be any large gap in the coverage. The sequence generating such a feature would need to be identified by amplifying across the affected region, identifying the resulting product as longer than expected from the reference sequence, and then sequencing the product to ascertain the nature of the insertion.

However, the more common variants identified have been where the Pl genome has a gap in the aligned sequence, but the genotroph sequence has a form that is identical to that of the reference sequence. Such an example is shown in Fig. 15.3, where the alignment for the reads from Pl are mostly missing, as is the case for C1 and C2, but the coverage for the genotrophs C3, S, and the callus line FT37 are all identical to each other and the Bethune reference genome at the expected read count for a single copy region. Since there are few reads within this region, it appears that

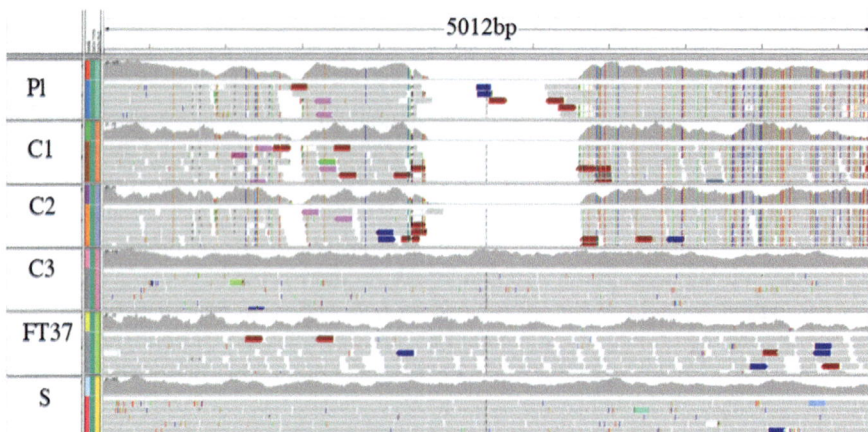

Fig. 15.3 An insertion event in the genotrophs C3, S, and the callus line FT37 that contains sequence absent from Pl. Note also the SNPs surrounding this insertion that may indicate that the region modified at this site is substantially larger than the simple insertion region

Pl (the progenitor line, see Fig. 15.1) does not contain these nucleotide sequences. Note that the differences between the two groups is not confined to the deletion event but extends on either side of it with multiple SNP. To ensure that this is not an issue with the sequence data, PCR amplification across this region, with primers designed within the conserved regions, has confirmed that the structure observed in the IGV visualization is, in fact, correct. Some of the regions that are of this nature can be up to 20kb in length.

The best described instance of this situation is that for the element LIS-1 (Chen et al. 2005, 2009; Cullis 2005). This insertion event has been shown to occur reproducibly at a specific single copy site in a number of the genotrophs, again with a short direct repeat of three base pairs (TCC) at the site of insertion. The original progenitor line was homozygous for the un-inserted site, while where LIS-1 was observed, the individuals were homozygous for the insertion. The alignment of the whole genome sequence reads from PL indicated that there were regions of LIS-1 that were present in Pl but neither in assembled form nor covering the whole of the LIS-1 sequence. This sequence was not present in the reference sequence, but the presence of LIS-1 has also been shown in other flax and linseed varieties. In particular, screening of a set of *Linum bienne* collected from Turkey identified two examples of the complete insertion of LIS-1, while the other individuals did not have LIS-1. Therefore, this sequence does appear to be part of the *Linum* pan genome.

For all of these instances of a region of the genome present in the genotrophs but absent from the original line, the question of their origin is seminal. The explanation of the deletions is where the activity of transposable elements or their transposases might be invoked for the structural variants. In the case of the insertions, the "new" sequence does not appear to be present intact within the progenitor line. However, as with LIS-1, the sequence does appear to be part of the larger repertoire of the *Linum* genome, being present intact in other varieties or in wild relatives. Of note here is that the region appears to be one of an alternative pair of structures at these genomic positions.

15.6 Regions of High SNP Polymorphisms

The structural variations between the Pl and genotroph genomes described above have included regions where the genotrophs are different from Pl but the same as Bethune and where the genotrophs are different from PL which is identical to Bethune. Another set of differences between the genotrophs and Pl is that where SNPs exist between the two groups. The SNP variants to be considered here are not confined to single-base variants, but regions where there are many differences between the lines over a relatively short stretch of the genome. The duality of structural variants is also observed in the distribution of SNPs between these lines, namely, there are regions where the genotrophs are different from Pl but have the same sequence as that in Bethune (Fig. 15.5a), while in other sites, Pl and Bethune are identical but some of the genotrophs are very different (Fig. 15.5b). The distribution of SNPs is not uniform across the genome, but as with the SNPs between Bethune and Pl, they are clustered. The distribution of SNPs common to the C3 and S genotrophs across chromosome 3 is shown in Fig. 15.6. Here are two regions, one in which Pl and the genotrophs are identical and very similar to the Bethune reference genome. The other region is where Pl is very similar to Bethune where S and C3 are identical but very different from Pl.

15.7 Integrity of the Flax Genome

The foregoing discussion has focused on the variation between Pl and the genotrophs. This focus can lead to the impression that the genome is unstable and that most of it is malleable. This is not generally the case. As noted for the SNP distribution in Fig. 15.4, the majority of the genome is identical between Pl and the genotrophs, with the region of variation being very localized.

What is the nature of the variants, be they insertions, deletions, or SNPs? The regions of the genome that vary include all those components expected; some have similarity to components of transposons, some include genes, and others have no identified function as determined from similarity searches for either sequence or protein homologies.

One of the gene families that might have been expected to vary considering the rapid evolution of these families is the set coding for disease resistance genes. The resistance gene analogs (RGAs) have been mapped on to the Bethune genome (You

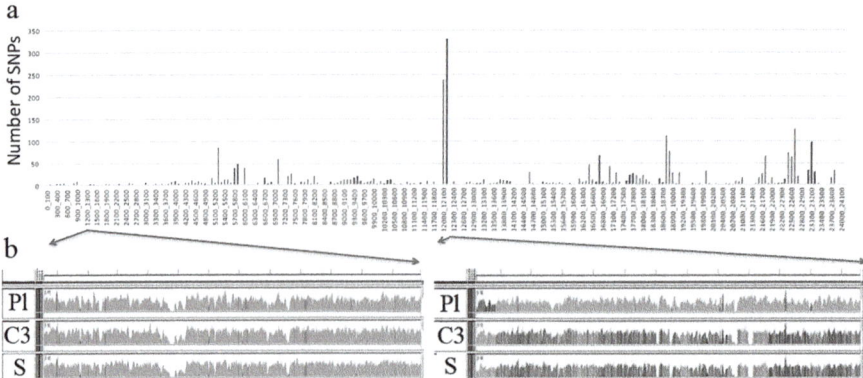

Fig. 15.4 (**a**) Distribution of SNPs that are common to S and C3 but different from Pl. (**b**) Expanded view of regions of low and high variability

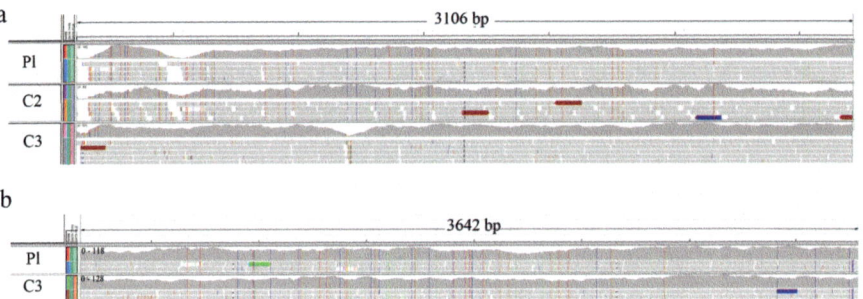

Fig. 15.5 Alignments across flax rust resistance genes (**a**) Alignment of PL, C2, and C3 across the flax N2 gene where Pl and C2 are identical but different from Bethune, while the region in C3 is very similar to that in Bethune. (**b**) Alignment of Pl and C3 across the flax N1 gene where these are identical but different from the equivalent region in Bethune

et al. 2018). Most of these genes have remained identical in Pl and the genotrophs, but some are variant (Fig. 15.5a, b).

15.8 Timing of Genomic Changes

The specific genomic variants described have been associated with inbred lines derived from the original Stormont cirus variety. As noted elsewhere (Cullis and Cullis 2019), possible trivial explanations for all of these observations would be that either the original Stormont cirus variety consisted of mixed seeds that were segregating or that the original seed mixture was heterozygous for all of these loci. The data on the nuclear DNA variation of responsive plants growing under inducing conditions and the variation in ribosomal RNA gene number both indicate that the variation takes place during the vegetative growth under inducing conditions (Cullis and Charlton 1981). This data has been supplemented using the variable genomic loci identified through the genome comparisons of Pl and the genotrophs.

15.8.1 Appearance of Genomic Variants During Growth of Pl Under Inducing Conditions

The sequence named Linum Insertion Sequence-1 (LIS-1) was identified in the S genotrophs at a specific site (Cullis 2005; Chen et al. 2005, 2009). The two ends of the insertion sequence and the un-inserted site provided markers to follow the appearance of this genomic fragment while Pl was growing under inducing conditions. Total DNA was extracted from leaves from various regions along the stem of Pl plants grown under inducing conditions. These DNAs were characterized as to the presence or absence of LIS-1. It was shown that this element was present in leaves at later stages of development but absent from those produced earlier (Fig. 15.6). Using primers that delineate many of the variants identified between Pl and the genotrophs has confirmed the pattern but also shown that each of these variants may be generated independently of any of the others. Thus, for the same leaf DNAs showing variation for LIS-1, both positions in this plant had become homozygous for a deletion (Fig. 15.6e, f) compared with the Pl parent. In both of these cases, and most frequently, the modified site became homozygous during vegetative growth, consistent with these genomic variants being homozygous in subsequent generations. These two examples also illustrate the independence of each of the genomic changes that are occurring during growth under an inducing environment, with one, in this case, occurring earlier in development than the other.

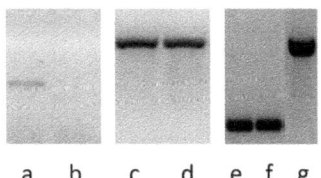

a b c d e f g

Fig. 15.6 Amplifications using DNA from leaves from the middle third and from the upper third of the stem of a Pl plant grown under inducing conditions (**a, c**) Amplification across the uninserted site, **a**, or the left border c, of LIS-1 using DNA from leaves from the middle third of the stem. (**b, d**) Amplification across the uninserted site, **b**, or the left border d, of LIS-1, using DNA from leaves from the upper third of the stem. (**e–g**) Amplifications across a region deleted in genotrophs compared to Pl. (**e**) Amplification using DNA from leaves from the middle third of the stem. (**f**) Amplification using DNA from leaves from the upper third of the stem. (**g**) Amplification using DNA from leaves from Pl growing under non-inducing conditions

The result of growing Pl in an inducing environment is a chimeric plant, namely, one in which the cells in older leaves can have different genomic constitutions from those of younger leaves. However, the growth of the plant does not change this leaf genomic constitution, that is, the genomic variants are not being generated in the leaf cells but in the apical meristem. Since the subsequent progeny are homozygous for the variant loci, the cells in the apical meristem that give rise to the gametes must have become homozygous for the specific variations. This is also seen for the uppermost leaves which have become homozygous for both the variants (Fig. 15.6).

These data clearly demonstrate that the genome is being modified during the vegetative growth prior to flowering. The modifications of the variant loci do not appear to be coordinated in terms of their timing during growth, as demonstrated in Fig. 15.6, where the insertion of LIS-1 is not complete in the first sample shown (the middle leaves), while for the deletion, it is already homozygous at this stage. However, it is possible that a hierarchy in terms of how labile a locus may be, that is, some sites always become modified after a shorter growth period under inducing conditions than others. An important question arising from these observations is the characteristics of those genomic locations that do become modified – what is being recognized to include these regions in whatever the mechanism by which the genomic changes occur?

15.9 Genetic Control of the Ability to Respond

Not all genotypes of flax respond to the growth environment by modifying their genome. However, there is genetic control of this characteristic (Durrant and Timmis 1973; Bickel et al. 2012). The control does not appear to be a single gene characteristic, and genetic mapping combined with whole genome sequencing should be able to identify the loci, and ultimately the genes, associated with the ability to modify the genome. An issue here is that Bethune, for which the reference sequence was developed, is one of the nonresponding varieties. Therefore, it may be necessary to assemble the Pl genome to identify those regions, present in Pl but absent from Bethune, which are responsible for the genome restructuring.

15.10 Mechanisms Underpinning the Flax Genome Restructuring

Any proposed mechanism for the genomic rearrangements needs to be consistent with the observations on the modified regions of the genome:

1. These variants are not usually single-base changes but can extend over thousands of base pairs.
2. The variations observed include one or more of the following:

    (a) Deletions (Fig. 15.2)
    (b) Insertions of new DNA sequence (Fig. 15.3)

(c) The presence of a high frequency of SNPs (Fig. 15.4)

3. Although each of the variants has a specific site within the genome, during the induction period, it is often possible to identify both forms.
4. Most variants become homozygous in the somatic generation.
5. In some cases, following induction, the progeny can remain stable when challenged with some, but not all growth environments, that is, some genomic plasticity can still remain in the "stable" genotrophs (Durrant and Jones 1971).

The new genomic arrangements that can be explained by previously identified mechanisms are those in which the genotrophs have a deletion compared to the original Pl line (Fig. 15.2). These frequently have a short direct repeat flanking the deleted region and therefore could have been generated by the activity of a transposase excising the region and not having it reinserted elsewhere in the genome. Such activity has been observed in plant cells, especially in those going through rounds of tissue culture.

Perhaps the more difficult question to answer with regard to these genomic changes is how the sites that are missing from Pl, either as a gap in the genomic alignments or due the high frequency of SNPS in the derived sequence, are reconstituted in the genotrophs in response to growth in stress environments? Where there is a gap in the alignment of whole genome reads from Pl, that is, present as a sequence in the genotrophs, the sequence present has usually been identical to that in the reference genome as shown in Fig. 15.3. Since this is an alignment of an approximately 100X sequence coverage, if the sequence was present, it should be aligned. As can be seen in Fig. 15.3, there are a small number of reads aligning from the Pl sequencing but much less than expected for a single copy sequence. However, these reads are not within this gap in Pl as the amplification of the region by PCR generally results in fragments of the expected size based on the alignments. In all the cases of this type of variant, the sequence present in the genotrophs is the same or very similar to that found in Bethune. Since the Bethune sequence is the basis of the reference sequence, this observation might simply be a consequence of only finding such variants due to the baseline used.

The third type of variation observed is where the region is covered by sequence, but there are many SNPs between the original Pl line and the genotrophs (Fig. 15.4). Again, two forms are apparent, one in which the sequence present in Pl is very different from that in Bethune, but the genotrophs that are variant have the same (or very similar) sequence to Bethune. The inverse of this is also observed, where Pl and Bethune are the same, but the genotroph site has many SNPs.

15.11 Proposed Model for the Generation of the Genomic Variation

The model is shown in Fig. 15.7. Since the genomic variants can cover many thousands of base pairs, it is difficult to envisage this level of variation occurring one base at a time. Also, identical variants have arisen repeatedly, so the mechanism needs to be one which can result in the same variant repeatedly. Therefore, it is

Pl Chromosome(s)

Extrachromosomal DNA fragments (via RNA intermediates?)

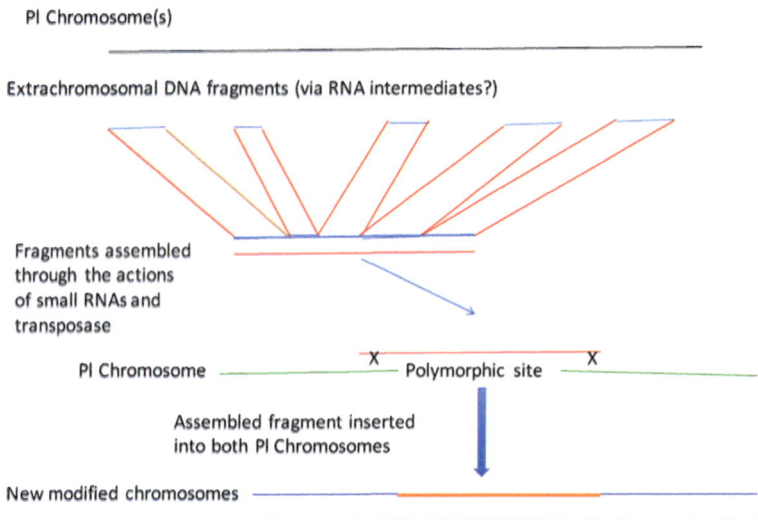

Fragments assembled
through the actions
of small RNAs and
transposase

Pl Chromosome ————————X———— Polymorphic site ——X——————————

Assembled fragment inserted
into both Pl Chromosomes

New modified chromosomes ——————————————————————————

Fig. 15.7 Model for the generation of genomic variants in flax in response to the growth environment

proposed that the alternative sequence is built extrachromosomally by assembling the regions of the new fragment. Once the alternative form has been assembled, it can then replace the extant sequence currently occupying that position on the chromosome. Since there does not appear to be any genomic rearrangement occurring in differentiated tissue, these events are proposed to occur in the apical meristem. Once a suite of variants has been assembled that result in an improved performance, those modified meristematic cells can grow out, take over the meristem, and result in an altered genome.

This model is consistent with all the observations described here. The presence of both variants at some stage of the growth would account for both forms being detectable during the induction of genomic changes. The replacement of a chromosomal region as a single event can explain the long-range multiple variants that have been observed. The replacement function would also explain the localization of the variants, rather than the process resulting in a complete scrambling of the genome. Finally, the replacement function would also result in the cells becoming homozygous (unless only a single assembled molecule was developed). The targeting of a chromosomal region would result in both homologues being modified, unless there was a mechanism to maintain heterozygosity. The failure to identify heterozygotes in the progeny of these induced individuals would support this replacement process.

The mechanism by which the new molecules are constructed and the template for their synthesis is still obscure. Here it is proposed that the assembly is mediated through the actions of transposases and small RNA molecules making it perhaps analogous to a eukaryotic genome-editing mechanism. The enzymes involved and the possible intermediates still need to be identified.

15.12 Summary

The response of certain fiber flax varieties to stress environments results in reproducible changes in the genome and apparent adaptations to the stress. These changes can be categorized as flipping between alternative states, with both alternatives also being observed in the wild progenitor of flax. Not all the possible genomic changes always occur; the suite of variants occurring is dependent on the environmental stimulus. The regions affected can contain genic sequences although many of them are not annotated as genes but are also frequently similar to sequences found in transposons. These observations are consistent with the hypothesis that the genome is restructured at a set of labile regions that can result in phenotypic variety without significant lethality. The genomic rearrangements occur in the apical meristem, and the environment then acts as a selective agent to discriminate among the new genome structures resulting in an alteration of the genome of the meristem. Those that confer growth advantage are then propagated in that (and possibly) subsequent generation.

References

Bickel CL, Lukacs M, Cullis CA (2012) The loci controlling plasticity in flax. Res Rep Biol 3:1–11
Chen Y, Schneeberger RG, Cullis CA (2005) A site-specific insertion sequence in flax genotrophs induced by environment. New Phytol 167:171–180
Chen Y, Lowenfeld R, Cullis CA (2009) An environmentally induced adaptive (?) insertion event in flax. J Genet Mol Biol 1:38–47
Cullis CA (1973) DNA differences between flax genotypes. Nature 243:515–516
Cullis CA (1976) Environmentally induced changes in ribosomal RNA cistron number in flax. Heredity 36:73–80
Cullis CA (1977) Molecular aspects of the environmental induction of heritable changes in flax. Heredity 38:129–154
Cullis CA (1979) Quantitative variation in the ribosomal RNA genes in flax genotrophs. Heredity 42(2):237–246
Cullis CA (1983) Environmentally induced DNA changes in plants. CRC Crit Rev Plant Sci 1:117–129
Cullis CA (2005) Mechanisms and control of rapid genomic changes in flax. Ann Bot 95:201–206
Cullis CA (2017) Mechanisms of induced inheritable genome variation in flax. Chapter 4. In: Li X-Q (ed) Somatic genome variation in animals, plants and microorganisms. Wiley, New York; 2016, pp 77–90. ISBN: 978-1-118-64706-6
Cullis CA, Charlton L (1981) The induction of ribosomal DNA changes in flax. Plant Sci Lett 20:213–217
Cullis CA, Cleary W (1986) Rapidly varying DNA sequences in flax. Can J Genet Cytol 28:252–259
Cullis CA, Cullis MA (2019) Origin of the flax genotrophs. In: Cullis CA (ed) Linum genetics and genomics. (in press)
Durrant A (1962) Induction, reversion and epitrophism of flax genotrophs. Nature 196:1302–1304
Durrant A, Jones TWA (1971) Reversion of induced changes in amount of nuclear DNA in Linum. Heredity 27:431–439
Durrant A, Timmis JN (1973) Genetic control of environmentally induced changes in Linum. Heredity 30:369–379

Evans GM (1968) Nuclear changes in flax. Heredity 23:25–38

Evans GM, Durrant A, Rees H (1966) Associated nuclear changes in the induction of flax genotrophs. Nature 212:697–699

Wang ZW, Hobson N, Galindo L, Zhu SL, Shi DH, McDill J, Yang LF, Hawkins S, Neutelings G, Datla R, Lambert G, Galbraith DW, Grassa CJ, Geraldes A, Cronk QC, Cullis C, Dash PK, Kumar PA, Cloutier S, Sharpe AG, Wong GKS, Wang J, Deyholos MK (2012) The genome of flax (Linum usitatissimum) assembled de novo from short shotgun sequence reads. Plant J 72(3):461–473. Article first published online: 14 Aug 2012. https://doi.org/10.1111/j.1365-313X.2012.05093.x

You FM, Xiao J, Li P, Yao Z, Jia G, He L, Zhu T, Luo MC, Wang X, Deyholos MK, Cloutier S (2018) Chromosome-scale pseudomolecules refined by optical, physical and genetic maps in flax. Plant J 95(2):371–384. https://doi.org/10.1111/tpj.13944

Chapter 16
Transgenic Flax and the Triffid Affair

Camille D. Ryan and Stuart J. Smyth

16.1 Introduction

Common flax (*Linum usitatissimum* L.) was one of the first crops cultivated by man used in food and linen cloth production in Europe and Asia, dating back to 5000–8000 BC (Berglund 2002). Today, there are two types of flax produced, seed flax for oil and fiber flax for the stem fiber. Fiber flax has an average annual global production of 500,000 hectares and is predominantly grown in China, the Russian Federation, and in Western Europe. Approximately 3.5 million hectares of oilseed cultivars (also referred to as linseed) are grown in Canada, India, China, the United States (US), and Argentina. Flax is used in the production of various industrial products including linen, fiber composites, paints, inks, and linoleum. Flax has grown in popularity as a nutritional supplement for its value as an essential omega-3 fatty acid for both human and animal consumption with the introduction of low linolenic acid flax varieties over the past several years (Galushko and Ryan 2012). It is often consumed raw or is used in whole-grain products such as cereals and breads. In 2014, Health Canada approved a health claim that linked consumption of ground whole flaxseed with lowered blood cholesterol levels, a major risk factor for heart disease (Health Canada 2014).

Other than a few brief intervals in the 1960s and 1970s, Canada has been the global leader in flax production, followed by the United States, China, and India.

C. D. Ryan
Bayer CropScience, St. Louis, MO, USA

S. J. Smyth (✉)
Department of Agriculture and Resource Economics, University of Saskatchewan, Saskatchewan, Canada
e-mail: stuart.smyth@usask.ca

© Springer Nature Switzerland AG 2019
C. A. Cullis (ed.), *Genetics and Genomics of Linum*, Plant Genetics and Genomics: Crops and Models 23, https://doi.org/10.1007/978-3-030-23964-0_16

Flax production in Canada ranged from a low of 500,000 tons to nearly 1.1 million tons in the first decade of this century, with an annual average of just over 700,000 tons between 2000 and 2015. Only a small portion of this is consumed domestically, averaging about 20% per year, with the export value averaging over C$200 million. Canada's export markets include the United States, China, Japan, Australia, Mexico, the United Kingdom, and Brazil. However, by far, Canada's largest export market for flax is Europe. On average 75% of Canada's production is exported to Europe each year, although in peak years, it has risen to well over 80%.

Given Canada's dominance as a global flax producer, private and public investments have been made into developing cultivars that enhance both productivity and oilseed quality. Transformation processes used in development include conventional breeding techniques as well as mutagenesis. Triffid flax was a flax variety that was genetically engineered for resistance to soil residues of sulfonylurea-based herbicides. Developed by the Crop Development Centre at the University of Saskatchewan over the period of the mid-1980s to the mid-1990s, the new variety was designed to provide growers with an alternative to continuous cropping of wheat and other cereals. It also provided a profitable oilseed option to the millions of hectares of canola production. This chapter tracks the history of Triffid, from its early development to its deregistration in 2001, to Triffid's discovery in the EU food supply chain in 2009 and the ongoing trade problems that have since ensued.

16.2 Background of Genetically Modified Flax to 2001

Researchers at the Crop Development Centre (CDC) at the University of Saskatchewan conducted research on several varieties of transgenic flax and after several years of field trials, selected variety FP967 to submit for registration to the Plant Biotechnology Office[1] (PBO) of Agriculture Canada in February 1994. This variety is more commonly known by its marketing brand name of CDC Triffid. In May 1996, the PBO gave partial approval to CDC Triffid. It was determined to be substantially equivalent to conventional flax varieties in terms of specific use and safety for the environment and for health. However, at that point in time, CDC Triffid only received approval for animal feed use. In a Canadian split-run decision, this new flax variety received approval for animal feed use only in 1996 (McHughen et al. 1997) and subsequently entered a seed multiplication program. Full variety release, including Health Canada's approval for Triffid to be marketed for human consumption, was not granted until 1998.

The implementation of the 1998 EU moratorium on GM crops and foods presented a formidable obstacle for the Canadian flax industry. Approximately three-quarters of the flax production in Canada at the time was exported to Europe, and

---

[1] The PBO was the forerunner to the Canadian Food Inspection Agency, which was the independent regulatory agency created when the PBO was removed from being an office within the Department of Agriculture and Agri-Food Canada in 1997.

the commercialization of a GM flax variety alarmed the European importing firms. While the canola industry had an effective identity preservation program in place to manage GM varieties and continued to supply the European and Japanese markets (1995–97) (Smyth and Phillips 2001), the flax industry did not have this option because of the dominant role of the European market. European importers were adamant that they would reject flax imports from Canada if GM flax was grown commercially.

Varietal registration (which allowed for seed multiplication but not commercial production) was granted in May 1996, and pedigreed seed production was initiated. Seed multiplication continued in 1997, and by the end of the 1997 harvest, there was an estimated 5000 tons of pedigreed CDC Triffid seed in existence in Canada. That year turned out to be the final year for multiplication of GM flaxseed. At that point, existing seed stocks of CDC Triffid were identified and contained in separate grain bins in compliance with pedigreed seed production regulations.

By late 1997 to early 1998, it became apparent to the Canadian flax industry that the proposed (at that time) EU moratorium would proceed. Discussions about how to handle the situation were initiated and led by the Flax Council of Canada (FCC). It was determined that all of the existing contained seed stocks would remain that way until a suitable location could be found to crush the flaxseed. Coordinated by the FCC, a CanAmera Foods crushing plant in Altona, Manitoba, was eventually contracted to crush the flax. The meal was subsequently mixed into livestock feed and fed to Canadian livestock, while the oil from the crush was diverted into industrial applications. All breeder seed stock held by the CDC was incinerated. This effectively removed all breeders' seed from pedigreed seed growers contracted to multiply the seed.

To ensure that CDC Triffid flax would not jeopardize future export markets, the developers of the variety applied to deregister the transgenic flax variety. This process was initiated in 2000 and by 2001 the Canadian Food Inspection Agency (CFIA) had officially deregistered CDC Triffid flax. This meant that it was illegal to produce or distribute this variety anywhere in Canada, but not illegal to grow it.[2] As agricultural history would write it, Triffid became the first transgenic crop technology to be withdrawn from the Canadian market. The Canadian flax industry believed that varietal deregistration in 2001 represented an end to Triffid and any trade issues associated with GM flax. Unfortunately, this was not to be the case.

---

[2] Many crop varieties are deregistered as they are replaced by new varieties. The new varieties have superior yield and disease resistance, and the old varieties become obsolete and are deregistered. In an industry where bin run seed is common and variety comingling can occur, making it illegal to grow a deregistered variety would be a logistical nightmare, so it is only illegal to sell and/or market a deregistered variety as that particular variety. It is not illegal to unintentionally grow a small percentage of a deregistered variety as part of another crop. When this happens, the shipment is downgraded to animal feed and enters that market (at a lower price) rather than the human food market.

16.3   The Re-emergence of GM Flax: 2009–2016

In July 2009, the EU reported that a Canadian shipment of flax had tested positive for the NPTII marker indicating a GM event. At this point, it was assumed that GM canola or some other GM crop had comingled in the shipment. However, by September 2009 the EU's Rapid Alert System for Food and Feed (RASFF) was notified by a German company that its bakery/cereal products had tested positive for Triffid.[3] Notification on the RASFF system is "equivalent to an air siren going on in the EU" (Ryan and Smyth 2012; Ryan and McHughen 2014). It is an incredibly effective communication tool. This notification in September was the first of more than 100 over the next several months that would report Triffid in bakery goods, cereals, and other products made by companies throughout the EU.[4]

With the EU having a zero-tolerance regulation for comingling of unapproved GM events, it meant that Canadian flax exports were in violation of this regulation and that all Canadian flax shipments would be rejected until Canada could guarantee that their flax shipments did not contain even trace amounts of GM flax. The Canadian seed trade industry responded swiftly to the initial notification. Stakeholders such as the Flax Council of Canada, the CFIA, and the Canadian Grain Commission (CGC) mobilized quickly in order to mitigate what threatened to shut market access for Canadian flax producers. With winter approaching and the looming closing of the St. Lawrence Seaway, stakeholders were highly motivated to ensure that markets opened before the winter freeze-up, usually sometime in December. This export option is not available for the three winter months of January, February, and March. Figure 16.1 shows the crop year export figures for Canadian flax to Europe, highlighting the impact of Europe's zero-tolerance policy during the height of Triffid concerns. Canadian flax exports to Europe were down by 51% when compared to the crop year just prior to the detection of GM flax. As is evident, the final quarter of the calendar year is the peak export period for flax to Europe.

The FCC and the CFIA met together in Brussels, Belgium, on October 19, 2009, to work with DG Sanco (Directorate General for Health and Consumer Affairs) and other EU stakeholders. The objective of the meeting was to develop a testing protocol to manage the situation in Canada (European Commission 2009). The protocol was immediately launched in the Canadian market, quickly establishing a system of sampling and testing of flax stores (one test for every 5000 bushels). Samples that tested positive at levels greater than or equal to 0.01% for Triffid[5] would not be accepted for

---

[3] Please refer to the initial (full) notification recorded as RASFF 2009.1171 at: https://webgate. ec.europa.eu/rasff-window/portal/index.cfm?event=notificationDetail&NOTIF_REFERENCE= 2009.1171.

[4] One hundred and eleven (111) notifications were filed with the RASFF between September 8, 2009, and January 18, 2011. A notification dated July 28, 2010, was qualified as a "border rejection" by Finland.

[5] No detection at a 0.01% level, 19 times out of 20.

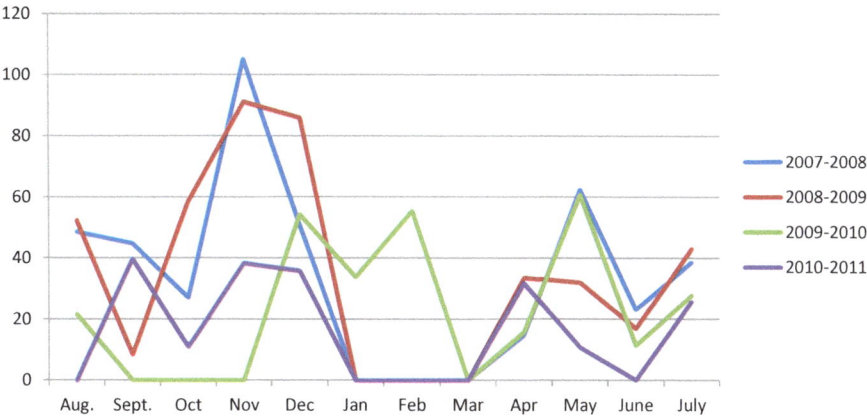

Fig. 16.1 Month over month flax exports to the EU 2007–2011 (000s tons) (Data Source: Canadian Grain Commission database, available at: http://www.grainscanada.gc.ca/statistics-statistiques/ecgwf-egcfb/ecgm-megc-eng.htm. NOTE: Some of the December 2009 exports were able to be shipped through the St. Lawrence Seaway, but much of this month's exports, as well as those of January and February 2010, were either railed to the port of Montreal and exported or exported through the West Coast of Canada, substantially adding to the export costs in Canada)

import into the EU market (FCC 2009). According to James (2010), Canadian flax averaged a failure rate of 20% at EU ports during the last 6 months of 2009.

By early 2010, following much speculation, the source of the Triffid contamination was identified by the Flax Council of Canada. It originated from two of the CDC pedigreed flax varieties – CDC Normandy and CDC Mons (Booker and Lamb 2012; Young et al. 2015).

Speculation did not end there. There were also allegations that some seed growers had held back seed after deregistration in the hopes that EU policy and consumer opinion would change and the market could take advantage of this valuable flax variety. It was widely communicated, although not verifiably proven that the subsequent farm testing revealed that one sample had tested positive for Triffid at a level of 100% purity. This would have been entirely possible as there were no tests capable of accurately testing flax at the time that it was deregistered, for the presence of Triffid at very low levels. Additionally, the widespread low-level presence of Triffid flax across the Canadian growing belt was speculatively due to pollen or gene transfer. The notion of "flow" and spread of flaxseed, however, is much more complex than that. Flax primarily self-pollinates so seed-mediated gene flow is probably a more important factor in terms of gene flow. There are two scenarios at work in seed-mediated gene flow. First, there is *persistence* of the cultivar (where it was previously grown). For growers that did not rotate for 3 years, volunteer flax may have *persisted* in fields, and this could potentially have included small amounts of volunteer Triffid flax. Secondly, there is seed mixing and movement by equipment (seeders, combines, trucks, etc.) that may account for *dispersal* of the GM flaxseed in trace amounts. Flaxseed sticks like glue when it is wet and exhibits static cling properties when it is dry. Both wet and dry, flaxseed will inevitably stick to farm equipment. While great care is taken by the seed growers to prevent this, a few seeds can get stuck in a combine or in seeding equipment. It only takes a handful of Triffid plants, combined with another flax variety, to produce as many as 1000 seeds at harvest. This could quite easily account for some of the widespread low-level presence of Triffid across the flax-growing region in Canada. Seed movement has many pathways, and it is quite likely that all pathways outlined above accounted for the widespread low-level presence of Triffid (Hall 2012).

In the end, the most likely cause of the low level of GM flaxseed comingling was the cross-pollination between different varieties of breeders's seed at the University of Saskatchewan. In the late 1990s, the official variety development standards for the planting distance between like varieties of crops in confined field trials were determined to be 3 feet. This meant that field plots of Triffid would have been grown a mere 3 feet from plots of other non-GM flax varieties during development. As cross-pollination occurred, the breeders's seed varieties in the neighboring plots would thus have a very low level of Triffid present in the breeders's seed of new flax varieties. Again, there were no tests available at this time that could have detected the presence of the GM flax in the breeders's seed of other flax varieties.

Testing was conducted on CDC Normandy and CDC Mons in January of 2010, and both tested positive for Triffid at 0.01%. These two varieties – both essentially deemed obsolete at the time – were withdrawn from the program, seed stores were destroyed, and varieties were deregistered on August 1, 2013. By early March of 2010, after extensive testing, extremely low indications of Triffid contamination were discovered in a low number of other breeder seed samples of four other varieties: CDC Bethune, CDC Sorrel, CDC Sanctuary, and CDC Glas (FP 2300). These varieties tested positive at trace levels well below the 0.01% detection level. Rather than withdrawing or destroying these varieties, they were reconstituted. First, individual plants were tested for the presence of Triffid, and those deemed Triffid-free were used as a seed source for the reconstitution process. The seed increase was then sent to New Zealand to a partnering organization where it was grown under confined conditions where flax has not been grown on the land for at least 5–10 years. Once cultivated, the seed was tested again to ensure that it was transgene-free. When the reconstitution process was completed in 2012, the seed was transported back to Canada for introduction into the foundation seed program and was distributed to seed growers and to seed companies and multiplied.

An event-specific assay has since been developed to detect FP967 T-DNA (Young et al. 2015). As part of the requirements for the EU and in keeping with the contractual obligations of the Protocol, testing of Canadian flax is ongoing. In fact, testing is conducted repeatedly all along the value chain, from farm-held stores of flax to the elevator and at ports where flax shipments await export.

16.4 Economic and Trade Impacts of Triffid Flax

Fully capturing the economic costs of Triffid flax is a difficult process. The social costs from farm gate to plate are even more difficult to pin down. That being said, however, we were able to estimate the costs associated with demurrage and quarantine, the testing costs, and other costs associated with segregation along the value chain. This is outlined in Table 16.1. In total, the costs incurred in Canada from 2009 to 2011 are estimated at almost $30 million.

The EU, too, sustained significant costs all along its value chain, estimated at over $50 million (CDN) (COCERAL/FEDIOL 2010; Babuscio et al. 2016). Together, the initial trade disruption costs on both sides of the Atlantic from the EU detection of GM flax were estimated to have reached C$80 million (Ryan and Smyth 2012; Babuscio et al. 2016). This cumulative figure includes the cargo ship quarantine costs, initial flax testing costs, and lost business opportunities within the European flax industry.

As part of the resolution of the import ban that the EU placed on Canadian flax in September 2009, Canada agreed to enact a Farm Stewardship Program for a 5-year period. The objective of this program was to ensure the Canadian flax industry and international flax import markets that Canadian shipments of flax would be free of GM flax.

In the winter of 2009–2010, protocols were developed, and an industry-wide testing program of flaxseed commenced. This 5-year testing program concluded in February 2014. The costs of this program are presented in Table 16.2. As the stewardship program was part of the Canadian Government's commitment to facilitate the lifting of the EU's import ban (enacted in 2009), funds to cover a portion of the testing costs were made available through Agriculture and Agri-Food Canada.

Table 16.1 Total estimated costs associated with the Triffid event in Canada (2009–2011)

Cost category		Notes	Source
Demurrage/quarantine costs	$12,000,000[a]	As of September 2010	Authors' calculations
Testing costs	$3,900,000[b]	2009–2011	Authors' calculations
Cost of segregation, other costs for breeders, certified seed suppliers, producers, grain companies, AAFC, and SaskFlax	$13,185,217	2009 to 2011	Dayananda 2011
Total estimated costs:	*$29,085,217 CDN*		

Source: Ryan and Smyth (2012)
[a]This cost estimate is calculated as follows: $30,000 (CDN) per day which is equivalent to $1 million (CDN) per month. We conservatively estimate a total of 12 months with this level of costs
[b]Based on the number of tests conducted (26,000) as reported by FCC and assuming a conservative (average) cost per test at $150, we estimate total testing costs (2009 to 2011) at almost $4 million (CDN)

Table 16.2 Canadian flaxseed testing costs ($CAD)

Time period	AAFC	Producers	Industry
Aug 2009–Feb 2014	$1,000,000	$1,000,000	$1,340,000
Mar 2014–Jun 2015	NA	$500,000	Unknown
	$1,000,000	$1,500,000	$1,340,000+

Source: Booker, Lamb, and Smyth (2015)

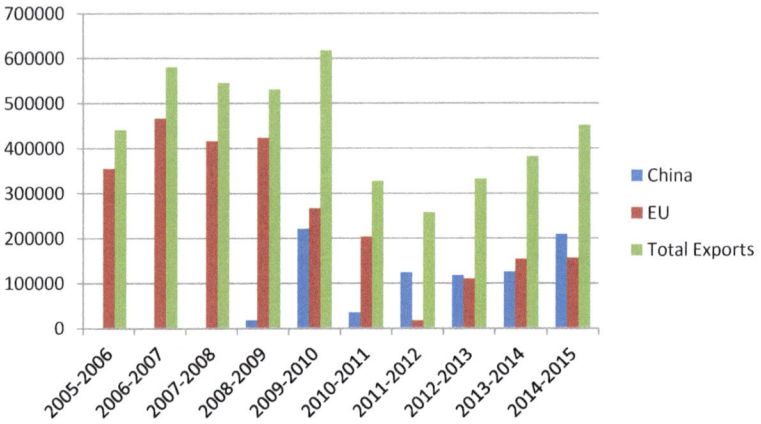

Fig. 16.2 Canadian flax exports to China and the EU, 2005–2015 (in 000 tons) (Source: Canadian Grain Commission 2015)

These funds were used to offset the cost of producer seed testing. Testing on all seed was required prior to seeding to ensure that any GM flax was removed from the crop production cycle. The flax export industry was responsible for testing costs at the railcar to point of export, with costs estimated at $1.34 million for the 5-year testing period. Cumulative testing costs in Canada from 2009 to 2014 are estimated to have been $3.34 million.

Following the conclusion of the 5-year stewardship program in 2014, the Federal Government's participation, through AAFC, ended, and testing costs have since been borne by producers and industry. The Flax Council of Canada estimated producer testing costs at $500,000; costs to the flax industry are not known. Given the cost ratio from previous years, it would not be expected that industry's cost would be lower than that of the producers. A conservative estimate of the cost of testing all flax exports to the EU for the 6 years, from initial detection to present, would be in the range of $4–5 million.

Additional opportunity costs were incurred from the periodic loss of the EU market (Fig. 16.2). The decline in EU flax exports was partially offset by the addition of China as an export destination. Chinese importers knew that Canadian flax exports to the EU had declined and that Canada had flax surpluses (despite declining flax production). This allowed China to import Canadian flax at a lower price than offered by EU flax importers.

While flax production has nearly fully recovered to pre-GM flax detection levels, the EU market has not recovered. As observed for other commodity markets that have been required to contend with coexistence challenges (rice and corn), export markets typically do not recover to previous levels. Prior to GM flax detection, Canada exported an average of 75% of the flax it produced to the EU. Over the past 3 years, EU flax imports have, on average, represented one-third of the flax produced in Canada. The Chinese export market does not fully account for the loss of 40% of Canada's flax export market.

What is not clear is when the testing of flax will end. The EU has a threshold of 0.1% for unapproved events and GM flax occurrence rates are well below this level. Therefore, all flax exports adhere to the EU standards for unapproved GM events. In theory, the Canadian flax industry should be able to end all flax testing and stewardship program requirements. And while this makes perfect sense from a logical commodity export market perspective, it does not from a politically driven one. There is no incentive for the EU to end their demand that every shipment of flax from Canada be tested to guarantee that it is free of GM flax. In fact, given the EU's recent policy change that allows Member States to individually decide whether they will allow the production of GM crops, it is anticipated that individual Member States will be averse to allowing an end to the testing protocol.

Conversely, what incentive is there for the Canadian flax industry to end its testing of flax? Historically, the EU has been the single most important export market, and the industry has a strong preference to see commodity trade return to pre-2009 levels. Ending the GM testing protocol could potentially make EU flax importers nervous about importing Canadian flax and drive them to seek alternative flax markets. The political sensitivity of testing has now become embedded within the flax trade between Canada and the EU and is likely to be viewed as entrenched. Therefore, this cost will likely have to be borne by producers and the flax industry going forward. The continued occurrence of false-positive tests presents an ongoing challenge to flax exporters.

With international flax trade between Canada and the EU still adversely affected by an incident from 2009, there is little indication that a resolution to the problem will be found in the recent Comprehensive and Economic Trade Agreement (CETA) between Canada and Europe. This trade agreement does mention the thorny issue of GM crops and other products of biotechnology but is incapable of providing a resolution. Article X.03: Bilateral Cooperation on Biotechnology states: "The Parties agree that cooperation and information exchange on issues related to biotechnology products are of mutual interest. The dialogue covers any relevant issues of mutual interest to Canada and the EU, including, … any trade impact related to asynchronous approvals of biotechnology products or the accidental release of unauthorized products, and any appropriate measures in this respect" (Government of Canada 2016). While clearly recognizing that international commodity trade has been disrupted, there is no mechanism for resolution identified, other than further ongoing discussions and dialogue.

16.5   Conclusions

The Triffid issue represents an unfortunate turn of events for Canada's flax industry. Triffid flax became the agricultural "poster child" for all that can go wrong with a transgenic crop variety in a zero-tolerance export market. While prices have recovered to some degree, the Triffid experience has left many flax growers discouraged. It was fortunate for the Canadian flax industry that China swooped in at the most vulnerable point and bought up a majority of the flax in early 2010. This helped to offset some of the economic losses that the industry incurred. Some stocks were shipped to the United States during that time as well. In the meantime, the Russian Federation and the Ukraine have leapt in and increased production to service short supplies in the EU flax market. It is not surprising that Canada has lost market share to these two countries. Linseed production in former Soviet Union countries increased by an estimated 45% to 480,000 metric tons in 2011 from the previous year (Ruitenberg 2011). By 2013, linseed production levels fell to just over 300,000 metric tons, which demonstrates a correction in the market. Still, that production level in the Russian Federation is triple what it was prior to 2009.

As time has passed, a certain amount of complacency around the Triffid issue has set in on the part of the EU. Despite this, there has been little to no resumption of exports into the food market as Canadian growers just do not want to take the risk. The EU's industrial market is being looked after, however, so the EU is relatively happy at this point.

Despite this, the Triffid issue has not been completely resolved. The Canadian industry is not where it needs to be. According to the Stewardship agreement with the EU, the industry still has to test, and that represents ongoing costs for the Canadian flax industry. According to the Flax Council of Canada (2011), however, "…the situation *is* workable…."

Economic impacts are very difficult to quantify. Some costs are explicit. Others less so and are time or opportunity cost related. This makes it difficult to attach a specific number to costs associated with the Triffid issue. One thing is for certain though, significant costs were incurred on both sides of the Atlantic.

When low level presence (LLP) events are detected or coexistence thresholds exceeded, agreements to their resolution clearly need to firmly establish "sunset clauses" or define the timeline for when the testing protocol will end; otherwise, as is evident from the situation described above, the likelihood of testing for the minute presence of GM material will continue indefinitely. Commodity export markets are not capable of providing market signals with respect to when testing protocols should end; therefore, it is crucial that those involved in the implementation of such protocols establish clear and concise timeframes regarding how long the testing period will last. Upon reaching the end of the agreed period, the testing should cease at point of export. A perfectly valid option is for the importer to incur the cost of continued testing, but the cost of such testing should no longer be borne by the exporting market.

The case of the low-level presence of GM flax in shipments to the EU demonstrates the vital importance of establishing clear definitions for the timelines of any commodity testing protocol at the point of protocol implementation. As shown above there are no economic incentives present in the Canada-EU flax trade market in the 2 years since the end of the Stewardship program. Alternatively, there has been no political incentive to find a resolution to this trade distortion, given the CETA negotiation process and agreement that paralleled the GM flax problems. As this case study demonstrates, when economic and political incentives are lacking, clear definition of timelines is required in the negotiation of any such testing protocols. Without clear and firm timelines, the costs of testing will be borne by exporters, even when the presence of LLP of GM commodities is below the rate of false positive.

References

Babuscio T, Hill W, Ryan CD, Smyth SJ (2016) The Canadian and Eurepean impacts from the detection of GM flax. Chapter 14. In: Kalaitzandonakes N, Phillips PWB, Smyth SJ, Wesseler J (eds) The coexistence of genetically modified, organic and conventional foods: government policies and market practices. Springer Publishers, New York

Berglund DR (2002) Flax: new uses and demands. In: Janick J, Whipkey A (eds) Trends in new crops and new uses. ASHS Press, Alexandria, VA, pp 358–360

Booker H, Lamb E (2012) Quantification of low-level GM seed presence in Canadian commercial flax stocks. AgBioforum 15(1):31–35

Booker HM, Lamb EG, Smyth SJ (2015) Ex-post Assessment of Genetically Modified, Low Level Presence of Canadian Flax. Transgenic Research 26(3):399–409

Canadian Grain Commission (2015) Canadian grain exports (Annual). CGC, Winnipeg, MB. Available online at: http://www.grainscanada.gc.ca/statistics-statistiques/cge-ecg/cgem-mecg-eng.htm. Accessed 28 Aug 2015

COCERAL/FEDIOL (2010) Low level presence of GMOs not authorized in Europe: the linseed CDC Triffid case. Report. July

Dayananda B (2011) The European union policy of zero tolerance: insights from the discovery of CDC Triffid. Thesis. Available online at: http://library2.usask.ca/theses/available/etd-06272011-111926/

European Commission (2009) Summary record of the standing committee on the food chain and animal health. Available online at: http://ec.europa.eu/food/committees/regulatory/scfcah/modif_genet/sum_19102009_en.pdf. Accessed on 12 Nov 2009

Flax Council of Canada (2009) GMO flax update 19th October 2009. Available online at: http://www.flaxcouncil.ca/files/web/GMO%20Flax%20Update%2019%20October%202009.pdf. Accessed on 20 Oct 2009

Flax Council of Canada (2011) 2010/2011 Flax sample testing program: approved labs and test submission forms. Available online at: http://www.flaxcouncil.ca/files/web/2010-2011%20Testing%20Program_Sept1_R26.pdf. Accessed on 28 Sept

Galushko V, Ryan CD (2012) Intellectual property rights (IPRs) and knowledge sharing in flax breeding. Int J Technol Glob 6(3):171–187

Government of Canada (2016) Canada-European union: comprehensive economic and trade agreement (CETA). Available online at: http://www.international.gc.ca/trade-agreements-accords-commerciaux/agr-acc/ceta-aecg/index.aspx?lang=eng. Accessed 24 May 2016

Hall L (2012) Personal communication with Linda Hall, Associate Professor, Department of Agricultural, Food and Nutritional Science, University of Alberta. April 2nd

Health Canada (2014) Summary of health Canada's assessment of a health claim about ground whole flaxseed and blood cholesterol lowering. http://www.hc-sc.gc.ca/fn-an/alt_formats/pdf/label-etiquet/claims-reclam/assess-evalu/flaxseed-graines-de-lin-eng.pdf

James T (2010) Presentation to the Canada Grains Council Meeting, Winnipeg

McHughen A, Rowland GG, Holm FA, Bhatty RS, Kenaschuk EO (1997) CDC Triffid transgenic flax. Can J Plant Sci 77:641–643

Ruitenberg R (2011) Russia to grow more linseed as Canada falters, oil world says. Bloomberg News. Available online at: http://www.bloomberg.com/news/2011-09-20/russia-to-grow-more-linseed-as-canada-falters-oil-world-says.html. September 20. Accessed on 27 Sept

Ryan CD, McHughen A (2014) Tomatoes, potatoes and flax: implications of withdrawn products. Chapter 51. In: Smyth SJ, Phillips PWB, Castle D (eds) Handbook on agriculture, biotechnology and development. Edward Elgar Publishing Ltd., Cheltenham, UK

Ryan CD, Smyth SJ (2012) Economic implications of low-level presence in a zero-tolerance European import market: the case of Canadian Triffid flax. AgBioforum 15(1):21–30

Smyth SJ, Phillips PWB (2001) Competitors co-operating: establishing a supply chain to manage genetically modified canola. Int Food Agribus Man Rev 4:51–66

Young L, Hammerlindl J, Babic V, McLeod J, Sharpe A, Matsalla C, Bekkaoui F, Marquess L, Booker H (2015) Genetics, structures, and prevalance of FP967 (CDC Triffid) T-DNA in flax. Springerplus 4:146

Chapter 17
Fundamental Insights into Plant Biology that Might Be Offered by Flax

Christopher A. Cullis

Flax is one of the oldest domesticated species. It is unusual in that it was domesticated for two very different phenotypes, namely, for an unbranched stem for long fibers for linen (flax) production and a small shorter bushy phenotype for high seed yield (linseed). These two contrasting commercial phenotypes have more recently been augmented by the increasing use of flax as an important source of fiber, of omega-3 fatty acids, of plant estrogens, and of lignans in the diets of both humans and animals. Canada was the first country to support a health claim for flax that provides the highest plant-based source of unsaturated fatty acids. New value-added markets for flax in the human health and animal nutrition realm, as well as the fiber and oil being used for an array of industrial products including linoleum flooring, car panels, industrial oils, and solvents, and a myriad of other composite materials are important in the revival of this important oilseed and fiber crop. In addition, it has commercial potential that is changing with new opportunities. The use of fibers in composite materials may be enhanced by the development of dual-purpose varieties, that is, those that have sufficient seed yield allied to high-quality fibers. Such a crop would have an environmental advantage as well as the straw, which would be useful without the need for disposal resulting in additional carbon sequestration.

The collections across the world contain representatives of many of the varieties, both old and new, as well as an extensive sampling of the related wild germplasm, as a consequence of this long history of commercial use. In addition to the close relatives of *Linum usitatissimum*, many related species have also been collected and documented. Along with these collections, multigenerational families segregating for useful traits have been developed and phenotyped. Therefore, the experimental material to understand the genetic underpinning and evolution of important plant characteristics is available.

C. A. Cullis (✉)
Department of Biology, Case Western Reserve University, Cleveland, OH, USA
e-mail: cac5@case.edu

© Springer Nature Switzerland AG 2019
C. A. Cullis (ed.), *Genetics and Genomics of Linum*, Plant Genetics and
Genomics: Crops and Models 23, https://doi.org/10.1007/978-3-030-23964-0_17

An example of the careful combinations of characterized germplasm was the development of the series of single gene tester lines for rust resistance, which were a necessary precursor to the isolation of the flax rust resistance genes. These genes were initially isolated using sophisticated genetic and cloning strategies prior to the availability of whole genome sequences and expression libraries. The goal of the Total Utilization of Flax GENomics (TUFGEN) project (http://www.genomeprairie.ca/project/previous/total-utilization-flax-genomics/) was to increase the benefits and versatility of flax by developing genomic-based tools to assist in crop breeding, to improve field performance, and to enhance seed and fiber traits. This project has laid the basis for the continued contribution of flax to the understanding of plant growth, development, and both biotic and abiotic interactions with the external environment, as well as the potential reinvigoration of the crop. The complete assembled genome sequence for the reference genome of the variety Bison was augmented with the resequencing of the many accessions of the Canadian flax core collection which yielded ~1.7M SNPs. The sequencing of multiple expression libraries, including those for the small RNAs that appear to be important for controlling many plant responses to stresses, has been the basis for gene prediction validation. These resources have allowed the construction of high-density genetic maps, and QTL for important agronomic and seed quality traits have been identified in both cultivated flax and its wild progenitor.

Flax is very amenable to modern molecular genetic techniques and so provides a model for investigating many of the fundamental plant processes. Gene targets involved in key processes like bast fiber and seed development have been identified through transcriptome analysis, but functional data for many genes is still lacking. However, major approaches to exploring gene function using virus-induced gene silencing, a TILLING (Targeting Induced Local Lesions in Genomes) platform and mutant collection, and CRISPR-cas9-based gene editing are available for flax and can be applied to facilitate the accelerated access by flax breeders to natural genetic variation in the flax gene pool. The recent demonstration that a combination of single-stranded oligonucleotides (ssODNs) and CRISPR/Cas9 could be used to generate herbicide-tolerant flax opens up new perspectives for both breeding purposes and the fundamental biology in this species. However, such genetic engineering approaches are currently unable to create large mutant populations displaying a range of different, and often novel, phenotypes and, therefore, need to be supplemented by the generation of mutant pools, such as those illustrated by the generation of the flax series of *lbf* mutants characterized by a fiber cell wall composition not present in wild-type populations. The creation of mutant populations will continue to be a powerful approach for improving our knowledge about flax biology and by extension plant biology in general.

The use of the biological and molecular resources is supported through the ability to transform flax. Therefore, any genes that are identified as possible underpinning a particular characteristic can be tested directly for their contribution to a particular phenotype. Of course, the transformability of flax had an unanticipated outcome with respect to the Triffid affair. This unintended release of an unapproved transgenic flax event that contaminated commercial harvests serves as a case study for the control of such experimental material.

The history of fiber flax varietal stability has indicated that, in spite of being almost completely self-fertilizing, replanting seed from the fiber crop resulted in a steady drift of phenotype. Subsequently, this change was shown to be associated with nuclear DNA variation in a specific subset of the genome, which was reproducibly modified when responding varieties were grown under nutrient deprivation. The application of next-generation sequencing of the genotrophs (stable lines derived from a known progenitor by a single-generation growth under nutrient stress conditions) has demonstrated that the genome can be partitioned into two compartments, one which is unaffected by the changes and the other undergoing structured, reproducible changes. Not all the possible genomic changes always occur; the suite of variants occurring is dependent on the environmental stimulus. The regions affected can contain genic sequences although many of them are not annotated as genes but are also frequently similar to sequences found in transposons. These changes can be categorized as flipping between alternative states, with both alternatives also being observed in the wild progenitor of flax. These observations are consistent with the hypothesis that the genome is restructured at a set of labile regions that can result in phenotypic variety without significant lethality.

Perhaps, the most perplexing facet of this, apparently unique, process is the appearance of sequence variants, possibly many thousands of bases long, which are not present intact in the original line. The source of these new sequences and the mechanism(s) of their assembly and integration into the genome might provide new avenues to identify and generate stress-tolerant genotypes of other important plant species.

Thus, flax exhibits a range of interesting characteristics, unavailable in other models. In addition to the characteristics already mentioned, flax produces a wide range of complex metabolites that have potential in the human health arena. It would be a useful model for understanding the pathways for synthesis of these molecules and for the production of sufficient quantities for further study.

Thus, the biological and molecular resources to identify the underlying genetics, mechanisms, and genes are all available in an amenable, manipulative model that is also a commercially relevant plant.

Index

© Springer Nature Switzerland AG 2019
C. A. Cullis (ed.), *Genetics and Genomics of Linum*, Plant Genetics and
Genomics: Crops and Models 23, https://doi.org/10.1007/978-3-030-23964-0

Printed by Printforce, the Netherlands